Enumerative Theory of Maps

Mathematics and Its Applications

Volume 468

Enumerative Theory Of Maps

by

Liu Yanpei

Department of Mathematics
Northern Jiaotong University
Beijing 100044, The People's Republic of China

Science Press
Beijing/New York

Springer Science+Business Media, B.V.

contraction of a specific edge properly chosen. Along this line one can see how different kinds of operations are constructed for enumerating a variety of types of maps. The way of decomposition is very closely related to the choice of parameters which the enumeration is according to. There are tricks which have to be exploited to avoid the complicatedness involved in deriving the functional equation from the decomposition.

As soon as a functional equation has been built up, the problem that follows is to find a suitable way to solve it, or to transform and simplify it for clarifying the solution. Here, we present a number of methods for solving the equations directly, or converting them into some special ones which are solvable in certain cases. The most interesting part is to try to find a way by which the Lagrangian inversion can be suitably applied for determining the coefficients in the power series form of the solution. Hopefully, many simpler formulae for enumerating a variety of maps have been obtained via a series of subtle treatments in this way.

In spite of whether the functional equation is completely solved or not, one is always allowed to investigate stochastic behaviors by estimating asymptotic properties of the solution as an enumerating function of certain kind of maps, when the order of maps is big enough, up to tending to infinity.

According to the basic theoretical idea as described above, the whole book is divided into three main parts. The first part, from Chapter 2 through Chapter 8, is on the ordinary theory of enumerating maps. The second, from Chapter 9 through Chapter 11, is on chromatic and dichromatic sums which can be seen as a kind of generalization of enumeration with much complication and much difficulty. And the third which consists of only one chapter, Chapter 12 is on the stochastic behaviors. Of course, Chapter 1 provides the necessary knowledge and basic techniques for the requirements of the whole book. In order to save space, the last section of each chapter is designed to be notes in which some historical remarks, new progress and unsolved problems with clues for possibly solving them are indicated in corresponding areas.

On this occasion, I should express my heartiest thanks to all those having made contributions themselves directly or indirectly to this book.

The initiation of this theory was established by Professor W. T. Tutte whose articles and directions were invoked for me to enter the field when I was working in the Department of Combinatorics and Optimization at the University of Waterloo, Canada, in the period of 1982–1984. Without these, I would be doing something else at the present time.

Many other friends of mine including Professors R. Cori, P.L. Hammer, D.M. Jackson, R. C. Mullin, R.C. Read, L.B. Richmond, P. Rosenstiehl, B. Simeone, T.T.S. Walsh, W. Xu and J. Yan are constantly concerned with me in material and spirit. For the final version, many people including J.L. Cai (PhD), Y.X. Chang (PhD), F.M. Dong (PhD), J.Q. Dong, R.X. Hao, Y.Q. Huang (PhD), S. Lawrencenko (PhD), A.P. Li (PhD), D.M. Li (PhD), X. Liu (PhD), Yi. Liu (PhD), T.Y. Liu, T.J. Lu (PhD), K. Ouyang (PhD), X.R. Sun (PhD), H. Ren, E.L. Wei, F.E. Wu (PhD) and M.L. Zheng (PhD) provide errata in part or whole.

The source Latex files for the whole book were typed and run on computers by my daughter Liu Ying.

Institutions including the Department of C & O, University of Waterloo, Canada; DIMACS and RUTCOR, Rutgers University, USA; the Department of Computer Science, the Department of Mathematics and the Department of Statistics, University of Rome 'La Sapienza', Italy; the Center of Applied Mathematics, E.H.E.S.S., France; the Department of Mathematics and Computer Science, University of Bordeaux I, France; and the Department of ECECS, University of Cincinnati, USA provided me the opportunities to visit with the hospitality. In particular, the Northern Jiaotong University where I am working now offers me favorable circumstances.

Last but not least, the partial financial support from the NSF in USA, the CNR in Italy and the NNSF in China should be especially acknowledged as well.

Y.P. Liu

Beijing, P.R. China.
May 1998

Contents

Chapter 1

Preliminaries

Throughout, for the sake of brevity, we adopt the following logical conventions: disjunction, conjunction, negation, implication, equivalence, universal quantification and existential quantification denoted by the symbols: $\vee$, $\wedge$, $\neg$, $\Rightarrow$, $\Leftrightarrow$, $\forall$ and $\exists$, respectively.

In the context, $(i.j.k)$ (or $i.j.k$) refers to item k of section j in chapter i for formulae (or theorems and the like).

A reference $[k]$ refers to item k in the bibliography where k consists of the first few letters(initials) of the last name(s) of the author(s) in alphabetical order followed by a number to distinguish publications of the same author(s).

Terminologies not explained in this book can be found in [Liu58], or probably in [GoJ1] or [Tut39].

§1.1 Maps

A *map*, denoted by M, is a mathematical concept which can, of course, be seen as a kind of abstraction from that appearing in geography, is defined to be a basic permutation $\mathcal{P}$ on a disjoint union $\mathcal{X}$ of quadricells with Axiom 1 and Axiom 2 bellow.

Let X be a finite set, and K the Klein group of four elements which are denoted by 1, α, β, and $\alpha\beta$. For $x \in X$, the set $Kx=\{x, \alpha x, \beta x, \alpha\beta x\}$ is said to be a *quadricell.*

We may write $\mathcal{X} = \sum_{x\in X} Kx$. Naturally, both α and β themselves are permutations on $\mathcal{X}$. A permutation $\mathcal{P}$ on $\mathcal{X}$ is said to be *basic* if for any $x \in \mathcal{X}$ there does not exist an integer k such that $\mathcal{P}^k x = \alpha x$.

Axiom 1 $\alpha\mathcal{P} = \mathcal{P}^{-1}\alpha$.

Axiom 2 *The group* Ψ_J *which is generated by* $J = \{\alpha, \beta, \mathcal{P}\}$ *is transitive on* $\mathcal{X}$.

Thus, we may write the map $M = (\mathcal{X}_{\alpha,\beta}(X), \mathcal{P})$. From Axiom 1, α and β are asymmetric, *i.e.*, $(\mathcal{X}_{\alpha,\beta}(X), \mathcal{P}) \neq (\mathcal{X}_{\beta,\alpha}(X), \mathcal{P})$ in general. Sometimes, α is called the *first operator* and β, the *second operator*. Generally, for a map $M = (\mathcal{X}_{\alpha,\beta}(X), \mathcal{P})$, it is not necessary that $(\mathcal{X}_{\beta,\alpha}(X), \mathcal{P})$ is also a map because it is not guaranteed to have Axiom 1 for β. In fact, Axiom 1 allows us to define the *vertices* of a map as the pairs of *conjugate* orbits of $\mathcal{P}$ on $\mathcal{X}$. We always write $\mathcal{P}$ as the product of the orbits (in cyclic order) obtained by choosing exactly one, which represents a vertex as well, in each *conjugate pair* determined by x and αx for $x \in \mathcal{X}$.

For a map $M = (\mathcal{X}_{\alpha,\beta}(X), \mathcal{P})$ given, from the definition it is easy to check that $M^* = (\mathcal{X}_{\beta,\alpha}(X), \mathcal{P}\alpha\beta)$ is also a map with β as the first operator and α, the second. We call M^* the *dual* (map) of M. From the duality, the *faces* of M are defined to be the vertices of M^*. Moreover, an *edge* of M is defined to be the quadricell $Kx = \{x, \alpha x, \beta x, \alpha\beta x\}$ for $x \in X$. An edge $\{x, \alpha x, \beta x, \alpha\beta x\}$ can be seen as a pair of *semiedges* $\{x, \alpha x\}$ and $\{\beta x, \alpha\beta x\}$, or $\{x, \beta x\}$ and $\{\alpha x, \alpha\beta x\}$ as well.

The graph whose vertices and edges are those of a map M is said to be the *underlying graph* of M and is denoted by $G(M)$. From Axiom 2, $G(M)$ has to be connected. Conversely, a map M whose vertices and edges are those of a graph G is said to be an *underlain map* of G and denoted by $M(G)$. Of course, M is an underlain map of $G(M)$. Although any map has a unique underlying graph, a graph is in general allowed to have many underlain maps.

In fact, any underlain map of a graph (connected of course) is an embedding on a surface. This enables us to denote a map M by (G, F) such that $G = (V, E) = G(M)$ where V, E and F are the vertex, edge and face sets respectively. Only one vertex without edge is always defined to be a map, which is called the *trivial map*, or the *vertex map*. If a map has a single edge, then it is called an *edge map* denoted by L. If an edge map has a loop, then it is called a *loop map*; otherwise, the *link map*. Apparently, only two possible loop maps exist. They are $L_1 = (\mathcal{X}, (x, \alpha\beta x))$ and $L_2 = (\mathcal{X}, (x, \beta x))$. The unique link map is $L = (\mathcal{X}, (x)(\alpha\beta x))$.

Let ν, ϵ and ϕ be the numbers of vertices, edges and faces of a map M respectively. The number

$$\mathrm{Eul}(M) = \nu - \epsilon + \phi \tag{1.1.1}$$

is said to be the *Euler characteristic* of M.

Further, if a map $M = (\mathcal{X}_{\alpha,\beta}(X), \mathcal{P})$ satisfies the following Axiom 3, then it is said to be *nonorientable*; otherwise, *orientable*.

Axiom 3 *The group Ψ_L generated by $L = \{\alpha\beta, \mathcal{P}\}$ is transitive on $\mathcal{X}_{\alpha,\beta}(X)$.*

Because it can be shown that if the group Ψ_L is not transitive on $\mathcal{X}_{\alpha,\beta}(X)$ then it has exactly two orbits one of which is conjugate of the other on $\mathcal{X}_{\alpha,\beta}(X)$, a map $(\mathcal{X}_{\alpha,\beta}(X), \mathcal{P})$ is orientable iff the group Ψ_L has two orbits on $\mathcal{X}_{\alpha,\beta}(X)$.

Let $M = (\mathcal{X}_{\alpha,\beta}(X), \mathcal{P})$ be a map and $e_x = \{x, \alpha x, \beta x, \alpha\beta x\}$ be the edge incident with $x \in \mathcal{X}_{\alpha,\beta}(X)$. For convenience, we always write $\mathcal{X}$ instead of $\mathcal{X}_{\alpha,\beta}(X)$ without specific indication and see that

$$\mathcal{X} = X + \alpha X + \beta X + \alpha\beta X \tag{1.1.2}$$

where $\gamma X = \{\gamma x | \forall x \in X\}$ for $\gamma = \alpha,\ \beta$, or $\alpha\beta$. Moreover, an edge e_x for $x \in X$ is simply denoted by e.

Now, we introduce two kinds of operations for an edge e on a map M. By the *deletion* of e on M, denoted by $M - e$, is meant that

$$M - e = (\mathcal{X} - e,\ \mathcal{P}\langle e\rangle), \tag{1.1.3}$$

where $\mathcal{P}\langle e\rangle$ is the restriction of $\mathcal{P}$ on $\mathcal{X} - e$. The other, called *contraction* of e on M and denoted by $M \bullet e$, is

$$M \bullet e = (\mathcal{X} - e,\ \mathcal{P}[e]) \tag{1.1.4}$$

where $\mathcal{P}[e]$ is obtained by composing the two vertices u and v incident to e as

$$\{\Lambda B,\ \alpha B^{-1} A^{-1}\}$$

when

$$u = \{xA, \alpha x\alpha A^{-1}\} \text{ and } v = \{\alpha\beta xB, \beta x\alpha B^{-1}\}$$

while all other vertices are in agreement with those for $\mathcal{P}$.

Theorem 1.1.1 *Any map M has* $\mathrm{Eul}(M) \le 2$.

Proof Because the deletion of an edge on the common boundary of two faces in a map M reduces one in the face number of M, we can always find a map M', $\nu(M') = \nu(M)$, such that M' has only one face and $\mathrm{Eul}(M') = \mathrm{Eul}(M)$ by a series of the operations. From the connectedness, $\nu(M') \le \epsilon(M') + 1$. Therefore,

$$\mathrm{Eul}(M) = \nu(M') - \epsilon(M') + 1 \le 2.$$

The theorem is proved. □

Two more operations which are often used have to be explained. Suppose $v = (AB)$ is a vertex of a map M. Let $\mathcal{P}'$ be obtained by substituting (Ax) and $(\alpha\beta xB)$ for (AB) in $\mathcal{P}$ where x is incident with the new edge. Then, the map $M' = (\mathcal{X}+Kx, \mathcal{P}')$ is said to be obtained by *splitting* the vertex v on M. If $v = (xy)$ is a vertex in a map M, then the map

$$M' = (\mathcal{X} - Kx - Ky + Kz, \mathcal{P}'),\ z = x = y,$$

where $\mathcal{P}'$ is the resultant one of deleting (xy) from $\mathcal{P}$, is said to be obtained by *missing* the vertex v on M. The inverse of deletion of an edge is called the *addition* of an edge and the inverse of missing a vertex, *subdividing* an edge.

Of course, the inverse of contraction of an edge is splitting a vertex as defined above. It is easily seen that the Euler characteristic is unchanged under the contraction of an edge, missing a vertex and their inverses: splitting a vertex, subdividing an edge.

However, the invariance of the Euler characteristic under edge deletion and its inverse, the edge addition, is only for an edge on the common boundary of two faces, or say, under *standard* deletion and addition.

Because any map can be transformed into another which has only one face such that the Euler characteristic is unchanged by virtue of what appears in the proof of Theorem 1.1.1, we are allowed to consider one face maps for the sake of finding the simplest one with a given Euler characteristic. For brevity, a map is represented by its faces with the convention: $x^{-1} = \alpha\beta x$ and hence $(\alpha x)^{-1} = \beta x$. On the whole, we are allowed to realize $x = \alpha x$ and hence $\beta x = \alpha\beta x$.

For orientable maps we have the following two properties: Orien.1 and Orien.2 which can be derived from the operations mentioned above.

Orien.1 If a one face map $M = (Rxx^{-1}Q)$, $R, Q \neq \emptyset$, then

$$\mathrm{Eul}(M) = \mathrm{Eul}(RQ).$$

Orien.2 If a one face map $M = (PxQyRx^{-1}Sy^{-1}T)$, then

$$\mathrm{Eul}(M) = \mathrm{Eul}(PSRQTxyx^{-1}y^{-1}).$$

For nonorientable maps, we have the following two properties: Norien.1 and Norien.2 which can be derived from the operations as well.

Norien.1 If a one face map $M = (PxQxR)$, then

$$\mathrm{Eul}(M) = \mathrm{Eul}(PQ^{-1}Rxx).$$

Norien.2 If a one face map $M = (Axxyzy^{-1}z^{-1})$, then

$$\mathrm{Eul}(M) = \mathrm{Eul}(Ax_1x_1x_2x_2x_3x_3).$$

Theorem 1.1.2 *If a map M is orientable, then we have* Eul(M) = 0(mod 2). *Moreover, M is on the surface of genus p (orientable), $p \geq 0$, iff*

$$\mathrm{Eul}(M) = 2 - 2p$$

where Eul(M) *is the Euler characteristic of M defined by* (1.1.1).

Proof From the orientability, each edge $e = Kx$ is only allowed to have $\{x, \alpha\beta x\} = \{x, x^{-1}\}$ (or $\{\alpha x, \beta x\}$ as well) in one of the two orbits of the group Ψ_L on $\mathcal{X}$. By using the properties: Orien.1 and Orien.2 as far as possible, we may finally find that Eul(M) is equal to either $\mathrm{Eul}(O_0)$, $O_0 = (xx^{-1})$, or

$$\mathrm{Eul}(O_p),\ O_p = \Big(\prod_{i=1}^{p} x_i y_i x_i^{-1} y_i^{-1}\Big)$$

for an integer $p \geq 1$. By counting the numbers of vertices, edges and faces in O_0 and O_p, the first statement of the theorem can be obtained. The second statement is a conclusion of the characterization of orientable surfaces. □

Theorem 1.1.3 *For a nonorientable map M, M is on the surface of genus q (nonorientable) iff M has*

$$\mathrm{Eul}(M) = 2 - q$$

where $q \geq 1$.

Proof From the nonorientability, there always is x in $\mathcal{X}$ such that both x and αx appear in the face of a one face map. Or in our words here, x appears twice. By using the properties Norien.1 and Norien.2 as far as possible, we may finally find that

$$\mathrm{Eul}(M) = \mathrm{Eul}(N_q),\ N_q = \Big(\prod_{i=1}^{q} x_i x_i\Big)$$

for an integer $q \geq 1$. Hence, from counting the numbers of vertices, edges, and faces in N_q, by virtue of the characterization of nonorientable surfaces the theorem is soon obtained. □

All O_p, $p \geq 0$ and N_q, $q \geq 1$, are called *standard maps* on the corresponding surface. If $\mathrm{Eul}(M) = 2$, *i.e.*, $p(M) = 0$, then M is said to be *planar*. The cases of $p(M) = 1$, $q(M) = 1$ and 2, which are often encountered, show that M is on the *torus*, the *projective plane* and the *Klein bottle* respectively.

For two maps $M_1 = (\mathcal{X}_{\alpha,\beta}(X_1), \mathcal{P}_1)$ and $M_2 = (\mathcal{X}_{\alpha,\beta}(X_2), \mathcal{P}_2)$, if there exists a bijection

$$\tau : \mathcal{X}_{\alpha,\beta}(X_1) \longrightarrow \mathcal{X}_{\alpha,\beta}(X_2)$$

such that the diagrams (1.1.5) as shown below are commutative for $\gamma_1 = \gamma_2 = \alpha$, for $\gamma_1 = \gamma_2 = \beta$ and for $\gamma_1 = \mathcal{P}_1$ and $\gamma_2 = \mathcal{P}_2$, then we say M_1 and M_2 are *isomorphic* while τ is called an *isomorphism* between them. An isomorphism of a map M to itself is called an *automorphism* of M. All automorphisms of a map M forms a group which is called the *automorphism group* of M and denoted by $\mathrm{Aut}(M)$. The order of $\mathrm{Aut}(M)$ is written as $\mathrm{aut}(M) = |\mathrm{Aut(M)}|$.

$$
\begin{array}{ccc}
\mathcal{X}_{\alpha,\beta}(X_1) & \xrightarrow{\ \tau\ } & \mathcal{X}_{\alpha,\beta}(X_2) \\
\gamma_1 \downarrow & & \downarrow \gamma_2 \\
\mathcal{X}_{\alpha,\beta}(X_1) & \xrightarrow[\ \tau\ ]{} & \mathcal{X}_{\alpha,\beta}(X_2)
\end{array}
\tag{1.1.5}
$$

If a map $M = (\mathcal{X}_{\alpha,\beta}(X), \mathcal{P})$ has an element, the *root* denoted by $r = r(M)$, in $\mathcal{X}_{\alpha,\beta}(X)$ marked beforehand, then M is called a *rooted map* and the marked edge, the *rooted* edge of M, which is usually denoted by $a = e_r(M)$. And, likewise the *rooted vertex* and the *rooted face*. Two rooted maps are said to be *isomorphic* if there is an isomorphism between them such that their roots are in correspondence.

Theorem 1.1.4 *For any rooted map its automorphism group is the trivial group.*

Proof Let τ be an automorphism of a map M with r being the root. Because $\tau(r) = r$, from (1.1.5) we see that

$$\tau(\alpha r) = \alpha r,\ \ \tau(\beta r) = \beta r \text{ and } \tau(\mathcal{P}r) = \mathcal{P}r.$$

Thus, for any $\psi \in \Psi_J$, the group generated by $J = \{\alpha, \beta, \mathcal{P}\}$, we have $\tau(\psi r) = \psi r$. From Axiom 2, the theorem follows. □

Based on this theorem, we may find

Theorem 1.1.5 *Let ν_i and ϕ_i be the respective number of vertices and faces of valency i, $i \geq 1$, on a map M. Then,*

$$\mathrm{aut}(M) \,\Big|\, (2i\nu_i,\ 2j\phi_j \mid \forall i,\ i \geq 1,\ \forall j,\ j \geq 1) \qquad (1.1.6)$$

where $(2i\nu_i,\ 2j\phi_j \mid \forall i,\ i \geq 1,\ \forall j,\ j \geq 1)$ is the greatest common divisor of all the numbers in the parentheses.

Proof From (1.1.5), an automorphism τ on M has to have the property that for $x \in \mathcal{X}$ which is incident to a vertex of valency i, $i \geq 1$, and with a face of valency j, $j \geq 1$, $\tau(x)$ has to be incident to a vertex of valency i and with a face of valency j as well. We may classify the elements which are incident to a vertex of valency i in $\mathcal{X}$ by the rule:

$$x \sim_{\mathrm{Aut}} y \Longleftrightarrow \exists \tau \in \mathrm{Aut}(M),\ x = \tau y.$$

And then, it is seen that all the classes obtained in this way have the same cardinality which is the order of the automorphism group $\mathrm{Aut}(M)$ from Theorem 1.1.4. Since the number of the elements incident to a vertex of valency i is $2i\nu_i$, we have $\mathrm{aut}(M)|2i\nu_i$. Similarly, we may also find $\mathrm{aut}(M)|2j\phi_j$. From the arbitrariness of the choice of i, $i \geq 1$, and j, $j \geq 1$, the theorem is obtained. □

From this theorem, one can soon find

$$\mathrm{aut}(M) \leq (2i\nu_i,\ 2j\phi_j \mid \forall i,\ i \geq 1,\ \forall j,\ j \geq 1). \qquad (1.1.7)$$

Because of the relation

$$4\epsilon = 2\sum_{i=1}^{\nu} i\nu_i = 2\sum_{j=1}^{\phi} j\phi_j,$$

and then

$$(2i\nu_i,\ 2j\phi_j \mid \forall i,\ i \geq 1,\ \forall j,\ j \geq 1) \,\Big|\, 4\epsilon, \qquad (1.1.8)$$

Theorem 1.1.5 implies that

$$\mathrm{aut}(M) \,\Big|\, 4\epsilon \text{ and hence } \mathrm{aut}(M) \leq 4\epsilon. \qquad (1.1.9)$$

Theorem 1.1.4 allows us to enumerate rooted maps without considering the symmetry. Meanwhile, Theorem 1.1.5 would have more advantages for enumerating general maps with the consideration of the symmetry from those obtained in the corresponding asymmetric case because an efficient algorithm for finding the automorphism group of a map can be designed on the basis.

§1.2 Polynomials on maps

From the two operations defined by (1.1.3) and (1.1.4), we are allowed to introduce a number of polynomials on maps. Here, only those related to the two kinds of polynomials, the chromatic polynomials and the dichromatic polynomials, are discussed. For generality, each edge e is assigned to have a weight: $w(e) = 0$ or 1 in other words, to be binary on a map M.

For a map M let V and E be the sets of vertices and edges of M respectively.

Suppose each edge $e \in E$ is with a weight $w(e)$ whose value is assigned to be binary. We may introduce a map function $\Phi(M)$ by the following recursion: for $e \in E$,

$$\Phi(M) = \begin{cases} A(e)\Phi(M-e) + B(e)\Phi(M \bullet e), \\ \qquad \text{if } e \in E \text{ is neither a loop} \\ \qquad\qquad \text{nor an isthmus;} \\ (X+Yz)^{\bar{w}(e)}(Xz+Y)^{w(e)}\Phi(M-e), \\ \qquad \text{if } e \text{ is a loop;} \\ (X+Yz)^{w(e)}(Xz+Y)^{\bar{w}(e)}\Phi(M \bullet e), \\ \qquad \text{if } e \text{ is an isthmus} \end{cases} \tag{1.2.1}$$

where

$$\begin{cases} A(e) = \bar{w}(e)X + w(e)Y; \\ B(e) = w(e)X + \bar{w}(e)Y \end{cases} \tag{1.2.2}$$

with the two conditions below being satisfied.

Cond1. If M is the vertex map, then

$$\Phi(M) = 1. \tag{1.2.3}$$

Cond2. If $M = M_1 + M_2$, or in other words, M is the union of M_1 and M_2 provided without common vertex of M_1 and M_2, then

$$\Phi(M) = z\Phi(M_1)\Phi(M_2). \tag{1.2.4}$$

Let $k_i = k_i(M)$ and $l_i = l_i(M)$ be the numbers of isthmuses and loops of weight i, $i = 0, 1$, respectively.

Theorem 1.2.1 *If the underlying graph of a map M has each of its edges either an isthmus or a loop, then we have*

$$\Phi(M) = (X + Yz)^{k_1+l_0}(Xz + Y)^{k_0+l_1}. \tag{1.2.5}$$

Proof When M has only one loop or isthmus, it is easy to check that (1.2.5) holds. In general, if e is a loop, then from (1.2.1), we should have

$$\Phi(M) = (X + Yz)^{\bar{w}(e)}(Xz + Y)^{w(e)} \times\Phi(M - e)$$

and hence (1.2.5) by the hypothesis of induction; otherwise, we may also obtain (1.2.5) in the similar way for e as an isthmus instead of a loop. □

In fact, we may further have

Theorem 1.2.2 *For any map M the function*

$$\Phi(M) = \Phi(M; X, Y, z)$$

is always a polynomial of X, Y, and z as the undeterminates.

Proof Because of (1.2.1–4), the function $\Phi(M)$ of any map M can always be expressed into a form in linear combination of the functions of those with all the edges either loop or isthmus such that the coefficients are all polynomials of X, Y and z.

Hence, from Theorem 1.2.1, the theorem follows. □

If two maps M_1 and M_2 are mutual dual, then it can be shown that the functions $\Phi(M_1)$ and $\Phi(M_2)$ have the same type.

Theorem 1.2.3 *For any map M, let M^* be its dual. Then, we have*

$$\Phi(M; X, Y, z) = \Phi(M^*; Y, X, z) \tag{1.2.6}$$

where Φ is defined by (1.2.1).

Proof Because of the duality between the two operations: deletion and contraction in the pair of M and M^*, the theorem can be soon derived. □

If two maps are isomorphic such that corresponding edges are of the same weight, then they are seen to be equivalent. By a *combinatorial invariant* of maps, we shall mean such a function of maps that has the same value when two maps are in the same equivalent class.

Theorem 1.2.4 *The function Φ is a combinatorial invariant on maps.*

Proof First, it is easily checked that if M has at most one edge which is neither a loop nor an isthmus, the theorem holds.

We are then allowed to suppose e_1 and e_2 are two edges, each of which is neither a loop nor an isthmus.

Further, we are also allowed to assume that none of e_1 and e_2 is a multiedge from (1.2.1–4) without loss of generality. Since

$$\begin{aligned}
&\Big(A(e_1)\Phi(M - e_1) + B(e_1)\Phi(M \bullet e_1)\Big)\\
&\qquad -\Big(A(e_2)\Phi(M - e_2) + B(e_2)\Phi(M \bullet e_2)\Big)\\
&= A(e_1)B(e_2)\Phi\Big((M - e_1) \bullet e_2\Big)\\
&\qquad +B(e_1)A(e_2)\Phi\Big((M \bullet e_1) - e_2\Big)\\
&\qquad -A(e_2)B(e_1)\Phi\Big((M - e_2) \bullet e_1\Big)\\
&\qquad -B(e_2)A(e_1)\Phi\Big((M \bullet e_2) - e_1\Big)\\
&= 0,
\end{aligned}$$

by considering that

$$\begin{cases} \Phi((M-e_1)\bullet e_2) &= \Phi((M\bullet e_2)-e_1); \\ \Phi((M\bullet e_1)-e_2) &= \Phi((M-e_2)\bullet e_1), \end{cases}$$

the theorem can be obtained recursively. □

If maps are restricted to planar maps, then it is known that the polynomial Φ is a kind of generalizations of the Jones polynomial and the bracket polynomial on knots in topology[Liu58–59].

For simplicity, we would prefer to concentrate on the polynomials of maps with constant weights which are always assumed to be 1 such that the coefficients in the recursion for an edge which is neither loop nor isthmus in (1.2.1) are absolute constants which are assumed to be 1 or -1.

By the *dichromatic polynomial* of a map M, denoted by $Q(M)$, we shall mean such a polynomial of M determined by the following recursion:

$$Q(M) = \begin{cases} Q(M-e)+Q(M\bullet e), & \text{if } e \text{ is neither a loop nor an isthmus;} \\ (1+\nu)Q(M-e), & \text{if } e \text{ is a loop;} \\ (1+\mu)Q(M\bullet e), & \text{if } e \text{ is an isthmus} \end{cases} \qquad (1.2.7)$$

in which the condition below are satisfied.

Q1. When M is the vertex map, then

$$Q(M) = \mu. \qquad (1.2.8)$$

Q2. When M is the union of two maps M_1 and M_2 which have only one vertex in common, then

$$Q(M) = \mu^{-1}Q(M_1)Q(M_2). \qquad (1.2.9)$$

Theorem 1.2.5 *For any map M its dichromatic polynomial has the following form:*

$$Q(M) = \sum_{S} \mu^{p_0(S)} \nu^{p_1(S)} \tag{1.2.10}$$

where S is over all spanning subgraphs of $G(M)$, $p_0(S)$ and $p_1(S)$ are the component number and the cyclomatic number of S respectively.

Proof Because the vertex map only has one spanning graph which is itself, from (1.2.10) the condition Q1 is satisfied.

For a map M which is the union of M_1 and M_2 with only one vertex in common, any spanning subgraph S of M is also a union of S_1 and S_2 which are spanning subgraphs of M_1 and M_2 respectively.

Since it is easily checked that

$$p_0(S) = p_0(S_1) + p_0(S_2) - 1$$

and

$$p_1(S) = p_1(S_1) + p_1(S_2),$$

the condition Q2 is also satisfied.

We then prove that $Q(M)$ expressed by (1.2.10) satisfies (1.2.7).

When e is neither a loop nor an isthmus. Because all the spanning subgraphs can be divided into two classes, those which contain e and those do not contain e. Both of them consist of all the spanning subgraphs of $M - e$ and $M \bullet e$, respectively. From

$$\begin{cases} p_0(S) = p_0(S; M - e), \\ \qquad p_1(S) = p_1(S; M - e), \\ \qquad\qquad \text{if } e \notin S; \\ p_0(S) = p_0(S; M \bullet e), \\ \qquad p_1(S) = p_1(S; M \bullet e), \\ \qquad\qquad \text{if } e \in S, \end{cases}$$

the form (1.2.10) leads to (1.2.7).

When e is a loop. Because for any spanning subgraph S of M, we have

$$\begin{cases} p_0(S) = p_0(S; M - e), \\ \qquad p_1(S) = p_1(S; M - e), \\ \qquad\qquad \text{if } e \notin S; \\ p_0(S) = p_0(S; M - e), \\ \qquad p_1(S) = p_1(S; M - e) + 1, \\ \qquad\qquad \text{if} e \in S, \end{cases}$$

the form (1.2.10) still leads to (1.2.7).

When e is an isthmus. Since for any S as mentioned above, we have

$$\begin{cases} p_0(S) = p_0(S; M \bullet e) + 1, \\ \qquad p_1(S) = p_1(S; M \bullet e), \\ \qquad\qquad \text{if } e \notin S; \\ p_0(S) = p_0(S; M \bullet e), \\ \qquad p_1(S) = p_1(S; M \bullet e), \\ \qquad\qquad \text{if } e \in S, \end{cases}$$

the form (1.2.10) leads (1.2.7) as well.

In consequence, from the well definedness suggested by Theorem 1.2.4 the theorem is obtained. □

From (1.2.9) the two undeterminates in the dichromatic polynomial do not have symmetry. However, whenever we introduce another polynomial

$$\chi(M;x,y) \;=\; \mu^{-1}Q(M;\mu,\nu), \tag{1.2.11}$$

it is easily verified that the two undeterminates $x=\mu+1$ and $y=\nu+1$ in χ have the symmetry.

In fact, from (1.2.7–9), we soon see that χ satisfies the following recursion:

$$\begin{cases} \chi(M) = \begin{cases} \chi(M-e)+\chi(M\bullet e), \\ \qquad \text{if } e \text{ is neither a loop} \\ \qquad\quad \text{nor an isthmus;} \\ x\chi(M\bullet e), \ \text{if } e \text{ is an isthmus;} \\ y\chi(M-e), \ \text{if } e \text{ is a loop;} \end{cases} \\ \chi(M)=1, \ \text{if } M \text{ is the vertex map;} \\ \chi(M)=\chi(M_1)\chi(M_2), \ \text{if } M=M_1\cup M_2 \\ \qquad \text{provided } M_1\cap M_2\in V, \ \text{a vertex.} \end{cases} \tag{1.2.12}$$

The polynomial χ is called the *dichromate* in Tutte's terminology, or *Tutte polynomial* in literature.

Suppose the edges in a map M are numbered $e_1, e_2, \cdots, e_\epsilon$. Let T be a spanning tree on M. If a tree edge e_i has the property that no cotree edge in the fundamental cocircuit formed in $\bar{T}+e_i$ has its suffix exceeding i, then e_i is said to be *internally active*. On the other hand, if a cotree edge e_j has the property that no tree edge in the fundamental circuit formed in $T+e_j$ has its suffix exceeding j, then e_j is said to be *externally active*. It is shown that the number of spanning trees T of M with given numbers $\mathrm{int}(T)$ and $\mathrm{ext}(T)$ of internally or externally active edges respectively is independent of the way of numbering edges.

Theorem 1.2.6 (Tutte, 1954) *The dichromate $\chi(M)$ of a map M has the following spanning tree expansion:*

$$\chi(M) = \sum_{\substack{0\leq i\leq\epsilon \\ 0\leq j\leq\epsilon}} \tau(M;i,j)x^i y^j \tag{1.2.13}$$

where $\tau(M;i,j)$ *is the number of spanning trees of* M *with* $\mathrm{int}(T)=i$ *and* $\mathrm{ext}(T)=j$, $0\le i\le\epsilon$, $0\le j\le\epsilon$.

Proof The crucial step is to show the case when e is neither a loop nor an isthmus because other cases can be easily obtained in a similar way. Let $T=T(M;i,j)$ be a spanning tree on M with $\mathrm{int}(T)$ $=i$ and $\mathrm{ext}(T)=j$. For e internally active in T, $M-e$ has exactly one

$$T_1=T(M-e;i,j)=T-e+e_l$$

corresponding to T where $l=\max\{s|e_s\in C_e^*,\ C_e^*$ is the fundamental cocircuit formed by $\bar{T}+e\}$. Moreover, $T\bullet e$ is a tree on $M\bullet e$ but not a $T(M\bullet e;i,j)$. If e is externally active for T, then $T-e$ is a tree on $M-e$ but not a $T(M-e;i,j)$ and $T\bullet e-e_t$ is the unique $T(M\bullet e;i,j)$ corresponding to T when $t=\max\{s|e_s\in C_e-e,\ C_e$ is the fundamental circuit formed by $T+e\}$. Therefore,

$$\tau(M;i,j)\ =\ \tau(M-e;i,j)\ +\ \tau(M\bullet e;i,j)$$

is satisfied. By virtue of (1.2.12), the theorem is soon obtained. □

The *chromatic polynomial*, denoted by $P(M)$, of a map M is defined by the recursions as follows:

$$\begin{cases} P(M)=\begin{cases} P(M-e)-P(M\bullet e), \\ \qquad\qquad \text{if } e \text{ is neither a loop} \\ \qquad\qquad\quad \text{nor an isthmus;} \\ (1-\lambda^{-1})P(M-e), \\ \qquad\qquad\quad \text{if } e \text{ is an isthmus;} \\ 0, \text{ if } e \text{ is a loop;} \end{cases} \\ P(M)=\lambda, \text{ if } M \text{ is the vertex map;} \\ P(M)=P(M_1)P(M_2), \text{ if } M \text{ is the} \\ \qquad\qquad \text{disjoint union of } M_1 \text{ and } M_2. \end{cases} \tag{1.2.14}$$

Theorem 1.2.7 (Whitney, 1932) *For any map M, its chromatic polynomial has the following expansion:*

$$P(M;\lambda) = \sum_{S}(-1)^{\alpha_1(S)}\lambda^{p_0(S)} \tag{1.2.15}$$

where S is over all spanning subgraphs of M, $\alpha_1(S)$ and $p_0(S)$ are the numbers of edges and components in S respectively.

Proof First, we see that for a loop map L,

$$P(L) = \lambda + (-1)^1\lambda = 0.$$

When M is the disjoint union of M_1 and M_2. For any spanning subgraph S of M, let S_1 and S_2 are the parts of S in M_1 and M_2 respectively. Since

$$\alpha_1(S) = \alpha_1(S_1) + \alpha_1(S_2);$$

$$p_0(S) = p_0(S_1) + p_0(S_2),$$

we have

$$\begin{aligned} P(M) &= \sum_{S_1}(-1)^{\alpha_1(S_1)}\lambda^{p_0(S_1)} \\ &\quad \times \sum_{S_2}(-1)^{\alpha_1(S_2)}\lambda^{p_0(S_2)} \\ &= P(M_1)P(M_2). \end{aligned}$$

From the two facts above, we soon find that for any map M with a loop, $P(M) = 0$.

Then, we prove that $P(M)$ defined by (1.2.15) satisfies the recursion (1.2.14) for e being not a loop. When e is an isthmus. Because all the spanning subgraphs can be divided into two parts. One is those without e, and the other those with e. The former is denoted by S_1, which is just a spanning subgraphs of $M - e$. The latter is denoted by S_2, which is a spanning subgraph of $M \bullet e$ when e is contracted. Since e is an isthmus we have

$$\alpha_1(S_2) = \alpha_1(S_2 - e) + 1;$$

$$p_0(S_2) = p_0(S_2 - e) - 1.$$

However, $S_2 - e$ is S_1 in this case. Hence, from (1.2.15), we have

$$P(M;\lambda) = \sum_{S_1}(-1)^{\alpha_1(S_1)}\lambda^{p_0(S_1)}$$

$$+\sum_{S_2}(-1)^{\alpha_1(S_2)}\lambda^{p_0(S_2)}$$

$$= (1-\lambda^{-1})\sum_{S_1}(-1)^{\alpha_1(S_1)}\lambda^{p_0(S_1)}$$

$$= P(M-e;\lambda).$$

When e is neither a loop nor an isthmus. The spanning subgraph S can also be distinguished to be S_1 or S_2 as mentioned above. From the relations:

$$\alpha_1(S_1) = \alpha_1(S_1 - e);$$

$$p_0(S_1) = p_0(S_1 - e);$$

$$\alpha_1(S_2) = \alpha_1(S_2 \bullet e) + 1;$$

$$p_0(S_2) = p_0(S_2 \bullet e),$$

we soon find that

$$P(M;\lambda) = \sum_{S_1}(-1)^{\alpha_1(S_1)}\lambda^{p_0(S_1)}$$

$$+(-1)\sum_{S_2\bullet e}(-1)^{\alpha_1(S_2\bullet e)}\lambda^{p_0(S_2\bullet e)}$$

$$= P(M-e;\lambda) - P(M\bullet e;\lambda).$$

Therefore, the polynomial $P(M;\lambda)$ defined by (1.2.15) is the chromatic polynomial of M given by (1.2.14). □

The *flow polynomial* of a map M denoted by $F(M)$ is defined by the following recursion:

$$\begin{cases} F(M) = \begin{cases} -F(M-e) + F(M \bullet e), & \text{if } e \text{ is neither a loop nor an isthmus;} \\ (\lambda - 1)F(M \bullet e), & \text{if } e \text{ is a loop;} \\ 0, & \text{if } e \text{ is an isthmus;} \end{cases} \\ F(M) = 1, \text{ if } M \text{ is the vertex map;} \\ F(M) = F(M_1)F(M_2), \text{ if } M \text{ is the edge disjoint union of } M_1 \text{ and } M_2 \text{ with at most one vertex in common.} \end{cases} \tag{1.2.16}$$

Theorem 1.2.8 *The flow polynomial of a map M has the following form:*

$$F(M;\lambda) = (-1)^{\alpha_1(M)} \sum_S (-1)^{\alpha_1(S)} \lambda^{p_1(S)} \tag{1.2.17}$$

where S is over all the spanning subgraphs of M, α_1 and p_1 are the edge number and the cyclomatic number respectively.

Proof First, we demonstrate that for the link map L

$$F(L;\lambda) = (-1)^1\left(1 + (-1)^1\right) = 0.$$

Of course, for the vertex map, it is easily checked. For any map M which is the edge disjoint union of M_1 and M_2 with at most one vertex in common, let S be a spanning subgraph of M with $S_1 = M_1 \cap S$ and $S_2 = M_2 \cap S$. Because

$$\alpha_1(M) = \alpha_1(M_1) + \alpha_1(M_2);$$

$$\alpha_1(S) = \alpha_1(S_1) + \alpha_1(S_2);$$

$$p_1(S) = p_1(S_1) + p_1(S_2),$$

from (1.2.17), we soon find that

$$F(M;\lambda) = (-1)^{\alpha_1(M_1)}\sum_{S_1}(-1)^{\alpha_1(S_1)}\lambda^{p_1(S_1)}$$

$$\times(-1)^{\alpha_1(M_2)}\sum_{S_2}(-1)^{\alpha_1(S_2)}\lambda^{p_1(S_2)}$$

$$= F(M_1;\lambda)F(M_2;\lambda).$$

From what is obtained, it is known that for M with an isthmus, $F(M;\lambda) = 0$.

Then, we discuss on M without isthmus. When e is a loop. Since

$$\alpha_1(M) = \alpha_1(M \bullet e) + 1;$$
$$\alpha_1(S_1) = \alpha_1(S_1 \bullet e),\ p_1(S_1) = p_1(S_1 \bullet e);$$
$$\alpha_1(S_2) = \alpha_1(S_2 \bullet e) + 1,\ p_1(S_2) = p_1(S_2 \bullet e) + 1,$$

where S_1 and S_2 are the respective spanning subgraphs without and with e, we have

$$F(M;\lambda) = (-1)(1-\lambda)F(M \bullet e;\lambda) = (\lambda - 1)F(M \bullet e;\lambda).$$

When e is neither a loop nor an isthmus. From the relations:

$$\alpha_1(M) = \alpha(M - e) + 1 = \alpha_1(M \bullet e) + 1;$$
$$\alpha_1(S_1) = \alpha_1(S_1 - e),\ p_1(S_1) = p_1(S_1 - e);$$
$$\alpha_1(S_2) = \alpha_1(S_2 \bullet e) + 1,\ p_1(S_2) = p_1(S_2 \bullet e),$$

the recursion (1.2.16) can be soon derived.

In consequence, according to the well definedness implied by Theorem 1.2.4, the theorem is proved. □

§1.3 Enufunctions

For a set $\mathcal{M}$ of maps, let $N = \{n_i | i \geq 1\}$ be a set of combinatorial invariants meant by that if M_1 and M_2 in $\mathcal{M}$ are isomorphic then $n_i(M_1) = n_i(M_2)$, $i \geq 1$. We may introduce a function as

$$g_{\mathcal{M}}(y_i | i \geq 1) = \sum_{M \in \mathcal{M}} \prod_{i \geq 1} y_i^{n_i(M)} \tag{1.3.1}$$

which is said to be the *enumerating function*, or in short, *enufunction* of $\mathcal{M}$ for N because it can be expressed as the form:

$$g_{\mathcal{M}}(y_i|i \geq 1) = \sum_{n_i \geq 0, i \geq 1} A_{\mathcal{M}}(n_i|i \geq 1) \prod_{i \geq 1} y_i^{n_i} \tag{1.3.2}$$

where $A_{\mathcal{M}}(n_i|i \geq 1)$ is the number of non-isomorphic maps $M \in \mathcal{M}$ with the conditions as

$$n_i(M) = n_i \geq 0, \ i \geq 1.$$

For two disjoint sets of maps $\mathcal{M}_1$ and $\mathcal{M}_2$, the *addition* of their enufunctions $g_{\mathcal{M}_1}(y_i|\ i \geq 1)$ and $g_{\mathcal{M}_2}(y_i|i \geq 1)$ is defined to be

$$(g_{\mathcal{M}_1} + g_{\mathcal{M}_2})(y_i|i \geq 1) = \sum_{n_i \geq 0, i \geq 1} (A_{\mathcal{M}_1}$$

$$+A_{\mathcal{M}_2})(n_i|i \geq 1) \prod_{i \geq 1} y_i^{n_i}. \tag{1.3.3}$$

Theorem 1.3.1 *For any two sets of maps $\mathcal{M}_1$ and $\mathcal{M}_2$ we have*

$$g_{\mathcal{M}}(y_i|i \geq 1) = (g_{\mathcal{M}_1} + g_{\mathcal{M}_2} - g_{\mathcal{M}_1 \cap \mathcal{M}_2})(y_i|i \geq 1) \tag{1.3.4}$$

where $\mathcal{M} = \mathcal{M}_1 \cup \mathcal{M}_2$.

Proof Because for $n_i \geq 0, \ i \geq 1$, given, it is easily shown that

$$A_{\mathcal{M}}(n_i|i \geq 1) = (A_{\mathcal{M}_1} + A_{\mathcal{M}_2} - A_{\mathcal{M}_1 \cap \mathcal{M}_2})(n_i|i \geq 1),$$

the theorem is soon obtained. □

Let $\mathcal{M}$ be a set of maps and $\mathcal{M}_1 \times \mathcal{M}_2$ be the *Cartesian product* of the two sets of maps $\mathcal{M}_1$ and $\mathcal{M}_2$, or in the form

$$\mathcal{M}_1 \times \mathcal{M}_2 = \{(M_1, M_2)|\forall M_1 \in \mathcal{M}_1, \forall M_2 \in \mathcal{M}_2\}. \tag{1.3.5}$$

If there exists a bijection $\tau: \ \mathcal{M} \to \mathcal{M}_1 \times \mathcal{M}_2$, then $\mathcal{M}$ is said to be *composed* by $\mathcal{M}_1$ and $\mathcal{M}_2$, or in other words, $\mathcal{M}$ is a *composition* of $\mathcal{M}_1$ and $\mathcal{M}_2$.

For two enufunctions $g_{\mathcal{M}_1}(y_i|i \geq 1)$ and $g_{\mathcal{M}_2}(y_i|i \geq 1)$, the *product* of them is defined by

$$(g_{\mathcal{M}_1}g_{\mathcal{M}_2})(y_i|i \geq 1) = \sum_{n_i, i\geq 1} A_{\mathcal{M}_1\mathcal{M}_2} \prod_{i\geq 1} y_i^{n_i} \tag{1.3.6}$$

where

$$A_{\mathcal{M}_1\mathcal{M}_2} = \sum_{\substack{n_i^1+n_i^2=n_i \\ n_i^1, n_i^2\geq 0, i\geq 1}} A_{\mathcal{M}_1}(n_i^1|i \geq 1)A_{\mathcal{M}_2}(n_i^2|i \geq 1) \tag{1.3.7}$$

for $n_i \geq 0,\ i \geq 1$.

Theorem 1.3.2 *For a set of maps $\mathcal{M}$,*

$$g_{\mathcal{M}}(y_i|i \geq 1) = g_{\mathcal{M}_1}g_{\mathcal{M}_2}(y_i|i \geq 1) \tag{1.3.8}$$

iff $\mathcal{M}$ is a composition of $\mathcal{M}_1$ and $\mathcal{M}_2$.

Proof From the definition of a composition of two sets of maps, we may see that the relation (1.3.7) is satisfied iff $\mathcal{M}$ is a composition of $\mathcal{M}_1$ and $\mathcal{M}_2$. This implies the theorem. □

If the invariants $n_i(M)$, $i \geq 1$, are treated as the number of vertices of valency i for $M \in \mathcal{M}$, then the enufunction $g_{\mathcal{M}}(y_i|i \geq 1)$ is said to have a *vertex partition*. Similarly, if $n_i(M)$, $i \geq 1$, is the number of faces of valency i, then the enufunction denoted by $f_{\mathcal{M}}(y_i|i \geq 1)$ is said to have a *face partition*.

In order to enumerate a set of maps $\mathcal{M}$, we should in general investigate the corresponding set of rooted maps $\mathcal{M}^{\mathrm{r}}$ each of whose elements is a map in $\mathcal{M}$ with a root as described in §1.1. As long as we find the number of non-isomorphic rooted maps with a given condition, the number of corresponding non-isomorphic non-rooted maps(or in brief, maps as usual) with the same condition can always be found, in principle, by considering the automorphism groups of the maps.

Let M^{r} be a map in a set of rooted maps $\mathcal{M}^{\mathrm{r}}$ with r being the root. The edge, vertex and face incident to r in M^{r} are denoted by e_r, v_r and f_r, *i.e.*, the *rooted edge*, *vertex* and *face* respectively. For a rooted map M^{r}, let $s(M^{\mathrm{r}})$ and $n_i(M^{\mathrm{r}})$, $i \geq 1$, be the respective valency of the rooted vertex and the number of non- rooted vertices of valency i.

In the same way we may write $t(M^{\mathrm{r}})$ and $m_j(M^{\mathrm{r}})$, $i \geq 1$, as the respective valency of the rooted face and the number of the non- rooted faces of valency j in M^{r}. Of course, all of them are combinatorial invariants on maps. We then have the following enumerating functions:

$$g^{\mathrm{r}}_{\mathcal{M}}(x; y_i|i \geq 1) = \sum_{M \in \mathcal{M}^{\mathrm{r}}} x^{s(M)} \prod_{i \geq 1} y_i^{n_i(M)};$$

$$f^{\mathrm{r}}_{\mathcal{M}}(x; y_j|j \geq 1) = \sum_{M \in \mathcal{M}^{\mathrm{r}}} x^{t(M)} \prod_{j \geq 1} y_j^{m_j(M)} \tag{1.3.9}$$

which are respectively called the *rooted enufunctions* with vertex; and with face partition.

For a set of rooted maps $\mathcal{N}$ the enufunction with vertex or face partition has three cases which are often met for the enumeration.

One is for all $y_i = y$, $i \geq 1$. That is

$$\begin{aligned} \Psi_{\mathcal{N}}(x, y) &= h_{\mathcal{N}}(x; y, y, \cdots) \\ &= \sum_{M \in \mathcal{N}} x^{m(M)} y^{n(M)}, \end{aligned} \tag{1.3.10}$$

where

$$n(M) = \begin{cases} \sum_{i \geq 1} n_i(M), \text{ if } h_{\mathcal{N}} = g^{\mathrm{r}}_{\mathcal{N}}; \\ \sum_{j \geq 1} m_j(M), \text{ if } h_{\mathcal{N}} = f^{\mathrm{r}}_{\mathcal{N}}, \end{cases} \tag{1.3.11}$$

in other words, $n(M)$ is the number of non-rooted vertices or non-rooted faces and $m(M)$ is the valency of the rooted vertex or the rooted face on M according as $h_{\mathcal{N}}$ is the enufunction $g^{\mathrm{r}}_{\mathcal{N}}$ with vertex partition or $f^{\mathrm{r}}_{\mathcal{N}}$ with face partition.

Another is for the substitution $y_i = y^i$, $i \geq 1$. That is

$$\begin{aligned} \Phi_{\mathcal{N}}(x, y) &= h_{\mathcal{N}}(x\sqrt{y}; \sqrt{y}, y, (\sqrt{y})^3, \cdots) \\ &= \sum_{M \in \mathcal{N}} x^{m(M)} y^{l(M)} \end{aligned} \tag{1.3.12}$$

where

$$l(M) = \begin{cases} \frac{1}{2}\Big(s(M) + \sum\limits_{i\geq 1} in_i(M)\Big), \\ \qquad\qquad \text{if } h_{\mathcal{N}} = g^{\mathrm{r}}_{\mathcal{N}}; \\ \frac{1}{2}\Big(t(M) + \sum\limits_{j\geq 1} jm_j(M)\Big), \\ \qquad\qquad \text{if } h_{\mathcal{N}} = f^{\mathrm{r}}_{\mathcal{N}} \end{cases} \tag{1.3.13}$$

is just the number of edges on M in the case with either vertex or face partition.

The other is for $y_i = y\delta_{ik}$, where δ_{ik} is the Kronecker delta, $i \geq 1$. That is

$$\begin{aligned} \Lambda^{(k)}_{\mathcal{N}}(x,y) &= h_{\mathcal{N}}(x; 0, \cdots, 0, y, 0, \cdots) \\ &= \sum_{M\in\mathcal{N}} x^{m(M)} y^{n(M)}, \end{aligned} \tag{1.3.14}$$

where $m(M) = s(M)$ or $t(M)$ as mentioned above, and $n(M)$ is the number of nonrooted vertices or nonrooted faces according as $h_{\mathcal{N}}$ is $g^{\mathrm{r}}_{\mathcal{N}}$ or $f^{\mathrm{r}}_{\mathcal{N}}$. Of course, all of them are k-valent. In this case M is called an almost k-regular map, because it is k-regular with the only possible exception of the rooted vertex or the rooted face.

§1.4 Polysum functions

For a set of maps $\mathcal{M}$ given, let $\Phi(M;\lambda)$ be a polynomial of λ for $M \in \mathcal{M}$. We may introduce a function

$$h_{\mathcal{M}}(\Phi; x|y_i, i \geq 1) = \sum_{M\in\mathcal{M}} \Phi(M) x^{m(M)} \prod_{i\geq 1} y_i^{l_i(m)}, \tag{1.4.1}$$

where $m(M)$ is the valency of the rooted vertex or the rooted face and $l_i(M)$ is the number of non-rooted vertices, or non-rooted faces according as the vertex partition or the face partition is considered. In any case, $\{m(M), l_i(M), i \geq 1\}$ can always be treated as a set of combinatorial invariants on maps in $\mathcal{M}$.

Here, $h_{\mathcal{M}}(\Phi) = h_{\mathcal{M}}(\Phi; x|y_i, i \geq 1)$ is called a Φ-*sum function*. When $\Phi = P(M)$ or $Q(M)$, the chromatic polynomial or the dichromatic polynomial on M, it is then called the *chrosum* or *dichrosum function*, respectively. Generally it is called the *polysum function*.

Of course, if the polynomial Φ is a constant, then the Φ-sum function becomes the enufunction $g^{\mathfrak{r}}_{\mathcal{M}}$ or $f^{\mathfrak{r}}_{\mathcal{M}}$ which appear in the last section.

The fundamental study for polysum functions on a set of maps is often transformed into those with one or two undeterminantes, in particular into those cases described on enufunctions as in §1.3.

Because polysum functions are all in the ring which consists of the power series with the coefficients from the integer ring or the polynomial ring, and those having a square root in the corresponding ring are specifically important for many cases of determining the polysum function or specially the enufunction for a given set of maps, we are asked to find the condition of a function having a square root in the same ring.

Let $\Re$ be any commutative ring. The rings of polynomials and power series, in an undeterminant x with the coefficients in $\Re$ are respectively denoted by $\Re[x]$ and $\Re\{x\}$. If more than one undeterminante is involved, the associativity to the left is always assumed. For instance, $\Re\{x\}[y]$ is the ring of polynomials in y whose coefficients are power series in $\Re\{x\}$.

Let $\mathcal{F}$ be a field of characteristic zero. By considering some basic principles of the summation form, we soon find the following relations:

$$\mathcal{F} \subset \left\langle \begin{array}{c} \mathcal{F}_1\langle x, y\rangle \\ \\ \mathcal{F}_2\langle x, y\rangle \end{array} \right\rangle \subset \mathcal{F}\{x\}\{y\} \tag{1.4.2}$$

where $\mathcal{F}_1\langle x, y\rangle; \mathcal{F}_2\langle x, y\rangle$ are respectively the following chains:

$$\begin{cases} \mathcal{F}[x] \subset \mathcal{F}\{x\} \subset \mathcal{F}\{x\}[y] \subset \mathcal{F}[x]\{y\}; \\ \mathcal{F}[y] \subset \mathcal{F}\{y\} \subset \mathcal{F}\{y\}[x] \subset \mathcal{F}[y]\{x\}. \end{cases}$$

In what follows, we usually write

$$\begin{cases} \mathcal{F}\{x\}[y] = \mathcal{F}\{x, y]; \\ \mathcal{F}[y]\{x\} = \mathcal{F}[y, x\}; \\ \mathcal{F}\{x\}\{y\} = \mathcal{F}\{x, y\}. \end{cases} \tag{1.4.3}$$

For $f \in \mathcal{F}\{x, y\}$ we denote the coefficient of $x^m y^n$ by

$$F(m, n) = [x^m y^n]f, \tag{1.4.4}$$

for $m, n \geq 0$.

Lemma 1.4.1 *Let $h \in 1 + y\mathcal{F}\{x,y\}$. Then,*
(i) $\exists h^{-1}, \sqrt{h} \in 1 + y\mathcal{F}\{x,y\}$;
(ii) h^{-1} *and* $\sqrt{h}$ *are both unique;*
(iii) $h \in 1 + xy\mathcal{F}[y,x\} \Rightarrow h^{-1}, \sqrt{h} \in 1 + xy\mathcal{F}[y,x\}$.

Proof Because $h \in 1 + y\mathcal{F}\{x,y\}$, we see that $h = 1 + yR, R \in \mathcal{F}\{x,y\}$. In fact, we have

$$h^{-1}(yR) = \sum_{i \geq 0} (-yR)^i;$$

$$\sqrt{h(yR)} = 1 + \sum_{i \geq 0} (-1)^i \frac{2^{-(2i+1)}(2i)!}{i!(i+1)!} (yR)^{i+1}, \tag{1.4.5}$$

based on the well definedness of the two power series which are both in $1 + y\mathcal{F}\{x,y\}$ in the way of comparing the coefficients of $x^i y^j, i, j \geq 0$, from the relation

$$\begin{cases} (1 + yR)h^{-1}(yR) = 1; \\ \left(\sqrt{h(yR)}\right)^2 = 1 + yR. \end{cases}$$

This is the first two statements of the lemma.

From $h \in 1 + xy\mathcal{F}[y,x\}$, h has the form $1 + xyR', R' \in \mathcal{F}[y,x\}$. Further, by (1.4.5) with respective xy and R' instead of y and R, we soon find that each coefficient of $x^i, i \geq 0$, in $h^{-1}(xyR')$ and $\sqrt{h(xyR')}$ is a finite sum of finite products of polynomials in $\mathcal{F}[y]$. This is the last statement of the lemma. □

Lemma 1.4.2 *Let $h \in 1 + y\mathcal{F}\{x,y]$. Then h has a factorization of the form*

$$h = \prod_{0 \leq i \leq k} \left((1 - a_i y)^{q_i} + xyH_i\right) \tag{1.4.6}$$

where k is a non-negative integer, $0 = a_0, a_1, \cdots, a_k$ are distinct elements of $\mathcal{F}, q_i$ is a positive integer and H_i is a polynomial in $\mathcal{F}\{x,y]$ of degree less than $q_i, i = 0, 1, \cdots, k$.

Proof It can be seen from a similar lemma in [Wae1]. □

Lemma 1.4.3 *Let $h \in 1 + y\mathcal{F}\{x, y]$ have a factorization* (1.4.6) *and let $g = (1 + xyH_0)^{-1}h$. Then the following statements are equivalent:*

(i) *h has a square root in $\mathcal{F}[y, x\}$;*
(ii) *g has a square root in $\mathcal{F}[y, x\}$;*
(iii) *g has a square root in $\mathcal{F}\{x, y]$;*
(iv) *h has a factorization of the form*

$$h = (1 + xyH_0) \prod_{1 \le i \le k} \left((1 - a_i y)^{r_i} + xyG_i\right)^2$$

where $k, H_0, a_1, \cdots, a_k$ satisfy the condition in Lemma 1.4.2, *r_i is a positive integer, G_i is a polynomial in $\mathcal{F}\{x, y]$ of degree less than $r_i, i = 1, 2, \cdots, k$, and if $k = 0$ the product Π is defined to be* 1.

Proof (iv)⇒(i). If $k = 0, g = 1$. It is obviously true. Assume $k > 0$. By Lemma 1.4.1, $1 + xyH_0$ has a square root in $1 + xy\mathcal{F}[y, x\}$ and hence it is also true.

(i)⇒(ii). Since the square root of $1 + xyH_0$ has an inverse in $1 + xy\mathcal{F}[y, x\}$ the assertion is obtained.

(ii)⇒(iii). Suppose $t^2 = g$ where $t \in \mathcal{F}[y, x\}$. Then we have

$$t(0, y)^2 = g(0, y) = \prod_{1 \le i \le k} (1 - a_i y)^{q_i}.$$

By the uniqueness of factorization in $\mathcal{F}$ and hence $\mathcal{F}[y]$, it follows that $r_i = \frac{q_i}{2}$ is an integer, $i = 1, 2, \cdots, k$. Thus,

$$t(0, y) = \pm \prod_{1 \le i \le k} (1 - a_i y)^{r_i}. \tag{1.4.7}$$

However, t must have the form $u + xy^{r+1}v$ where

$$r = \sum_{1 \le i < k} r_i, \quad u \in \mathcal{F}\{x, y]$$

is of degree r, and $v \in \mathcal{F}[y, x\}$. Let

$$v = \sum_{m \ge 0} V_m y^m$$

where $V_m \in \mathcal{F}\{x\}, m \ge 0$. Then

$$g = t^2 = u^2 + 2xy^{r+1}uv + x^2y^{2r+2}v^2. \tag{1.4.8}$$

By (1.4.7), the coefficient U_r of y^r in u is nonzero. Equating the coefficients of y^{2r+1} for U_r in (1.4.8), we have

$$0 = 0 + 2xU_rV_0 + 0.$$

Because $\mathcal{F}\{x\}$ is an integral domain, it is only possible that $V_0 = 0$. Further, by equating the coefficients of $y^l, l \geq 2r+2$, we may soon find that $V_m = 0, m \geq 1$. Therefore, g has a square root in $\mathcal{F}\{x, y]$.

(iii)⇒(iv). Since

$$t(x,0)^2 = g(x,0) = h(x,0) = 1$$

and $t = u \in \mathcal{F}\{x,y]$, by Lemma 1.4.2 we have that t must be in a factorization of the form

$$t = \pm \prod_{1\leq i\leq k} \left((1-a_iy)^{r_i} + xyG_i\right)$$

where $G_i \in \mathcal{F}\{x,y]$ is of degree less than $r_i, i = 1, 2, \cdots, k$, and hence (iv) is satisfied. □

Further, the following theorem provides a characterization of a polynomial $h \in \mathcal{F}\{x,y]$ being the square of a power series in $\mathcal{F}\{x,y\}$ or $\mathcal{F}[y,x\}$.

Theorem 1.4.1 (Brown, 1965) *Let $h \in \mathcal{F}\{x,y]$. Then, h is the square of a power series in $\mathcal{F}\{x,y\}$(or $\mathcal{F}[y,x\}$) iff it can be written as a product f^2g such that $f \in \mathcal{F}\{x,y]$(or $\mathcal{F}[y,x\}$) and $g \in 1 + y\mathcal{F}\{x,y]$(or $1 + xy\mathcal{F}\{x,y]$).*

Proof See [Bro1]. □

Now, let us investigate the equation of the following quadratic form:

$$z^2 + P_1z + P_2 = 0 \tag{1.4.9}$$

with $P_1, P_2 \in \mathcal{F}\{x,y]$ (or $\mathcal{F}\{x,y\}$) are known. And let us write

$$\Delta = P_1^2 - 4P_2, \tag{1.4.10}$$

which is called the *discriminant* of the equation.

Theorem 1.4.2 *Equation* (1.4.9) *has a solution* z *in* $\mathcal{F}\{x,y\}$ (or $\mathcal{F}[y,x\}$) *iff* Δ *has a square root in* $\mathcal{F}\{x,y\}$(or $\mathcal{F}[y, x\}$).

Proof Because the equation (1.4.9) can be rewritten as

$$\left(z+\tfrac{1}{2}P_1\right)^2+P_2-\tfrac{1}{4}P_1^2=0.$$

or equivalently,

$$(z+\tfrac{1}{2}P_1)^2=\tfrac{1}{4}\Delta. \tag{1.4.11}$$

This implies that $z \in \mathcal{F}\{x,y\}$ (or $\mathcal{F}[y,x\}$) iff

$$\sqrt{\Delta} \in \mathcal{F}\{x,y\}(\text{ or } \mathcal{F}[y,x\}).$$

The theorem is obtained. □

Combining the two theorems above, we soon find that the equation (1.4.9) has a solution $z \in \mathcal{F}\{x,y\}$(or $\mathcal{F}[y,x\}$) iff $\Delta = f^2g$ such that $f \in \mathcal{F}\{x,y]$(or $\mathcal{F}[y,x\}$) and $g \in 1+y\mathcal{F}\{x,y]$(or $1+xy\mathcal{F}\{x,y]$).

§1.5 The Lagrangian inversion

For a ring $\Re$ of characteristic zero, let $\mathcal{L}\{\Re;x\}$ be the ring of Laurent series, or in other words, $\forall f \in \mathcal{L}\{\Re;x\}$ the number of terms with x^i, $i<0$, is finite, and let $V(f)=\min\{i|[x^i]f \neq 0\}$. Usually, it is restricted to the case that $V(f) \geq 0$. Occasionally, to the general case that $V(f) < \infty$.

For the sake of brevity, let

$$\partial_x^i = \begin{cases} \dfrac{1}{i!}\dfrac{\mathrm{d}^i}{\mathrm{d}x^i}\Big|_{x=0}, & i \geq 0; \\ x^{-i}|_{x=\infty}, & i<0 \end{cases} \tag{1.5.1}$$

which is called the *coefficient operator*, or the *partial*. In other words, for $f \in \mathcal{L}\{\Re;x\}$, we have

$$\partial_x^i f = [x^i]f. \tag{1.5.2}$$

Lemma 1.5.1 *For any $f \in \mathcal{L}\{\Re; x\}$, $V(f) < \infty$, we have*

$$\partial_x^{-1}\left(\frac{\mathrm{d}f}{\mathrm{d}x}\right) = 0 \tag{1.5.3}$$

where ∂_x^{-1} is defined in (1.5.1) *for $i = -1$.*

Proof Because the derivative of any power of x is never x^{-1}, the lemma follows. □

Lemma 1.5.2 *For any $f, g \in \mathcal{F}\{x\}$, $V(f)$, $V(g) < \infty$, we have*

$$\partial_x^{-1}\left(\frac{\mathrm{d}f}{\mathrm{d}x}\right)g = -\partial_x^{-1} f\frac{\mathrm{d}g}{\mathrm{d}x}. \tag{1.5.4}$$

Proof Since $\frac{\mathrm{d}}{\mathrm{d}x}(fg) = \frac{\mathrm{d}f}{\mathrm{d}x}g + f\frac{\mathrm{d}g}{\mathrm{d}x}$, the lemma is obtained. □

Lemma 1.5.3 *For any $f, r \in \mathcal{L}\{\Re; x\}$, $V(f) = k < \infty$, $V(r) = \alpha > 0$, we have*

$$\alpha\partial_x^{-1} f = \partial_z^{-1}\left(f(r)\frac{\mathrm{d}r}{\mathrm{d}z}\right). \tag{1.5.5}$$

Proof It is only necessary to discuss the case that $f = x^n$. For this, we have

$$\begin{aligned}
\partial_z^{-1}\left(f(r)\frac{\mathrm{d}r}{\mathrm{d}z}\right) &= \partial_z^{-1}\big(r(z)\big)^n\frac{\mathrm{d}r}{\mathrm{d}z} \\
&= \begin{cases} \partial_z^{-1}\left((n+1)^{-1}\dfrac{\mathrm{d}r^{n+1}}{\mathrm{d}z}\right) = 0,\ n \neq -1, \\ \qquad \text{by Lemma 1.5.1;} \\ \alpha + \partial_z^{-1}\left(\dfrac{\mathrm{d}}{\mathrm{d}z}\log h\right) = \alpha,\ n = -1, \\ \qquad \text{by Lemma 1.5.1} \end{cases} \\
&= \alpha\partial_x^{-1}x^n = \alpha\partial_x^{-1} f
\end{aligned}$$

where $r(z) = z^\alpha h(z)$, $h(0) \neq 0$.

In general, from the linearity of the partial, the lemma is soon found. □

Theorem 1.5.1 *Suppose $x = t\phi(x)$, $\phi(x) \in \Re\{x\}$, $\phi(0) \neq 0$. Then for any $f \in \mathcal{L}\{\Re; x\}$, $V(f) \geq 0$, we have*

$$\partial_t^n f = \frac{1}{n}\partial_\lambda^{n-1}\left(\phi^n(\lambda)\frac{\mathrm{d}f}{\mathrm{d}\lambda}\right) \tag{1.5.6}$$

where ∂_t^n is as defined as in (1.5.1).

Proof Since $\phi(0) \neq 0$, $\phi^{-1}(x)$ exists and hence $t = x\phi^{-1}(x)$. However, we may see that

$$\partial_t^n f = \partial_t^{-1}(t^{-(n+1)}f)$$

by $V(t) = 1$ and Lemma 1.5.3,

$$= \partial_x^{-1}\left(t^{-(n+1)}f\frac{\mathrm{d}t}{\mathrm{d}x}\right)$$

$$= -\frac{1}{n}\partial_x^{-1}\left(f\frac{\mathrm{d}t^{-n}}{\mathrm{d}x}\right)$$

$$= -\frac{1}{n}\partial_x^{-1}\left(f\frac{\mathrm{d}}{\mathrm{d}x}(x^{-n}\phi^n)\right)$$

$$= \frac{1}{n}\partial_x^{n-1}\left(\phi^n\frac{\mathrm{d}f}{\mathrm{d}x}\right),$$

which implies (1.5.6). □

From the theorem the following corollaries deduced are often employed in the enumeration of maps.

Corollary 1.5.1 *If $x = t\phi(x)$, $\phi(0) \neq 0$, then for any function $f \in \mathcal{L}\{\Re; x\}$, $V(f) \geq 0$, we have*

$$f(x) = f(0) + \sum_{n\geq 1}\frac{t^n}{n!}\frac{\mathrm{d}^{n-1}}{\mathrm{d}x^{n-1}}\left(\phi^n\frac{\mathrm{d}f}{\mathrm{d}x}\right)\Big|_{x=0}. \tag{1.5.7}$$

Lemma 1.5.4 *Let $x = a + t\phi(x)$, $\phi(a) \neq 0$. Then for any function $f(x) \in \mathcal{L}\{\Re; x\}$, we have*

$$\partial_t^n f = \frac{1}{n}\partial_{x=a}^{n-1}\left(\phi^n \frac{\mathrm{d}f}{\mathrm{d}x}\right). \tag{1.5.8}$$

Proof Let $y = x - a$. Then, $y = t\phi(y+a)$. From (1.5.6), we find that

$$\begin{aligned}\partial_t^n f &= \frac{1}{n}\frac{\mathrm{d}^{n-1}}{\mathrm{d}y^{n-1}}\left(\phi^n(y+a)\frac{\mathrm{d}f}{\mathrm{d}y}\right)\bigg|_{y=0} \\ &= \frac{1}{n}\frac{\mathrm{d}^{n-1}}{\mathrm{d}x^{n-1}}\left(\phi^n(x)\frac{\mathrm{d}f}{\mathrm{d}x}\right)\bigg|_{x=a}.\end{aligned}$$

This is the formula (1.5.8). □

Corollary 1.5.2 *Let $x = a + t\phi(x)$, $\phi(a) \neq 0$. Then, we have*

$$f(x) = f(a) + \sum_{n\geq 1}\frac{t^n}{n!}\frac{\mathrm{d}^{n-1}}{\mathrm{d}x^{n-1}}\left(\phi^n \frac{\mathrm{d}f}{\mathrm{d}x}\right)\bigg|_{x=a} \tag{1.5.9}$$

for any function $f(x) \in \mathcal{L}\{\Re; x\}$.

In order to generalize those mentioned above for the multivariate case, we have to explain some notations. Let

$$\underline{x} = (x_1, x_2, \cdots)$$

be the vector each of whose components is an undeterminante and

$$\underline{k} = (k_1, k_2, \cdots)$$

be the powers of $x_1, x_2, \cdots$ in the monoid $\underline{x}^{\underline{k}}$. For $f(\underline{x})$, $g_1(\underline{x})$, $\cdots$, $g_m(\underline{x}) \in \mathcal{L}\{\Re; \underline{x}\}$, let $V(g_i) = (V(g_i, x_1), \cdots, V(g_i, x_m))$ where

$$V(g_i, x_j) = \min\{l | \partial_{x_j}^l g_i \neq 0\}, \tag{1.5.10}$$

$j = 1, 2, \cdots, m$. Further, we are allowed to employ

$$\partial_{\underline{x}}^{\underline{i}} = \begin{cases} \dfrac{1}{\underline{i}!}\dfrac{\mathrm{d}^{\underline{i}}}{\mathrm{d}\underline{x}^{\underline{i}}}\bigg|_{\underline{x}=0}, & \underline{i} \geq 0; \\ \\ \underline{x}^{-\underline{i}}\big|_{\underline{x}=\infty}, & \underline{i} < 0 \end{cases} \tag{1.5.11}$$

where $\underline{i}! = i_1! i_2! \cdots$ and

$$\frac{\mathrm{d}^{\underline{i}}}{\mathrm{d}\underline{x}^{\underline{i}}} = \frac{\partial^{i_1}}{\partial x_1^{i_1}} \frac{\partial^{i_2}}{\partial x_2{}^{i_2}} \cdots$$

whence $\underline{i} = (i_1, i_2, \cdots)$. From the conventions above, the coefficient operator is still allowed to be written in the form

$$\partial_{\underline{x}}^{\underline{i}} f = [\underline{x}^{\underline{i}}] f \tag{1.5.12}$$

for $f \in \mathcal{L}\{\Re; \underline{x}\}$.

For $\underline{g}(\underline{z}) \in \mathcal{L}^m\{\Re; \underline{z}\}$ where $\underline{g} = (g_1, g_2, \cdots, g_m)$ and $\underline{z} = (z_1, z_2, \cdots, z_m)$, the determinant

$$\det\left(\frac{\partial \underline{g}}{\partial \underline{z}}\right) = \det\left(\frac{\partial g_i}{\partial z_j}\right)_{1 \leq i,j < m}$$

is said to be the *Jacobian* of $\underline{g}$, which is denoted by $J(\underline{g})$.

Lemma 1.5.5 *For a function $g(\underline{z}) \in \mathcal{L}\{\Re; \underline{z}\}$, let $V(g) = (p_1, p_2, \cdots, p_m) = \underline{p}$ and let $g(\underline{z}) = a\underline{z}^{\underline{p}} h(\underline{z})$ such that $h(\underline{0}) = 1$ and $a \neq 0$. Then, we have*

$$g^k(\underline{z}) \frac{\partial}{\partial z_j} g(\underline{z}) = \begin{cases} \dfrac{\partial}{\partial z_j}\left(\dfrac{1}{k+1} g^{k+1}(\underline{z})\right), & \text{if } k \neq -1; \\[2ex] \dfrac{p_i}{z_j} + \dfrac{\partial}{\partial z_j} \log h(\underline{z}), & \text{if } k = -1 \end{cases} \tag{1.5.13}$$

for any integer k.

Proof When $k \neq -1$. It is easy to see that

$$\begin{aligned} \frac{\partial}{\partial z_j}\left(\frac{1}{k+1} g^{k+1}(\underline{z})\right) &= \frac{1}{k+1} \frac{\partial}{\partial z_j} g^{k+1}(\underline{z}) \\ &= g^k(\underline{z}) \frac{\partial}{\partial z_j} g(\underline{z}). \end{aligned}$$

Otherwise, *i.e.*, $k = -1$. Because

$$g^{-1}(\underline{z})\frac{\partial}{\partial z_j}g(\underline{z}) = a^{-1}\underline{z}^{-\underline{p}}h^{-1}(\underline{z})\frac{\partial}{\partial z_j}a\underline{z}^{\underline{p}}h(\underline{z})$$

$$= \underline{z}^{-\underline{p}}h^{-1}(\underline{z})\left(\underline{z}^{\underline{p}}\frac{p_j}{z_j}h(\underline{z}) + \underline{z}^{\underline{p}}\frac{\partial h}{\partial z_j}\right)$$

$$= \frac{p_j}{z_j} + h^{-1}\frac{\partial h}{\partial z_j}$$

$$= \frac{p_j}{z_j} + \frac{\partial}{\partial z_j}\log h,$$

the lemma is soon obtained. □

Lemma 1.5.6 *Let* $\underline{k}$, $\underline{l}_i, \underline{p}_i$, $1 \le i \le m$, *be all integral* m*-vectors. For the determinant*

$$G_{\underline{k}}(\underline{z}) = \det \Delta_1,$$

$$\Delta_1 = \left(\frac{p_{ij}}{z_j}\delta_{k_i,-1} + \sum_{\underline{l}_i}\frac{l_{ij}}{z_j}a_i(\underline{l}_i, k_i)\underline{z}^{\underline{l}_i}\right), \tag{1.5.14}$$

we have the following expansion as

$$G_{\underline{k}}(\underline{z}) = \sum_{0\le t\le m}\sum_{\aleph_t\subseteq N_m}\sum_{\underline{l}_{\alpha_1},\cdots,\underline{l}_{\alpha_t}}\Delta_2,$$

$$\Delta_2 = G(t; P, L, \underline{k})\underline{z}^{\underline{l}_{\alpha_1}+\cdots,+\underline{l}_{\alpha_t}-\underline{1}}, \tag{1.5.15}$$

where $P = (p_{ij})$, $L = (l_{ij})$, *and*

$$G(t; P, L, \underline{k}) = (\det B)\prod_{s=1}^{t} a_{\alpha_s}(\underline{l}_{\alpha_s}, k_{\alpha_s}) \tag{1.5.16}$$

in which $B = (b_{ij})$,

$$b_{ij} = \begin{cases} p_{ij}\delta_{k_i,-1}, & \text{if } i \in N_m - \aleph; \\ l_{ij}, & \text{if } i \in \aleph \end{cases} \tag{1.5.17}$$

with $N_m = \{1, 2, \cdots, m\}$ *and* $\aleph = \aleph_t = \{\alpha_1, \alpha_2, \cdots, \alpha_t\}$.

Proof From the multilinearity on each row of $G_{\underline{k}}(\underline{z})$, we may write

$$G_{\underline{k}}(\underline{z}) = \sum_{\aleph \subseteq N_m} \det\left(A_{ij}(\alpha)\right)$$

where

$$A_{ij}(\alpha) = \begin{cases} \dfrac{p_{ij}}{z_j}\delta_{k_i,-1}, & \text{if } i \in N_m - \aleph; \\ \displaystyle\sum_{\underline{l}_i} \frac{l_{ij}}{z_j} a_i(\underline{l}_i, k_i)\underline{z}^{\underline{l}_i}, & \text{if } i \in \aleph. \end{cases}$$

Further, by the multilinearity to expand on row i, for each $i \in \aleph$, we find that

$$G_{\underline{k}}(\underline{z}) = \sum_{t=0}^{m} \sum_{\aleph_t \subseteq N_m} \sum_{\underline{l}_{\alpha_1}, \cdots, \underline{l}_{\alpha_t}} \underline{z}^{\underline{l}_{\alpha_1} + \cdots + \underline{l}_{\alpha_t} - \underline{1}} \times (\det B) \prod_{s=1}^{t} a_{\alpha_s}(\underline{l}_{\alpha_s}, k_{\alpha_s})$$

with B defined by (1.5.17). That is (1.5.15). □

Lemma 1.5.7 *Let* $f(\underline{x}) \in \mathcal{L}\{\Re; \underline{x}\}$, $\underline{g}(\underline{x}) \in \mathcal{L}^m\{\Re; \underline{x}\}$, *and let*

$$V(g_i) = \underline{p}_i = (p_{i1}, \cdots, p_{im}) \geq \underline{0}$$

be finite with $\underline{p}_i \neq \underline{0}$ *for* $1 \leq i \leq m$. *Then we have*

$$\det P[\underline{x}^{-\underline{1}}]f(\underline{x}) = [\underline{z}^{-\underline{1}}]\left(f(\underline{g})J(\underline{g})\right) \tag{1.5.18}$$

where $J(\underline{g})$ *is the Jacobian of* g.

Proof Because $V(g_i) = \underline{p}_i \geq \underline{0}$ and $\underline{p}_i \neq 0$, for $1 \leq i \leq m$, we know that $f(\underline{g}(\underline{x}))$ exists in $\mathcal{L}\{\Re; \underline{x}\}$. Let $f(\underline{x}) = \sum_{\underline{k}} F(\underline{k})\underline{x}^{\underline{k}} \in \mathcal{L}\{\Re; \underline{x}\}$. Then,

$$\begin{aligned} [\underline{z}^{-\underline{1}}]\left(f(\underline{g})J(\underline{g})\right) &= [\underline{z}^{-\underline{1}}]\sum_{\underline{k}} F(\underline{k})\underline{g}^{\underline{k}}(\underline{z})J(\underline{g}) \\ &= \sum_{\underline{k}} F(\underline{k})[\underline{z}^{-\underline{1}}]G_{\underline{k}}(\underline{z}), \end{aligned} \tag{1.5.19}$$

where

$$G_{\underline{k}}(\underline{z}) = \det\left(g_i^{k_i}(\underline{z})\frac{\partial}{\partial z_j}g_i(\underline{z})\right).$$

Let $g_i(\underline{z}) = c_i\underline{z}^{\underline{p}_i}h_i(\underline{z})$ such that $h_i(\underline{0}) = 1$ and $c_i \neq 0$, for $i = 1, 2, \cdots, m$. Thus from Lemma 1.5.5, we find that $G_{\underline{k}}(\underline{z})$ has the form as (1.5.14) such that

$$\sum_{\underline{l}_i} a_i(\underline{l}_i, k_i)\underline{z}^{\underline{l}_i} = \begin{cases} \dfrac{1}{k_i+1}g_i^{k_i+1}(\underline{z}), & k_i \neq -1; \\ \\ \log h_i(\underline{z}), & k_i = -1 \end{cases}$$

and $\underline{l}_i = (l_{i1}, \cdots, l_{im})$. By Lemma 1.5.6, we see that for a nonzero contribution to $[\underline{z}^{\underline{1}}]G_{\underline{k}}(\underline{z})$, $\underline{l}_{\alpha_1} + \underline{l}_{\alpha_2} + \cdots + \underline{l}_{\alpha_t} = \underline{0}$ with the only possible exception of $t = 0$. However, if $t = 0$,then

$$\begin{aligned} \det B &= (\det P)\prod_{1\leq s\leq m}\delta_{k_s,-1} \\ &= [\underline{z}^{-\underline{1}}]G_{\underline{k}}(\underline{z}). \end{aligned}$$

By(1.5.19) the lemma follows. □

Based on the Lemmas 1.5.5–7, we are allowed to derive the Lagrangian inversion for multivariables.

Theorem 1.5.2 *Let $f \in \mathcal{L}\{\Re; \underline{x}\}$ and $\phi_1(\underline{x}), \phi_2(\underline{x}), \cdots, \phi_m(\underline{x}) \in \Re\{\underline{x}\}$, $\phi_i(\underline{0}) \neq 0$, $i = 1, \cdots, m$. Suppose that $x_i = t_i\phi_i(\underline{x}), i = 1, 2, \cdots, m$. Then, we have*

$$\partial_{\underline{t}}^{\underline{k}} f = \partial_{\underline{x}}^{\underline{k}}\left(f(\underline{x})\underline{\phi}^{\underline{k}}(\underline{x})\Delta(\underline{x})\right) \tag{1.5.20}$$

where

$$\Delta(\underline{x}) = \det\left(\delta_{ij} - \frac{x_j}{\phi_i(\underline{x})}\frac{\partial\phi_i}{\partial x_j}\right). \tag{1.5.21}$$

Proof By an argument similar to that for the univariate, there exists a unique series $x_i(\underline{t}) \in \Re\{\underline{t}\}$ such that $x_i = t_i\phi(\underline{x})$ for $i = 1, 2, \cdots, m$. Because $[\underline{t}^{\underline{k}}]f(\underline{x}) = [\underline{t}^{-\underline{1}}]\underline{t}^{-(\underline{k}-\underline{1})}f(\underline{x})$, by substituting $t_i = x_i\phi_i^{-1}(\underline{x})$ we have, from Lemma 1.5.7 in which

$$\det P = \det(V(t_1)^{\mathrm{T}}, \cdots, V(t_m)^{\mathrm{T}})$$

$$= \det(I_m) = 1$$

(where "T" represents the transpose) is considered here, that

$$\partial_{\underline{t}}^{\underline{k}} f = [\underline{x}^{-\underline{1}}] f(\underline{x}) \underline{\phi}^{\underline{k}+\underline{1}} \underline{x}^{-(\underline{k}+\underline{1})} J(\underline{t}).$$

However

$$J(\underline{t}) = \det \left(\delta_{ij} \phi_i^{-1}(\underline{x}) - x_i \phi_i^{-2}(\underline{x}) \frac{\partial \phi_i}{\partial x_j} \right)$$

by extracting the factor ϕ_i^{-1} from row i for $i = 1, 2, \cdots, m$,

$$= \underline{\phi}^{-\underline{1}}(\underline{x}) \det \left(\delta_{ij} - x_j \phi_i^{-1}(\underline{x}) \frac{\partial \phi_i}{\partial x_j} \right). \tag{1.5.22}$$

By combining (1.5.21) and (1.5.22), we soon find that

$$\begin{aligned} \partial_{\underline{t}}^{\underline{k}} f &= [\underline{x}^{\underline{k}}] f(\underline{x}) \underline{\phi}^{\underline{k}} \det \left(\delta_{ij} - x_j \phi_i^{-1}(\underline{x}) \frac{\partial \phi_i}{\partial x_j} \right) \\ &= \partial_{\underline{x}}^{\underline{k}} \left(f(\underline{x}) \underline{\phi}^{\underline{k}} \Delta(\underline{x}) \right). \end{aligned}$$

This is the conclusion of the theorem. □

Naturally, by simply specifying the theorem to the univariate case, Theorem 1.5.1 is just treated as a corollary. However, we have isolated the result of Theorem 1.5.1 for clarity and ease of reference.

Corollary 1.5.3 *Let $g(\underline{x}) \in \Re\{\underline{x}\}$, and $\phi_i(\underline{x}) \in \Re\{\underline{x}\}, \phi_i(\underline{0}) \neq 0$ for $i = 1, 2, , \cdots, m$, where $\underline{x} = (x_1, \cdots, x_m)$. Suppose $x_i = t_i \phi_i(\underline{x})$, $1 \leq i \leq m$. Then, we have*

$$\partial_{\underline{t}}^{\underline{k}} \frac{g(\underline{x})}{\Delta(\underline{x})} = \partial_{\underline{x}}^{\underline{k}} g(\underline{x}) \underline{\phi}^{\underline{k}}(\underline{x}) \tag{1.5.23}$$

where $\Delta(\underline{x})$ is the same appearing in (1.5.21).

Proof Because for $t_i = x_i \phi_i^{-1}(\underline{x})$, $1 \leq i \leq m$,

$$\begin{aligned} \Delta(\underline{x}) &= \det \left(\delta_{ij} - x_j \phi_i^{-1}(\underline{x}) \frac{\partial \phi_i}{\partial x_j} \right) \\ &= \det \left(\delta_{ij} - x_i \phi_i^{-1}(\underline{x}) \frac{\partial \phi_i}{\partial x_j} \right), \end{aligned}$$

which is not zero, the function $f(\underline{x}) = g(\underline{x})\Delta^{-1}(\underline{x})$ is well defined. Then, $f(\underline{x}) \in \Re\{\underline{x}\} \subset \mathcal{L}\{\Re; \underline{x}\}$. From Theorem 1.5.2 the corollary is soon obtained. □

§1.6 The shadow functional

Let $f(\underline{y}) \in \Re\{\underline{y}\}$ where $\underline{y} = (y_1, y_2, \cdots)$ be a function. Of course,

$$V(f, y_i) \geq 0, \ i = 1, 2, \cdots.$$

We may introduce a transformation $\int_y : y^i \mapsto y_i$ for $i = 1, 2, \cdots$, with the convention that $y^0 = 1 \mapsto y_0$.

Since $\int_y$ is a functional from function space $\mathbf{F}$ with $\{1, y^1, y^2, \cdots\}$ as a basis to vector space $\mathbf{V}$ with $\{y_0, y_1, y_2, \cdots\}$ as a basis, it may be called the *shadow functional.*

For any

$$v_i = \sum_{j \geq 0} a_{ij} y^j \in \mathbf{V}, \ i = 1, 2,$$

it is easily seen that

$$\begin{aligned} \int_y (v_1 + v_2) &= \sum_{j \geq 0} (a_{1j} + a_{2j}) \int_y y^j \\ &= \sum_{j \geq 0} a_{1j} y_j + \sum_{j \geq 0} a_{2j} y_j \\ &= \int_y v_1 + \int_y v_2 \end{aligned}$$

and hence the shadow functional is a linear transformation.

The *inverse* of the shadow functional $\int_y$ is defined to be $\int_y^{-1} : y_j \mapsto y^j$, for $j = 1, 2, \cdots$ with the convention that

$$\int_y^{-1} y_0 = y^0 = 1.$$

Occasionally, we are also allowed to take $y_0 = 1$ in which 1 is, however, treated as a specific vector in $\mathbf{V}$.

Now, two linear operators called the *left* and the *right projections* and denoted by $\mathcal{L}_y$ and $\mathcal{R}_y$ on the space $\mathbf{V}$ with the basis $\{y_0, y_1, y_2, \cdots\}$ are respectively defined in what follows.

Let us write

$$v = \sum_{j \geq 0} a_j y_j \in \mathbf{V}.$$

Then,

$$\begin{cases} \mathcal{L}_y v = \sum\limits_{j \geq 0} (j+1) a_{j+1} y_j; \\ \mathcal{R}_y v = \sum\limits_{j \geq 1} \dfrac{1}{j} a_{j-1} y_j. \end{cases} \tag{1.6.1}$$

In other words, if y_i is treated as the vector of all components being zero with the only exception of the i- th which is y_i, $i \geq 0$, then the matrices corresponding to $\mathcal{L}_y$ and $\mathcal{R}_y$ are the respective matrices, denoted by L and R, which are

$$L = \begin{pmatrix} 0 & 1 & 0 & \cdot & \cdot & \cdot & \cdot & \cdot & 0 \\ \cdot & 0 & 2 & \cdot & \cdot & \cdot & \cdot & \cdot & \cdot \\ \cdot & \cdot & 0 & \cdot & \cdot & \cdot & \cdot & \cdot & \cdot \\ \cdot & \cdot & \cdot & \cdot & \cdot & \cdot & \cdot & \cdot & \cdot \\ \cdot & \cdot & \cdot & \cdot & \cdot & \cdot & 0 & \cdot & \cdot \\ \cdot & \cdot & \cdot & \cdot & \cdot & \cdot & k & \cdot & \cdot \\ \cdot & \cdot & \cdot & \cdot & \cdot & \cdot & 0 & \cdot & \cdot \\ \cdot & \cdot & \cdot & \cdot & \cdot & \cdot & \cdot & \cdot & \cdot \\ 0 & \cdot & \cdot & \cdot & \cdot & \cdot & \cdot & \cdot & \cdot \end{pmatrix},$$

$$R = \begin{pmatrix} 0 & \cdot & \cdot & \cdot & \cdot & \cdot & \cdot & \cdot & 0 \\ 1 & 0 & \cdot & \cdot & \cdot & \cdot & \cdot & \cdot & \cdot \\ 0 & \frac{1}{2} & 0 & \cdot & \cdot & \cdot & \cdot & \cdot & \cdot \\ \cdot & \cdot & \cdot & \cdot & \cdot & \cdot & \cdot & \cdot & \cdot \\ \cdot & \cdot & \cdot & \cdot & \cdot & \cdot & \cdot & \cdot & \cdot \\ \cdot & \cdot & \cdot & \cdot & \cdot & \cdot & \cdot & \cdot & \cdot \\ \cdot & \cdot & \cdot & \cdot & 0 & \frac{1}{k} & 0 & \cdot & \cdot \\ \cdot & \cdot & \cdot & \cdot & \cdot & \cdot & \cdot & \cdot & \cdot \\ 0 & \cdot & \cdot & \cdot & \cdot & \cdot & \cdot & \cdot & \cdot \end{pmatrix}. \tag{1.6.2}$$

It is easy to check that

$$LR = \begin{pmatrix} I & \underline{0}^{\mathrm{T}} \\ \underline{0} & 0 \end{pmatrix}, \quad RL = \begin{pmatrix} 0 & \underline{0} \\ \underline{0}^{\mathrm{T}} & I \end{pmatrix} \tag{1.6.3}$$

in which, I is the unity matrix and "T" is the transpose.

Theorem 1.6.1 *For any* $v = v(y_0, y_1, y_2, \cdots) \in \mathbf{V}$, *let* $f(y) = \int_y^{-1} v$. *Then we have*

$$\begin{cases} \dfrac{\mathrm{d}}{\mathrm{d}y} f(y) = \displaystyle\int_y^{-1} \mathcal{L}_y v; \\ \displaystyle\int f(y)\mathrm{d}y = \int_y^{-1} \mathcal{R}_y v. \end{cases} \tag{1.6.4}$$

Proof By comparing the coefficients of the same terms on the two sides in each formula of (1.6.4), the theorem is soon found. □

For a function $f(\underline{x}, \underline{y})$ of two kinds of variables, suppose $f(\underline{x}, \underline{y}) \in \mathbf{V}\{\underline{x}, \underline{y}\}$, a bilinear space. It is then easily verified that

$$\int_x^{-1} \int_y^{-1} f(\underline{x}, \underline{y}) = \int_y^{-1} \int_x^{-1} f(\underline{x}, \underline{y}). \tag{1.6.5}$$

This allows us to denote by $F(x, y)$ the function determined from (1.6.5). Conversely, for $F(x, y) \in \Re\{x, y\}$, we also have

$$f(\underline{x}, \underline{y}) = \int_x \int_y F(x, y) \tag{1.6.6}$$

since $\int_x$ and $\int_y$ are commutative as well.

Let $f(z) \in \Re\{z\}$. Then, we may introduce two operators on f for z, which are fundamental for the map enumerations. Let $f = f(z)$. Then, one is defined by

$$\delta_{x,y} f = \frac{f(x) - f(y)}{x - y} \tag{1.6.7}$$

which is called the (x, y)-*difference* of f. The other is defined by

$$\partial_{x,y} f = \frac{y f(x) - x f(y)}{x - y} \tag{1.6.8}$$

which is called the $\langle x, y\rangle$-*difference*.

Theorem 1.6.2 *For any function* $f(z) \in \Re\{z\}$, *let* $f = f(z)$. *Then, we have*

$$\partial_{x,y}(zf) = xy\delta_{x,y} f. \tag{1.6.9}$$

Proof Because of the linearity of the two operators $\partial_{x,y}$ and $\delta_{x,y}$, it is only necessary to discuss the case $f(z) = z^n$, $n > 0$. However, for this case, it is easily seen that

$$\begin{aligned}\partial_{x,y} zf &= \partial_{x,y} z^{n+1} \\ &= \frac{yx^{n+1} - xy^{n+1}}{x - y} \\ &= xy\frac{x^n - y^n}{x - y} \\ &= xy\delta_{x,y} z^n \\ &= xy\delta_{x,y} f.\end{aligned}$$

This is the theorem. □

If the function f is treated as the enufunction $h_{\mathcal{N}}(x, \underline{y})$, $\underline{y} = (y_1, y_2, \cdots)$ defined in §1.3 for a set $\mathcal{N}$ of rooted maps, we may, in difference from the inverse, write

$$\int_{\underline{y}}^{-1} h_{\mathcal{N}}(x, \underline{y}) = \Phi_{\mathcal{N}}(x, y) \tag{1.6.10}$$

which is defined by (1.3.12).

§1.7 Asymptotic estimation

Many known results for the estimation of some values of special functions can be employed to investigate the asymptotic behavior in the map enumeration.

Only two kinds of the most important special functions for us here are briefly mentioned. One of them is the gamma function $\Gamma(z)$, $z \in \mathbf{C}$, the complex field. That is

$$\Gamma(z) = \int_0^{\infty} t^{z-1} e^{-t} dt \tag{1.7.1}$$

for z which has its real part $\mathrm{Re}(z) > 0$. The form is said to be in the *Euler's integral.*

Because it can be shown that the gamma function satisfies the recursion

$$\Gamma(z + 1) = z\Gamma(z) \tag{1.7.2}$$

with $\Gamma(1) = 1$, which is checked by (1.7.1), we may soon find that

$$\Gamma(z+1) = z!\,. \tag{1.7.3}$$

If z is a real number $x > 0$, then

$$\Gamma(x) = \sqrt{2\pi}\, x^{x+\frac{1}{2}} \exp\left(-x + \frac{\theta}{12x}\right) \tag{1.7.4}$$

where $0 < \theta < 1$.

The following asymptotic estimation of the value of gamma function is usually called *Stirling's formula.* That is

$$\Gamma(az+b) \sim \sqrt{2\pi}\, e^{-az}(az)^{az+b-\frac{1}{2}} \tag{1.7.5}$$

for Arg $z < \pi$ and both of a and b are real numbers with $a > 0$. From (1.7.5), we soon find that

$$n! \sim \sqrt{2\pi} e^{-n} n^{n-\frac{1}{2}} \tag{1.7.6}$$

which is often applied in the asymptotic estimation.

The other is called the *hypergeometric function*, which is defined by

$$\begin{aligned} F(a,b;c;z) &= {}_2F_1(a,b;c;z) \\ &= \frac{\Gamma(c)}{\Gamma(a)\Gamma(b)} \sum_{n\geq 0} \frac{\Gamma(a+n)\Gamma(b+n)}{\Gamma(c+n)} \frac{z^n}{n!} \end{aligned} \tag{1.7.7}$$

for complex numbers a, b, c and z with $|z| = 1$ as the circle of convergence.

From (1.7.7), we see that

$$F(a,b;c;z) = F(b,a;c;z). \tag{1.7.8}$$

And further, the behavior of this series on its circle of convergence is as follows:

1. Divergence when $\mathrm{Re}(c-a-b) \leq -1$;

2. Absolute convergence when $\mathrm{Re}(c-a-b) > 0$;

3. Conditional convergence when $-1 < \mathrm{Re}(c-a-b) \leq 0$ with the point $z = 1$ is excluded.

Two cases have to be notified.

A. When a or b is equal to $-n, n = 0, 1, 2, \cdots$, the series (1.7.7) becomes a polynomial of degree n in z. More concretely, we have

$$F(-m, b; c; z) = \sum_{0 \le n \le m} \frac{(-m)_n (b)_n}{(c)_n} \frac{z^n}{n!}, \tag{1.7.9}$$

where $(x)_n = x(x+1)\cdots(x+n-1)$, for $x = -m, b$, or c.

B. When c is equal to $-m, m = 0, 1, 2, \cdots$, provided a or b is not a negative integer n with $n < m$, the series (1.7.7) is not defined. And, for $c = -m$, we have

$$\lim_{c \to -m} \frac{1}{\Gamma(c)} F(a, b; c; z) = \frac{(a)_{m+1}(b)_{m+1}}{(m+1)!}$$
$$\times z^{m+1} F(a+m+1, b+m+1; m+2; z). \tag{1.7.10}$$

Now, we list some elementary cases of the Gauss series.

$$\begin{cases} F(1, 1; 2; z) = z^{-1} \ln(1-z); \\ F(\frac{1}{2}, 1; \frac{3}{2}; z^2) = \frac{1}{2} z^{-1} \ln\left(\frac{1+z}{1-z}\right); \\ F(a, b; b; z) = (1-z)^{-a}; \\ F(-a, a; \frac{1}{2}; \sin^2 z) = \cos(2az). \end{cases} \tag{1.7.11}$$

For some special values of z, the series has simpler forms which are easily applied.

$$F(a, b; c; 1) = \frac{\Gamma(c)\Gamma(c-a-b)}{\Gamma(c-a)\Gamma(c-b)} \tag{1.7.12}$$

when $c \ne 0, -1, -2, \cdots$, and $\mathrm{Re}(c-a-b) > 0$;

$$F(a, b; a-b+1; -1) = 2^{-a}\pi^{\frac{1}{2}}$$
$$\times \frac{\Gamma(1+a-b)}{\Gamma(1+\frac{1}{2}a-b)\Gamma(\frac{1}{2}+\frac{1}{2}a)} \tag{1.7.13}$$

when $1 + a - b \ne 0, -1, -2, \cdots$;

$$F(a, b; a-b+2; -1) = 2^{-a}\pi^{\frac{1}{2}}(b-1)^{-1}$$

$$\times\Gamma(a-b+2)\Big(\frac{1}{\Gamma(\frac{1}{2}a)\Gamma(\frac{3}{2}+\frac{1}{2}a-b)}$$
$$-\frac{1}{\Gamma(\frac{1}{2}+\frac{1}{2}a)\Gamma(1+\frac{1}{2}(a-b))}\Big) \tag{1.7.14}$$

when $a-b+2\neq 0,-1,-2,\cdots$;

$$F\Big(a,b;\frac{1}{2}a+\frac{1}{2}b+\frac{1}{2};\frac{1}{2}\Big)=\pi^{\frac{1}{2}}\frac{\Gamma(\frac{1}{2}+\frac{1}{2}a+\frac{1}{2}b)}{\Gamma(\frac{1}{2}+\frac{1}{2}a)\Gamma(\frac{1}{2}+\frac{1}{2}b)} \tag{1.7.15}$$

when $\frac{1}{2}a+\frac{1}{2}b+\frac{1}{2}\neq 0,-1,-2,\cdots$;

$$F\Big(a,1-a;b;\frac{1}{2}\Big)=\frac{2^{1-b}\pi^{\frac{1}{2}}\Gamma(b)}{\Gamma(\frac{1}{2}a+\frac{1}{2}b)\Gamma(\frac{1}{2}+\frac{1}{2}b-\frac{1}{2}a)} \tag{1.7.16}$$

when $b\neq 0,-1,-2,\cdots$;

$$F\Big(a,\frac{1}{2}+a;\frac{3}{2}-2a;-\frac{1}{3}\Big)=\Big(\frac{8}{9}\Big)^{-2a}\frac{\Gamma(\frac{4}{3})\Gamma(\frac{3}{2}-2a)}{\Gamma(\frac{3}{2})\Gamma(\frac{4}{3}-2a)} \tag{1.7.17}$$

when $\frac{3}{2}-2a\neq 0,-1,-2,\cdots$;

$$F\Big(a,\frac{1}{2}+a;\frac{5}{6}+a;\frac{1}{9}\Big)=\Big(\frac{3}{4}\Big)^{a}\pi^{\frac{1}{2}}\frac{\Gamma(\frac{5}{6}+\frac{2}{3}-2a)}{\Gamma(\frac{1}{2}+\frac{1}{3}a)\Gamma(\frac{5}{6}+\frac{1}{3}a)} \tag{1.7.18}$$

when $\frac{5}{6}+\frac{2}{3}a\neq 0,-1,-2,\cdots$;

$$F\Big(a,\frac{1}{3}a+\frac{1}{3};\frac{2}{3}a+\frac{2}{3};e^{\frac{i\pi}{3}}\Big)=2^{\frac{2}{3}a+\frac{2}{3}}\pi^{\frac{1}{2}}$$
$$\times 3^{-\frac{1}{2}(a+1)}e^{\frac{i\pi a}{6}}\frac{\Gamma(\frac{1}{3}a+\frac{5}{6})}{\Gamma(\frac{1}{3}a+\frac{2}{3})\Gamma(\frac{2}{3})} \tag{1.7.19}$$

when $\frac{1}{3}a\neq -\frac{5}{6},-\frac{11}{6},-\frac{17}{6},\cdots$.

Finally, we explain the combinatorial numbers which are called *Stirling numbers* of *first kind* and *second kind* with results of the asymptotic estimation. Let $\underline{St}$ and $\underline{\underline{St}}$ be the first and the second Stirling numbers. Then they are determined by the equations:

$$_n(x)=\sum_{0\leq m\leq n}\underline{\mathrm{St}}_m^{(n)}x^m,$$

$$x^n = \sum_{0 \le m \le n} \underline{\underline{\mathrm{St}}}^{(n)}_{m}\, {}_m(x), \tag{1.7.20}$$

where ${}_l(x) = x(x-1)\cdots(x-l+1)$, for $l = n, m$. From (1.7.20), we may find that

$$\underline{\mathrm{St}}^{(n)}_m = \sum_{k=0}^{n-m} (-1)^k \begin{pmatrix} n-1+k \\ n-m+k \end{pmatrix}$$

$$\times \begin{pmatrix} 2n-m \\ n-m-k \end{pmatrix} \underline{\underline{\mathrm{St}}}^{(m-n+k)}_k,$$

$$\underline{\underline{\mathrm{St}}}^{(n)}_m = \frac{1}{m!} \sum_{k=0}^{m} (-1)^{m-k} \begin{pmatrix} m \\ k \end{pmatrix} k^n. \tag{1.7.21}$$

The asymptotic behaviors of them are expressed by

$$|\underline{\mathrm{St}}^{(n)}_m| \sim \frac{(n-1)!(r+\ln n)^{m-1}}{(m-1)!}, \text{ for } m = o(\ln n),$$

$$\underline{\underline{\mathrm{St}}}^{(m+n)}_m \sim \frac{m^{2n}}{2^n n!}, \text{ for } n = o(m^{\frac{1}{2}}) \tag{1.7.22}$$

where r is the Euler constant.

§1.8 Notes

1.8.1 As we know, the maps in geography as a mathematical concept firstly appeared in the four color problem in the 19th century [Liu2], [Liu6]. A combinatorial representation of polyhedra which are treated as maps was firstly discovered by Heffler in the 19th century as well [Hef1]. Its dual form was described by Edmonds in 1960 [Edm1]. However, the algebraic representation as described in §1.1 has been used only in the most recent decades [Lit1], [Tut35], [Tut39], [Tut25]. Based on the development of topology, the theory of maps has become quite nature. In fact, the classification of maps is equivalent to that of surfaces. Nevertheless, the main part of map theory provides a new insight into the development of graph theory as well. That is why enumerative theory is alone already more than enough to form the present book.

1.8.2 Although all polynomials on maps appearing in §1.2 have the corresponding polynomials on graphs, we only concentrate on those for maps because they are specifically important in map enumeration [Liu40], [Liu42], [Liu44], [Liu48–9], [Liu56], [Tut21–4], [Tut36–7]. In addition, the more general polynomial determined by (1.2.1–4) [Liu59] is also a generalization of Jones polynomial [Jon1] and bracket polynomial [Kau1] for links, particularly for knots in topology. They have been shown to have a very close relationship with the theory of quantum fields [Wit1] and with statistical mechanics [Bax1].

1.8.3 For formal power series, or particularly for the enufunctions and the polysum functions, readers might find helpful the book [GoJ1], in which many types of applications are provided, whence the theoretical accounts are comprehensively explained.

1.8.4 The generalization of Lagrangian inversion was firstly clarified by Good [Goo1], in which the applications to enumeration and to the queuing theory were also mentioned. However, an algebraically rigid foundation was founded by Tutte [Tut30].

1.8.5 Formulae in §1.7 can been seen in [AbS1]. The theoretical account of special functions especially with the relationship to the group representation can be found in [Waw1].

Chapter 2

Outerplanar Maps

§2.1 Plane trees

A *plane tree* is a planar map whose underlying graph is a tree.

Theorem 2.1.1 *A planar map is a plane tree iff it has exactly one face.*

Proof For the necessity suppose M is a plane tree. Because the underlying graph $G(M) = (V, E)$ of M is a tree, the size $\epsilon = |E|$ of $G(M)$ is $\nu - 1$ where $\nu = |V|$ is the order of $G(M)$. By Euler formula for planar maps, M has exactly $\nu - (\nu - 1) = 1$ face.

For the sufficiency let M which is a planar map have exactly one face. By Euler formula for planar maps, the underlying graph $G(M)$ of M has $\nu - 1$ edges. Therefore, from the connectedness of $G(M)$, $G(M)$ is a tree. □

If a plane tree is rooted then it is called a *rooted plane tree*. A rooted plane tree such that the root vertex is of valency 1 is said to be a *planted tree*.

Let $\mathcal{T}$ be the set of all rooted plane trees. For $T \in \mathcal{T}$, the root, the root-vertex, the root-edge, and the root-face of T are denoted respectively by $r(T)$, $v_r(T)$, $e_r(T)$ and $f_r(T)$.

For two maps M_1 and M_2 with their respective roots $r_1 = r(M_1)$ and $r_2 = r(M_2)$, the map

$$M = M_1 \bigcup M_2 \tag{2.1.1}$$

provided $M_1 \cap M_2 = \{v\}$ such that

$$v = v_{r1} = v_{r2} \tag{2.1.2}$$

is defined to have that its root, root-vertex and root-edge are as the same as those of M_1 but the root-face is the composition of $f_r(M_1)$ and $f_r(M_2)$. The operation for getting M from M_1 and M_2 is called the 1-*addition* and the map M is written as

$$M = M_1 \dot{+} M_2. \tag{2.1.3}$$

Further, for any two sets of rooted maps $\mathcal{M}_1$ and $\mathcal{M}_2$, the set of maps

$$\begin{aligned} \mathcal{M}_1 \odot \mathcal{M}_2 = \{M_1 \dot{+} M_2 | \forall M_1 \in \mathcal{M}_1, \\ \forall M_2 \in \mathcal{M}_2\} \end{aligned} \tag{2.1.4}$$

is said to be the 1-*production* of $\mathcal{M}_1$ and $\mathcal{M}_2$.

If $\mathcal{M}_1 = \mathcal{M}_2 = \mathcal{M}$, then we are allowed to write

$$\mathcal{M} \odot \mathcal{M} = \mathcal{M}^{\odot 2} \tag{2.1.5}$$

and further,

$$\mathcal{M}^{\odot k} = \mathcal{M}^{\odot (k-1)} \odot \mathcal{M}. \tag{2.1.6}$$

Lemma 2.1.1 *Let $\mathcal{T}_i$ be the set of all the plane trees whose root-vertices are of valency i. Then, we have*

$$\mathcal{T}_i = \mathcal{T}_1^{\odot i} \tag{2.1.7}$$

for $i \geq 1$.

Proof For any $T \in \mathcal{T}_i$, let $T_1, T_2, \cdots, T_i$ be all the maximum planted subtrees of T, which have the same root-vertex as T does. Then, it is seen that

$$T = T_1 \dot{+} T_2 \dot{+} \cdots \dot{+} T_i.$$

Hence, T is also a member of the set on the right hand side of (2.1.7).

Conversely, it is obvious that any member T of $\mathcal{T}_1^{\odot i}$ is also a member of $\mathcal{T}_i$ by considering the definitions of 1- addition and 1- production above. □

Because the set $\mathcal{T}$ of all plane trees has the expression as

$$\mathcal{T} = \sum_{i\geq 0} \mathcal{T}_i$$

where $\mathcal{T}_0$ consists of only the vertex map, Lemma 2.1.1 implies that

$$\mathcal{T} = \mathcal{T}_0 + \sum_{i\geq 1} \mathcal{T}^{\odot i}. \tag{2.1.8}$$

For a map M let $a = e_r(M)$ be the root-edge of M. The map which is obtained by contracting a into the root-vertex with the edge corresponding to the successive one on the root-face of M being the root-edge is denoted by

$$M' = M \bullet a$$

as described in (1.1.4).

Let $\mathcal{T}_{\geq 1} = \mathcal{T} - \mathcal{T}_0$ and

$$\mathcal{T}' = \{T \bullet a | \forall T \in \mathcal{T}_{\geq 1}\}. \tag{2.1.9}$$

Lemma 2.1.2

$$\mathcal{T}' = \mathcal{T} \odot \mathcal{T}. \tag{2.1.10}$$

Proof For $T' \in \mathcal{T}$, T' may be the vertex map ϑ which is obtained from the link map, the tree with only one edge, by contracting the edge into a vertex. It can of course be seen to be the map $\vartheta \dot{+} \vartheta$ which is in the set on the right hand side of (2.1.10). Otherwise, suppose that $T' = T \bullet a$, $T \in \mathcal{T}_{\geq 1}$. Let

$$T - a = T_1 + T_2.$$

Then it can be shown that $T' = T_1 \dot{+} T_2$ and hence a member of the set on the right hand side of (2.1.10).

Conversely, let $T' \in \mathcal{T} \odot \mathcal{T}$ and

$$T' = T_1 \dot{+} T_2.$$

Then $T' = T \bullet a$ where T is the tree obtained by adding an edge a from the root-vertex of T_2 to the one of T_1 as the root-edge $e_r(T)$ of T such that $e_r(T_1)$ is the successive edge of a on the boundary of the root-face $f_r(T)$. It is easily seen that T is a member of $\mathcal{T}_{\geq 1}$. Hence T' is a member of $\mathcal{T}'$ as well. □

Lemma 2.1.3 *There is a bijection between the set $\mathcal{T}(n-1)$ of plane trees of order $n-1$ and the set $\mathcal{T}_1(n)$ of planted trees of order n and hence we have*

$$|\mathcal{T}(n-1)| = |\mathcal{T}_1(n)| \tag{2.1.11}$$

for $n \geq 2$.

Proof For $T \in \mathcal{T}_1(n)$, let $T - o$ be the resultant map of deleting the root-vertex $o = v_r(T)$ such that the edge which corresponds to the successive one on the face boundary of T is the root-edge of $T - o$. It is easy to see that $T - o \in \mathcal{T}(n-1)$ and to check that this relation determines a bijection between $\mathcal{T}(n-1)$ and $\mathcal{T}_1(n)$. The lemma is obtained. □

From Lemma 2.1.3, we are allowed only to discuss planted trees instead of general plane trees for the enumeration without loss of generality.

For a general plane tree T, let $m(T)$ be the valency of the root-vertex and $n_i(T)$, $i \geq 1$, be the number of non-rooted vertices of valency i. Write

$$\mathcal{T}(m; \underline{n}) = \{T | m(T) = m, \ \underline{n}(T) = \underline{n}, T \in \mathcal{T}\} \tag{2.1.12}$$

where

$$\underline{n}(T) = (n_1(T), n_2(T), \cdots)$$

and $\underline{n} = (n_1, n_2, \cdots)$.

Lemma 2.1.4 *For integers $m \geq 1$ and $n_i(T) \geq 0$, $i \geq 1$, we have*

$$\left|\mathcal{T}_{\geq 1}(m; \underline{n})\right| = \sum_{i \geq 1} \left|\mathcal{T}(m + i - 2; \underline{n} - e_i)\right| \tag{2.1.13}$$

where

$$e_i = (e_{1i}, e_{2i}, \cdots, e_{ji}, \cdots)$$

and $e_{ji} = 1, i = j$; 0, otherwise.

Proof For any $T \in \mathcal{T}_{\geq 1}(m; \underline{n})$ we may find exactly one plane tree $T' = T \bullet a$ by contracting the root-edge a. Because the valency of the root-vertex of T' is $m + i - 2$ and T' has exact one vertex less than T does, which is of valency i if the non-rooted end of the root-edge of T has valency i, we see that T' has exactly one contribution to the term on the right hand side of (2.1.13).

Conversely, for $T' \in \mathcal{T}(m + i - 2; \underline{n})$ with given m and i, it may happen that $m = 1$ and hence $i = 1$. In this case T' must be the vertex map which exactly corresponds to the link map in $\mathcal{T}_{\geq 1}$ by splitting the vertex of the vertex map into a link. Otherwise, by splitting the root-vertex of T' into an edge (a link of course) which is chosen to be the root-edge such that the non-rooted end is incident with the edges which correspond to the first $i - 1$ edges from the root-edge of T' in the rotation at the root-vertex, we find exact one tree T in $\mathcal{T}_{\geq 1}$ and further $T \in \mathcal{T}_{\geq 1}(m; \underline{n})$. □

For $\mathcal{T}$ write the onufunction

$$t = t_{\mathcal{T}}(x; \underline{y}) = \sum_{T \in \mathcal{T}} x^{m(T)} \underline{y}^{\underline{n}(T)}$$

$$= \sum_{m \geq 0, \underline{n} \geq \underline{0}} c_{\mathcal{T}}(m; \underline{n}) x^m \underline{y}^{\underline{n}} \tag{2.1.14}$$

and let

$$[t]_{i-1} = \sum_{m \geq i, \underline{n} \geq \underline{0}} c_{\mathcal{T}}(m; \underline{n}) x^m \underline{y}^{\underline{n}} \tag{2.1.15}$$

which is called a *truncation* of t for $i \geq 0$.

Theorem 2.1.2 *The enufunction t satisfies the following functional equation:*

$$t = 1 + \sum_{i \geq 1} x^{2-i} y_i [t]_{i-2}. \tag{2.1.16}$$

Proof Since $\mathcal{T} = \mathcal{T}_0 + \mathcal{T}_{\geq 1}$, we have that $t = t_{\mathcal{T}_0} + t_{\mathcal{T}_{\geq 1}}$.

Because $\mathcal{T}_0$ consists of only the vertex map, it is easy to evaluate that $t_{\mathcal{T}_0} = 1$.

From (2.1.13–15), by noticing that

$$c_{\mathcal{T}}(m; \underline{n}) = \left| \mathcal{T}(m; \underline{n}) \right|, \tag{2.1.17}$$

we may deduce that

$$t_{\mathcal{T}_{\geq 1}} = \sum_{i\geq 1} x^{2-i} y_i [t]_{i-2}. \tag{2.1.18}$$

Therefore, (2.1.16) is true. The theorem is obtained. □

If the enufunction is written as

$$t = 1 + \sum_{m\geq 1} \tau_m x^m, \tag{2.1.19}$$

then it is determined by the vector $\tau = (\tau_1, \tau_2, \cdots)$. Let

$$e_i = (0, \cdots, 1_i, \cdots)$$

be the vector with all components 0 but the i-th which is 1 and $I = [1, 1, \cdots]$, the matrix with all entrances 0 but those being 1 on the diagonal. According to Theorem 2.1.2, by comparing the coefficients on both sides of (2.1.16), we may obtain the following linear equation system:

$$(I - Y)\tau^{\mathrm{T}} = y_1 e_1^{\mathrm{T}} \tag{2.1.20}$$

where the letter "T" represents the transpose of a vector and

$$Y = \begin{pmatrix} y_2 & y_3 & y_4 & y_5 & \cdot & \cdot & \cdot \\ y_1 & y_2 & y_3 & y_4 & \cdot & \cdot & \cdot \\ 0 & y_1 & y_2 & y_3 & \cdot & \cdot & \cdot \\ 0 & 0 & y_1 & y_2 & \cdot & \cdot & \cdot \\ & & & & & & \\ \cdot & \cdot & \cdot & \cdot & \cdot & \cdot & \cdot \end{pmatrix}. \tag{2.1.21}$$

From (2.1.20) we may find an explicit expression of τ and hence of t as follows:

$$\tau^{\mathrm{T}} = \sum_{i\geq 0} y_1 Y^i e_1^{\mathrm{T}}. \tag{2.1.22}$$

On the other hand, from Lemma 2.1.3 let j be the valency of the nonrooted end of the root-edge of a planted tree. Then we have

$$\tau_1 = y_1 + \sum_{j\geq 2} y_j \tau_1^{j-1}$$

$$= y_1 + \sum_{j\geq 1} y_{j+1}\tau_1^j. \tag{2.1.23}$$

Now, we are allowed to find an explicit expression of τ_1 by employing Corollary 1.5.2.

Theorem 2.1.3 *The enufunction of planted trees with vertex partition has the following form:*

$$\tau_1 = \sum_{n\geq 1}\sum_{\underline{j}\in J} \frac{(n-1)!}{\underline{j}!} \underline{y}^{\underline{j}} \tag{2.1.24}$$

where $\underline{j} = (j_1, j_2, \cdots)$ *and*

$$J = \Big\{\underline{j}\Big| \sum_{i\geq 1} j_i = n, \sum_{i\geq 1} ij_i = 2n-1\Big\}.$$

Of course, n *is the number of nonrooted vertices.*

Proof By using (1.5.7) in Corollary 1.5.1 with $\tau_1 = \tau$, $1 = t$,

$$\phi(x) = y_1 + \sum_{j\geq 1} y_{j+1}\tau_1^j$$

and $f(x) = x = \tau_1$ with (2.1.23), we may see that

$$\tau_1 = \sum_{n\geq 1} \frac{1}{n!} \frac{\mathrm{d}^{n-1}}{\mathrm{d}\xi^{n-1}} \phi(\xi)^n \Big|_{\xi=0}. \tag{2.1.25}$$

Since $\phi(\xi) = \xi^{-1}\psi(\xi)$, where

$$\psi(\xi) = \sum_{j\geq 1} y_j \xi^j$$

with (2.1.23), we find that

$$\frac{\mathrm{d}^{n-1}}{\mathrm{d}\xi^{n-1}} \phi(\xi)^n \Big|_{\xi=0} = (n-1)! \partial_\xi^{2n-1} \psi(\xi)^n \tag{2.1.26}$$

in which ∂_ξ^{2n-1} is the coefficient operator described in §1.5.

Since it is seen that

$$\partial_\xi^{2n-1}\psi(\xi)^n \begin{cases} > 0, \text{ if } \underline{j} \in J; \\ = 0, \text{ otherwise,} \end{cases}$$

and further

$$\partial_{\xi}^{2n-1}\psi(\xi)^n = \sum_{\underline{j}\in J}\frac{n!}{\underline{j}!}\underline{y}^{\underline{j}}, \tag{2.1.27}$$

we may soon derive (2.1.24) from (2.1.25–26). □

From this theorem, by considering the root-vertex and the *root-semiedge* as imagined with careful calculation, we can easily find

Corollary 2.1.1 *For a sequence of positive integers* $n_1, n_2, \cdots$, *such that*

$$\sum_{i\geq 1} n_i = n,$$

there exists a tree of order n *with the number of vertices of valency* i *being* n_i *iff it satisfies that*

$$\sum_{i\geq 1} in_i = 2(n-1).$$

□

If the shadow functional $\int_y$ defined in §1.6 is employed here, then the following theorem can also be obtained.

Theorem 2.2.4 *The enufunction* τ_1 *of planted trees with vertex partition satisfies the following functional equation:*

$$\tau_1 = y_1 + \int_y \frac{y^2\tau_1}{1-y\tau_1}. \tag{2.1.28}$$

Proof From (2.1.23) we have

$$\begin{aligned}\tau_1 - y_1 &= \sum_{i\geq 1} y_{i+1}\tau_1^i \\ &= \int_y \left(y^2\tau_1 \sum_{i\geq 0}(y\tau_1)^i\right) \\ &= \int_y \frac{y^2\tau_1}{1-y\tau_1}\end{aligned}$$

and hence (2.1.28). □

In fact, (2.1.24) provides a solution of Eqs.(2.1.23) and (2.1.28) in its own right. Further, from Lemma 2.1.3 we may also find a solution of Eq.(2.1.20) and an equation generalized from Eq.(2.1.28) for general plane trees as well.

§2.2 Wintersweets

A *wintersweet* is a rooted planar map such that the root-vertex is not on a circuit and that it would be a tree if the circuits were contracted into vertices to be *articulate*, or in other words, to be of valency 1. Of course, loops are treated to be circuits as well. From the definition, we can see that the root-edge is a cut edge. If a wintersweet has no circuit, then it becomes a tree in its own right.

Let $\mathcal{W}$ be the set of all rooted wintersweets. We may classify wintersweets into three categories: $\mathcal{W}^{(\mathrm{I})}$, $\mathcal{W}^{(\mathrm{II})}$ and $\mathcal{W}^{(\mathrm{III})}$. Here, $\mathcal{W}^{(\mathrm{I})}$ consists of only one map, the vertex map which is treated as the degenerate case. $\mathcal{W}^{(\mathrm{II})}$ contains all the wintersweets each of which has its root- edge incident to a circuit.

Let $\mathcal{W}_i$, $i \geq 0$, be the set of wintersweets with the root- vertices of valency i. Of course, it is easily seen that $\mathcal{W}^{(\mathrm{I})} = \mathcal{W}_0$. Further, we have

$$\mathcal{W} = \sum_{i \geq 0} \mathcal{W}_i \tag{2.2.1}$$

and

$$\mathcal{W}_i = \mathcal{W}_1^{\odot i} \tag{2.2.2}$$

for $i \geq 2$.

Lemma 2.2.1 *Let* $\mathcal{W}_1' = \{W - o | \forall W \in \mathcal{W}_1\}$ *where* o *is the root-vertex of* W. *Then, we have*

$$\mathcal{W}_1^{(\mathrm{III})'} = \mathcal{W} \tag{2.2.3}$$

where $\mathcal{W}_1^{(\mathrm{III})} = \mathcal{W}_1 \cap \mathcal{W}^{(\mathrm{III})}$.

Proof First, we may see that any member W' of $\mathcal{W}_1^{(\mathrm{III})'}$ is a member of $\mathcal{W}$ because the nonrooted end of the root-edge of W' is not incident on a circuit.

Then, for any $W \in \mathcal{W}$ we may construct a wintersweet W_1 by adding an edge $(o, v_r(W))$ which is chosen to be the root-edge of W_1. However, the vertex $v_r(W_1)$ is of course not incident to a circuit. This implies that $W_1 \in \mathcal{W}_1^{(\mathrm{III})}$ and hence $W = W_1 - o \in \mathcal{W}_1^{(\mathrm{III})'}$. □

For a map $W_1 \in \mathcal{W}_1^{(\mathrm{III})}$ with $\langle o_1, o_2 \rangle$ being the root-edge, let W_{-1} be the map obtained from W_1 by choosing $\langle o_2, o_1 \rangle$ as the root-edge instead of $\langle o_1, o_2 \rangle$. Apparently, it can be checked that $W_{-1} \in \mathcal{W}$.

Lemma 2.2.2 *Let* $\mathcal{W}'_{-1} = \{W_{-1} - o_1 | \forall W_1 \in \mathcal{W}_1^{(\mathrm{III})}\}$. *Then, we have*

$$\mathcal{W}'_{-1} = \mathcal{W} \tag{2.2.4}$$

whenever noticing that the root-edge of $W_{-1} - o_1$ *is chosen to be the successor of* $\langle o_1, o_2 \rangle$ *instead of* $\langle o_2, o_1 \rangle$ *on the boundary of the root-face of* $W_{-1} - o_1$.

Proof Because it is easily verified that

$$\mathcal{W}'_{-1} = \mathcal{W}_1^{(\mathrm{III})'} \tag{2.2.5}$$

from Lemma 2.2.1, the lemma is derived. □

Lemma 2.2.3 *Let* $\mathcal{C}$ *be the set of all circuits among which the loop map is the smallest. Then, we have*

$$\left|\mathcal{W}^{(\mathrm{II})}\right| = \left|\mathcal{W} \times \mathcal{C}\right| \tag{2.2.6}$$

where $\times$ *is the Cartesian product.*

Proof For any $W \in \mathcal{W}^{(\mathrm{II})}$, write $W = C \dot{+} W_{-1}$ where the common vertex of C and W_{-1} is o_1 which is the root- vertex of W_1. Since $W_1 \in \mathcal{W}_1^{(\mathrm{III})}$, from Lemma 2.2.1 it corresponds to exact one member of the set on the right hand of (2.2.6).

Conversely, for any $W \in \mathcal{W}$ and $C \in \mathcal{C}$, we may always construct the unique map $W_1 \in \mathcal{W}_1^{(\mathrm{III})}$ by adding an edge $\langle o, v_r(W) \rangle$ as the root-edge of W_1. Then, $C \dot{+} W_{-1} \in \mathcal{W}^{(\mathrm{II})}$. □

Now we investigate the relationship between $\mathcal{W}^{(\mathrm{III})}$ and $\mathcal{W}$.

Lemma 2.2.4 *For $\mathcal{W}^{(\mathrm{III})}$, we have*

$$\mathcal{W}^{(\mathrm{III})'} = \mathcal{W} \tag{2.2.7}$$

where $\mathcal{W}^{(\mathrm{III})'} = \{W \bullet a | \forall W \in \mathcal{W}^{(\mathrm{III})}\}$ *with* a *being the root-edge of* W.

Proof For any $W \in \mathcal{W}^{(\mathrm{III})'}$, since the nonrooted vertex of a is not incident to a circuit, we know that $W \in \mathcal{W}$.

Conversely, for any $W \in \mathcal{W}$, because the root-vertex is a cut vertex, we are always allowed to find a map $\tilde{W}$ with the root-edge being a cut edge by splitting the root-vertex into an edge which is chosen to be the root-edge of $\tilde{W}$. Since the root-vertex of W is not incident to a circuit, we see that $\tilde{W} \in \mathcal{W}^{(\mathrm{III})}$. □

For integers m, n_1, n_2, $\cdots$, let $\mathcal{W}(m; n_1, n_2, \cdots)$ be the set of all the wintersweets which have the root-vertices of valency m, and the number of nonrooted vertices of valency i be n_i for $i \geq 1$.

Lemma 2.2.5 *For integers $m \geq 0$ and $\underline{n} = (n_1, n_2, \cdots) \geq \underline{0}$, we have*

$$\left|\mathcal{W}^{(\mathrm{III})}(m; \underline{n})\right| = \sum_{i \geq 1} \left|\mathcal{W}(m + i - 2; \underline{n} - e_i)\right| \tag{2.2.8}$$

where e_i is the vector of all components 0 but only the i-th which is just 1.

Proof For $W \in \mathcal{W}^{(\mathrm{III})}(m; \underline{n})$, suppose the nonrooted end of the root-edge is of valency i. By contracting the root-edge of W, we obtain exactly one map W'. From Lemma 2.2.4, it is a wintersweet. Since W' has one vertex of valency i less than W does and only the root-vertex of W' is of valency $m+i-2$ which is different from that of W. Therefore,

$$W' \in \mathcal{W}(m + i - 2; \underline{n} - e_i).$$

On the other hand, for any $W \in \mathcal{W}$ such that the valency of root-vertex is $m + i - 2$ and the number of nonrooted vertices of valency j is n_j when $j \neq i$; $n_i - 1$ when $j = i$, we can uniquely find $\tilde{W}$ for m and i given by splitting the root-vertex of W into an edge as the root-edge of $\tilde{W}$ such that the nonrooted end is only incident to the first $i - 1$-st from the root-edge in the rotation at the root-vertex of W. Because the root-vertex of W is a cut vertex, we see that $\tilde{W} \in \mathcal{W}$. Further, since the root-vertex of W is not on a circuit, we have $\tilde{W} \in \mathcal{W}^{(\mathrm{III})}$. By virtue of calculating the valency of the root-vertex and the number of nonrooted vertices of valency j, $j \geq 1$, it can be seen that $\tilde{W} \in \mathcal{W}^{(\mathrm{III})}(m; \underline{n})$. □

In view of the convention that for vectors $\underline{y} = (y_1, y_2, \cdots)$ and $\underline{n} = (n_1, n_2, \cdots)$, write

$$\underline{y}^{\underline{n}} = \prod_{i \geq 1} y_i^{n_i} \tag{2.2.9}$$

and let

$$w = w_{\mathcal{W}}(x; \underline{y}) = \sum_{W \in \mathcal{W}} x^{m(W)} \underline{y}^{\underline{n}(W)}$$

$$= \sum_{m \geq 0} \sum_{\underline{n} \geq 0} c_{\mathcal{W}}(m; \underline{n}) x^m \underline{y}^{\underline{n}} \tag{2.2.10}$$

be the enufunction of the wintersweets.

Theorem 2.2.1 *The enufunction w satisfies the equation as*

$$w = 1 + \frac{x y_3 w}{1 - y_2} + \sum_{i \geq 1} x^{2-i} y_i [w]_{i-2} \tag{2.2.11}$$

where $[w]_{i-2}$ is the truncation of w by getting off all the terms of x^j for $j \leq i - 2$.

Proof Because $\mathcal{W} = \mathcal{W}^{(\mathrm{I})} + \mathcal{W}^{(\mathrm{II})} + \mathcal{W}^{(\mathrm{III})}$, the only thing we should do is to evaluate $w_1 = w_{\mathcal{W}^{(\mathrm{I})}}$, $w_2 = w_{\mathcal{W}^{(\mathrm{II})}}$, and $w_3 = w_{\mathcal{W}^{(\mathrm{III})}}$. Since $\mathcal{W}^{(\mathrm{I})}$ consists of only one element, the vertex map, it is easy to see $w_1 = 1$. Since the enufunction of the set of circuits $\mathcal{C}$ is

$$\frac{1}{1 - y_2},$$

from (2.2.6) we derive

$$w_2 = \frac{x y_3}{1 - y_2} w$$

by considering that a new nonrooted 3-valent vertex appears. Finally, from lemma 2.2.5, we have

$$w_3 = \sum_{m \geq 1} \sum_{\underline{n} \geq 0} c_{\mathcal{W}^{(\mathrm{III})}}(m; \underline{n}) x^m \underline{y}^{\underline{n}}$$

$$= \sum_{i\geq 1} x^{2-i} y_i \sum_{m\geq i-1} \sum_{\underline{n}\geq 0} c_{\mathcal{W}}(m;\underline{n}) x^m \underline{y}^{\underline{n}}$$

$$= \sum_{i\geq 1} x^{2-i} y_i [w]_{i-2}.$$

In consequence the theorem is obtained. □

If the function is treated to be a power series of x as

$$w = \sum_{m\geq 0} \omega_m x^m, \tag{2.2.12}$$

then it is determined by the vector $\underline{w} = (\omega_1, \omega_2, \cdots)$.

Theorem 2.2.2 *The following linear equation:*

$$(I - Y_{\mathcal{W}})\xi^{\mathrm{T}} = c e_1^{\mathrm{T}}, \tag{2.2.13}$$

where

$$c = y_1 + \frac{y_3}{1 - y_2} \tag{2.2.14}$$

and

$$Y_{\mathcal{W}} = \begin{pmatrix} y_2 & y_3 & y_4 & y_5 & \cdot & \cdot & \cdot \\ c & y_2 & y_3 & y_4 & \cdot & \cdot & \cdot \\ 0 & c & y_2 & y_3 & \cdot & \cdot & \cdot \\ 0 & 0 & c & y_2 & \cdot & \cdot & \cdot \\ \cdot & \cdot & \cdot & \cdot & \cdot & \cdot & \cdot \end{pmatrix}, \tag{2.2.15}$$

is well defined and the solution is $\xi = \underline{w}$.

Proof According to Theorem 2.2.1, by substituting (2.2.12) for (2.2.13) and by writing the right hand side of (2.2.11) as a power series of x, we may uniquely find the equation (2.2.13) through equating the correspondent coefficients. The theorem is obtained. □

Lemma 2.2.6 *Let $\mathcal{W}_i$ be the set of all the wintersweets with the valency of root-vertex being i for $i \geq 1$. Then, we have*

$$\mathcal{W}_i = \mathcal{W}_1^{\odot i} \tag{2.2.16}$$

for $i \geq 2$.

Proof For any $W \in \mathcal{W}_i$, since the root-edge is a cut edge, we have the unique expression as

$$W = \dot{\sum_{1 \leq j \leq i}} W_j$$

such that $W_j \in \mathcal{W}_1, 1 \leq j \leq i$, and hence W is a member of the set on the right hand side of (2.2.16).

Conversely, it is easily checked that any map on the right hand side is a member on the left hand side of (2.2.16). □

From Lemma 2.2.6 we may soon find that

$$\omega_i = \omega_1^i \tag{2.2.17}$$

for $i \geq 2$. This allows us only to discuss ω_1 instead of $\underline{w}$ for the enumeration of general wintersweets mentioned above.

Lemma 2.2.7 *The following relation:*

$$\omega_1 = \left(y_1 + \frac{y_3}{1 - y_2}\right)\left(1 - \sum_{i \geq 2} y_i \omega_1^{i-2}\right)^{-1} \tag{2.2.18}$$

is satisfied by ω_1.

Proof From the first equation of (2.2.13), we see that

$$\omega_1 = \left(y_1 + \frac{y_3}{1 - y_2}\right) + \sum_{i \geq 2} y_i \omega^{i-1}$$

by Lemma 2.2.6,

$$= \left(y_1 + \frac{y_3}{1 - y_2}\right) + \omega_1 \sum_{i \geq 2} y_i \omega_1^{i-2}.$$

Therefore, the lemma is immediately obtained. □

A wintersweet which has its root-vertex of valency 1 is also called *planted.*

Theorem 2.2.3 *The enufunction of planted wintersweets with vertex partition has the following explicit expression:*

$$\omega_1 = \sum_{n\geq 1}\sum_{l\geq 0}\sum_{\underline{j}\in J(n,l)} \frac{B_n(l,j_2,j_3)}{\underline{j}!}\underline{y}^{\underline{j}}, \tag{2.2.19}$$

where

$$J(n,l) = \Big\{\underline{j}\Big|\sum_{i\geq 1} j_i = n, \sum_{i\geq 1} ij_i = 2(n+l)-1\Big\},$$

with n and l being the respective numbers of nonrooted vertices and circuits in a planted wintersweet such that

$$B_n(l,j_2,j_3) = \sum_{s\geq 0} A_n(s,l,j_2,j_3) \tag{2.2.20}$$

and

$$A_n(s,l,j_2,j_3) = \frac{(n-s-1)!(l+s-1)!j_2!j_3!}{(j_2-s)!(j_3-l)!l!(l-1)!s!} \tag{2.2.21}$$

for $s,l,j_2,j_3 \geq 0$, and $n\geq 1$.

Proof By comparing (2.2.18) in Lemma 2.2.7 with (2.1.24), from theorem 2.1.3 we find that

$$\omega_1 = \sum_{n\geq 1}\sum_{\underline{j}\in J} \frac{(n-1)!}{\underline{j}!}\underline{y}^{\underline{j}}\Big|_{y_1\to y_1+\frac{y_3}{1-y_2}}$$

where $\underline{j} = (j_1, j_2, \cdots)$ and

$$J = \Big\{\underline{j}\Big|\sum_{i\geq 1} j_i = n, \sum_{i\geq 1} ij_i = 2n-1\Big\}.$$

Of course, n is the number of nonrooted vertices in corresponding trees.

Since

$$\Big(y_1 + \frac{y_3}{1-y_2}\Big)^{j_1} = \sum_{l=0}^{j_1} \frac{j_1!}{l!(j_1-l)!} {y_1}^{j_1-l}{y_3}^{l} \sum_{s\geq 0} \frac{(l+s-1)!}{(s-1)!l!} y_2^s$$

$$= \sum_{s\geq 0}\sum_{l=0}^{j_1} \frac{j_1!(l+s-1)!}{l!(j_1-l)!(l-1)!s!} y_1^{j_1-l}y_2^s y_3^l,$$

we have

$$\omega_1 = \sum_{\substack{n\geq 1,\ s\geq 0\\ \underline{j}\in J}} \sum_{l=0}^{j_1} \frac{(n-1)!j_1!(l+s-1)!}{\underline{j}!l!(j_1-l)!(l-1)!s!} y_1^{j_1-l} y_2^{j_2+s} y_3^{j_3+l} \prod_{i\geq 4} y_i^{j_i}.$$

By employing the following substitutions

$$\begin{cases} j_1' = j_1 - l, & j_1 = j_1' + l; \\ j_2' = j_2 + s, & j_2 = j_2' - s; \\ j_3' = j_3 + l, & j_3 = j_3' - l; \\ j_i' = j_i, & \text{for } i \geq 4; \\ n' = n + s; & n = n' - s, \end{cases}$$

we find

$$\omega_1 = \sum_{\substack{n'\geq 1,\ s\geq 0\\ l\geq 0}} \sum_{\underline{j}'\in J(n',l)} \frac{A_{n'}(s,l,j_2',j_3')}{\underline{j}'!} \underline{y}^{\underline{j}'}$$

where

$$A_{n'}(s,l,j_2',j_3') = \frac{(n'-s-1)!(l+s-1)!j_2'!j_3'!}{(j_2'-s)!(j_3'-l)!l!(l-1)!s!}$$

and

$$J(n',l) = \Big\{\underline{j}'\Big|\sum_{i\geq 1} j_i' = n', \sum_{i\geq 1} ij_i' = 2(n'+l)-1\Big\}.$$

Of course, n', s, and l are the numbers of nonrooted vertices, 2- valent vertices on circuits, and circuits in a wintersweet.

By ignoring the primes which are with the variables n and j_i for $i \geq 1$, we may soon obtain the theorem after rearrangements. □

Further, we investigate what a functional equation involved with the shadow functional is satisfied by the enufunction w of general wintersweets.

Theorem 2.2.4 *The functional equation*

$$\Big(1 - \frac{xy_3}{1-y_2}\Big)w = 1 + x\int_y \Big(y\delta_{x,y}\big(zw(z)\big)\Big) \tag{2.2.22}$$

is well defined in power series with all coefficients non-negative and the solution is

$$w = 1 + \underline{x}\Big(\sum_{i\geq 0}\Big(y_1 + \frac{y_3}{1-y_2}\Big)Y_{\mathcal{W}}^i e_1^{\mathrm{T}}\Big) \tag{2.2.23}$$

where $\underline{x} = (x, x^2, x^3, \cdots)$ *and* $\delta_{x,y}$ *is the* (x, y)*-difference defined by* (1.6.7).

Proof From Lemma 2.2.3, we see that

$$\begin{aligned} w_2 &= \sum_{W \in \mathcal{W} \times \mathcal{C}} x^{m(W)} \underline{y}^{\underline{n}(W)} \\ &= \Big(\sum_{W_1 \in \mathcal{W}} x^{m(W_1)} \underline{y}^{\underline{n}(W_1)} \Big) \\ &\quad \times \Big(x y_3 \sum_{C \in \mathcal{C}} y_2^{n_2(C)} \Big) \\ &= \frac{x y_3}{1 - y_2} w. \end{aligned}$$

For any $W \in \mathcal{W}$ let z_i represent the new nonrooted vertex obtained by splitting the root-vertex of W, $i = 1, 2, \cdots, m(W)+1$. From Lemma 2.2.5 we have

$$\begin{aligned} w_3 &= \sum_{W \in \mathcal{W}} \sum_{i=1}^{m(W)+1} x^i z_{m(W)-i+2} \underline{y}^{\underline{n}(W)} \\ &= \int_z \Big(\sum_{W \in \mathcal{W}} xz \frac{x^{m(W)+1} - z^{m(W)+1}}{x - z} \underline{y}^{\underline{n}(W)} \Big) \\ &= x \int_z \Big(z \delta_{x,z} \big(uw(u) \big) \Big). \end{aligned}$$

According to the expressions of w_2, w_3, and $w_1 = 1$ obtained above, from $w = w_1 + w_2 + w_3$, (2.2.22) is soon derived with the substitutions of y and z for z and u respectively.

By equating the correspondent coefficients on the two sides of (2.2.22), the expression (2.2.23) can also be deduced. □

By virtue of (2.2.23) we may soon find the following

Corollary 2.2.1 *For a sequence of positive integers* $l, j_1, j_2, \cdots$, *there exists a wintersweet with* l *circuits and* j_i *nonrooted vertices of valency* i, $i \geq 1$, *iff* $\underline{j} \in J(n, l)$ *where* n *is the number of nonrooted vertices.* □

In view of (2.2.7) and (2.2.19), we may find the explicit solutions of Eqs. (2.2.11), (2.2.13) and (2.2.22).

§2.3 Unicyclic maps

A *unicyclic map* is a rooted planar map such that there is only one circuit which is the boundary of the unique inner face and that the root-edge is on the circuit. Of course, loops are considered to be circuits. Let $\mathcal{U}$ be the set of all unicyclic maps. Then, $\mathcal{U}$ can be classified into two categories: $\mathcal{U}_{\mathrm{I}}$ and $\mathcal{U}_{\mathrm{II}}$. The first consists of all those with the root-edges being on loops and the last contains all the others.

Lemma 2.3.1 *Let* $\mathcal{U}'_{\mathrm{I}} = \{U \bullet a | \forall U \in \mathcal{U}_{\mathrm{I}}\}$ *where* a *is the root-edge of* U. *Then we have*

$$\mathcal{U}'_{\mathrm{I}} = \mathcal{T} \tag{2.3.1}$$

which is the set of all plane trees.

Proof First, it is apparent that any $U \in \mathcal{U}'_{\mathrm{I}}$ is a tree. Then, any tree can be seen as that resulting from contracting the root-edge which is a loop in the map obtained by adding a loop which is chosen to be the root-edge at the root-vertex of the tree, and hence a member of the set on the left hand side of (2.3.1). □

Let $\mathcal{U}_{\mathrm{II}}(m : \underline{n})$, $\underline{n} = (n_1, n_2, \cdots)$, be the subset of $\mathcal{U}_{\mathrm{II}}$ in which each map is with the valency m of the root-vertex and the number n_i, $i \geq 1$, of nonrooted vertices of valency i.

Lemma 2.3.2 *For integers* $m \geq 2$, $n_i \geq 0$, $i \geq 1$, *we have*

$$\left|\mathcal{U}_{\mathrm{II}}(m; \underline{n})\right| = \sum_{i \geq 2} \left|\mathcal{U}(m + i - 2; \underline{n} - e_i)\right| \tag{2.3.2}$$

where e_i *is the vector with all its components being* 0 *but only the* i*-th being* 1.

Proof For any $U \in \mathcal{U}_{\mathrm{II}}(m; \underline{n})$, suppose the valency of the non-rooted end of the root-edge is i. Because the root-edge is always on the circuit which is not the loop, i is not less than 2. By contracting the root-edge which is not a loop, we obtain a unique unicyclic map

U' which has the valency of the root-vertex $m+i-2$ and the number $n_j(U')$ of nonrooted vertices of valency j all the same as U except only for $j=i$ with one less than $n_i(U)$. That implies

$$U' \in \mathcal{U}(m+i-2; \underline{n}-e_i).$$

Conversely, for any integers $i \geq 2$, $m \geq 2$, and a map $U' \in \mathcal{U}$ with the valency of the root-vertex being $m+i-2$ given, we may also find a unique map U obtained by splitting the root-vertex into an edge such that the nonrooted end is only incident to those which corresponds to the first $i-1$ edges around the rotation starting from the root-edge at the root-vertex of U'. Because the root-edge of U is not a loop and U is the unicyclic map with the valency of the root-vertex being m in its own right, it leads that $U \in \mathcal{U}_{\mathrm{II}}(m; \underline{n})$. □

Now, let us write

$$\begin{aligned} u = u(x) &= u(x : \underline{y}) \\ &= \sum_{m \geq 2} \mu_m x^m \end{aligned} \tag{2.3.3}$$

which is the enufunction of $\mathcal{U}$ with vertex partition.

Theorem 2.3.1 *Let $[u]_i$ be the truncation of u by getting off all the terms with the powers of x less than or equal to i, $i \geq 1$. Then, the enufunction u satisfies the following relation:*

$$u = x^2 t + \sum_{i \geq 2} x^{2-i} y_i [u]_{i-1} \tag{2.3.4}$$

where t is the enufunction of plane trees described in §2.1.

Proof Let u_1 and u_2 be the enufunctions of $\mathcal{U}_{\mathrm{I}}$ and $\mathcal{U}_{\mathrm{II}}$. Then, we have $u = u_1 + u_2$.

For u_1, it is easy to check that

$$u_1 = x^2 t \tag{2.3.5}$$

from Lemma 2.3.1 by considering the loop having two contributions to the valency of the root-vertex of a map in $\mathcal{U}_{\mathrm{I}}$.

From Lemma 2.3.2, we see that

$$\begin{aligned} u_2 &= \sum_{m\geq 2}\sum_{\underline{n}\geq 0} c_{\mathcal{U}_{\text{II}}}(m;\underline{n})x^m\underline{y}^{\underline{n}} \\ &= \sum_{i\geq 2} x^{2-i}y_i \sum_{m\geq 2}\sum_{\underline{n}\geq 0} c_{\mathcal{U}}(m+i-2;\underline{n}-e_i) \\ &\qquad\qquad \times x^{m+i-2}\underline{y}^{\underline{n}-e_i} \\ &= \sum_{i\geq 2} x^{2-i}y_i \sum_{m\geq i}\sum_{\underline{n}\geq 0} c_{\mathcal{U}}(m;\underline{n})x^m\underline{y}^{\underline{n}} \\ &= \sum_{i\geq 2} x^{2-i}y_i[u]_{i-1}. \end{aligned} \tag{2.3.6}$$

In view of (2.3.5–6), (2.3.4) is soon obtained. □

If $\mu = (\mu_2, \mu_3, \cdots)$ is considered, then u is determined by the vector μ from (2.3.3).

Theorem 2.3.2 *The vector μ satisfies the following linear equation system:*

$$(I - Y_{\mathcal{U}})\mu^{\mathrm{T}} = \underline{\tau}_0^{\mathrm{T}} \tag{2.3.7}$$

where $\underline{\tau}_0 = (1, \tau)$, $\tau = (\tau_1, \tau_2, \cdots)$ is the vector for plane trees known in §2.1 and

$$Y_{\mathcal{U}} = \begin{pmatrix} y_2 & y_3 & y_4 & y_5 & \cdot & \cdot & \cdot \\ 0 & y_2 & y_3 & y_4 & \cdot & \cdot & \cdot \\ 0 & 0 & y_2 & y_3 & \cdot & \cdot & \cdot \\ 0 & 0 & 0 & y_2 & \cdot & \cdot & \cdot \\ \cdot & \cdot & \cdot & \cdot & \cdot & \cdot & \cdot \end{pmatrix}. \tag{2.3.8}$$

Further, u has the following expression:

$$u = \underline{x}\Big(\sum_{i\geq 0} Y_{\mathcal{U}}^i \underline{\tau}_0^{\mathrm{T}}\Big) \tag{2.3.9}$$

where $\underline{x} = (x^2, x^3, x^4, \cdots)$.

Proof From Theorem 2.3.1, by equating the coefficients of the same powers of x between the two sides in (2.3.4), the equation system

(2.3.7) is soon obtained. Then by expanding $(I - Y_{\mathcal{U}})^{-1}$, we may find (2.3.9). □

In what follows, we investigate the relationship among the components of the vector μ. Write

$$\mathcal{U}_k = \{U \mid \forall U \in \mathcal{U}, m(U) = k\}. \tag{2.3.10}$$

Of course, $m(U)$ is the valency of root-vertex on U.

Lemma 2.3.3 *For $k \geq 3$, we have*

$$\mathcal{U}_k = \mathcal{U}_{k-1} \odot \mathcal{T}_1 \tag{2.3.11}$$

where $\mathcal{T}_1$ is the set of all planted trees appearing in §2.1.

Proof For any $U \in \mathcal{U}_k$, since $k \geq 3$ there is an edge incident with the root-vertex of U. From the unicyclicity the edge is a cut edge. Because all the edges which can be visited by a path from the root-vertex through the edge induces a submap of U which is a planted tree and all other edges forms a unicyclic map which has the valency of the root-vertex $k-1$, we see that U is a member of the set on the right hand side of (2.3.11).

Conversely, for any $T \in \mathcal{T}_1$ and $U_1 \in \mathcal{U}_{k-1}$, since $U = T \dot{+} U_1$ is a unicyclic map with the valency of the root-vertex being k, we see that $U \in \mathcal{U}_k$.

In consequence, the lemma holds. □

As for $\mathcal{U}_2$, it can be classified into two categories: $\mathcal{U}_2^{(\mathrm{I})}$ and $\mathcal{U}_2^{(\mathrm{II})}$ such that $\mathcal{U}_2^{(\mathrm{I})} \subseteq \mathcal{U}_{\mathrm{I}}$ and $\mathcal{U}_2^{(\mathrm{II})} \subseteq \mathcal{U}_{\mathrm{II}}$. Of course, $\mathcal{U}_2^{(\mathrm{I})}$ consists of only one map which is the loop map.

Lemma 2.3.4 *Let $\tilde{\mathcal{U}} = \{U \bullet a \mid \forall U \in \mathcal{U}_2^{(\mathrm{II})}\}$ where a is the root-edge of U. Then, we have*

$$\tilde{\mathcal{U}} = \mathcal{U} \tag{2.3.12}$$

for $\mathcal{U}$ defined as in the beginning of this section.

Proof For any $\tilde{U} \in \tilde{\mathcal{U}}$ there is a $U \in \mathcal{U}_2^{(\mathrm{II})}$ such that $\tilde{U} = U \bullet a$. Since a is not a loop we know that $\tilde{U} \in \mathcal{U}$.

On the other hand, for any $U \in \mathcal{U}$, since the root-edge is on the circuit, we may always find a map U_1 by splitting the root-vertex into an edge as the root-edge such that the rooted one of its two ends is of valency 2. From the unicyclicity of U, U_1 is unicyclic as well. In view of that the root-edge of U_1 is not a loop, we see that $U_1 - a_1 \in \tilde{\mathcal{U}}$.

Therefore, the lemma is true. □

According to Lemma 2.3.3, it is easily seen that

$$\mu_k = \tau_1 \mu_{k-1} \tag{2.3.13}$$

for $k \geq 3$.

Further, from Lemma 2.3.4 according to the uniqueness of correspondence between $\mathcal{U}_2^{(\mathrm{II})}$ and $\mathcal{U}$, we find that

$$\mu_2^{(\mathrm{II})} = \sum_{i \geq 2} y_i \mu_i \tag{2.3.14}$$

where i represents the valency of the nonrooted end of the root- edge of a map in $\mathcal{U}_2^{(\mathrm{II})}$.

Lemma 2.3.5 *For μ_2, we have the following expression:*

$$\mu_2 = \frac{1}{1 - \sum_{i \geq 0} y_{i+2} \tau_1^i} \tag{2.3.15}$$

where τ_1 is the enufuction of planted trees.

Proof From (2.3.13–14) we obtain that

$$\begin{aligned} \mu_2^{(\mathrm{II})} &= \sum_{i \geq 2} y_i \tau_1^{i-2} \mu_2 \\ &= \sum_{i \geq 0} y_{i+2} \tau_1^i \mu_2. \end{aligned}$$

By considering that $\mu_2^{(\mathrm{I})} = 1$, the lemma is soon deduced. □

From (2.1.23), we are allowed to find an expression of τ_1^s by employing the Lagrangian inversion (1.5.7) in the case of $\tau_1 = x, 1 = t$,

$$\phi(x) = \sum_{j \geq 0} y_{j+1} \tau_1^j$$

and $f(x) = x^s = \tau_1^s$. That is

$$\tau_1^s = \sum_{n\geq 1} \frac{1}{n!} \frac{\mathrm{d}^{n-1}}{\mathrm{d}\xi^{n-1}} \left(\phi(\xi)^n \frac{\mathrm{d}\xi^s}{\mathrm{d}\xi} \right)\Big|_{\xi=0}$$

$$= \sum_{n\geq 1} \sum_{\underline{j}\in J(n,s)} \frac{s(n-1)!}{\underline{j}!} \underline{y}^{\underline{j}} \tag{2.3.16}$$

where

$$J(n,s) = \left\{ \underline{j} \geq 0 \Big| \sum_{i\geq 1} j_i = n \sum_{i\geq 1} i j_i = 2n - s \right\} \tag{2.3.17}$$

for $n \geq 1$ and $s \geq 1$.

From (2.3.16–17), we may soon find an explicit expression of the enufunction t of plane trees mentioned in §2.1.

Theorem 2.3.3 *For unicyclic maps of order $n+1$ with the valency of root-vertex being 2, let*

$$J_n = \left\{ \underline{j} \geq 0 \Big| \sum_{i\geq 1} j_i = n, \sum_{i\geq 1} i j_i = 2n \right\}. \tag{2.3.18}$$

Then, we have

$$\mu_2 = \frac{1}{1-y_2} + \sum_{k\geq 1} \sum_{n\geq k+1} \sum_{\underline{j}\in J_n} \mathcal{U}_2(k,n;\underline{j}) \underline{y}^{\underline{j}} \tag{2.3.19}$$

where

$$\mathcal{U}_2(k,n;\underline{j}) = \sum_{s\geq 1} \sum_{\underline{i}\in J_{\geq 2}(k,s)} \frac{sk!(n-k-1)!}{\underline{i}!(\underline{j}-\underline{i})!};$$

$$J_{\geq 2}(k,s) = \left\{ \underline{i} \geq 0 \Big| \sum_{j\geq 2} i_j = k, \sum_{j\geq 2} j i_j = s + 2k \right\}. \tag{2.3.20}$$

Proof From (2.3.15), we have

$$\mu_2 = 1 + \sum_{k\geq 1} \sum_{s\geq 0} \sum_{\underline{i}\in J_{\geq 2}(k,s)} \frac{k!}{\underline{i}!} \underline{y}^{\underline{i}} \tau_1^s.$$

By considering that

$$\sum_{j\geq 2} i_j = k; \ \sum_{j\geq 2} j i_j = 2k$$

has the only solution as $i_2 = k;\ i_j = 0,\ j > 2$, we obtain that

$$\mu_2 = \frac{1}{1-y_2} + \sum_{k\geq 1}\sum_{s\geq 1}\sum_{\underline{i}\in J_{\geq 2}(k,s)} \frac{k!}{\underline{i}!} \underline{y}^{\underline{i}} \tau_1^s$$

by (2.3.16) within interchange of the summations and substitutions of subscripts,

$$= \frac{1}{1-y_2} + \sum_{k\geq 1}\sum_{n\geq k+1}\sum_{s\geq 1}\sum_{\substack{\underline{i}\in J_{\geq 2}(k,s)\\ \underline{j}\in J_n}}$$

$$\times \frac{sk!(n-k-1)!}{\underline{i}!(\underline{j}-\underline{i})!} \underline{y}^{\underline{j}},$$

and hence the theorem. □

Further, from (2.3.13) all μ_k for $k \geq 3$ and then u is determined by an explicit expression.

Lemma 2.3.6 *For $U \in \mathcal{U}$, let $U_{(i,m(U)+2-i)}$ be the map obtained by splitting the root-vertex into a root-edge $\langle o_1, o_2\rangle$ such that the valency of o_1 is i and hence the valency of o_2 is $m(U)+2-i$ for $1 \leq i \leq m(U)+1$. Then, we have*

$$\mathcal{U}_{\mathrm{II}} = \sum_{U\in\mathcal{U}} \left\{U_{(i,m(U)+2-i)} \middle|\ 2 \leq i \leq m(U)\right\} \tag{2.3.21}$$

where $\mathcal{U}$ is the set of all unicyclic maps.

Proof Because for any $U \in \mathcal{U}$, among all the resultant maps of splitting the root-vertex of U, only those for $2 \leq i \leq m(U)$ are members of $\mathcal{U}_{\mathrm{II}}$. From Lemma 2.3.4 the lemma is found. □

Theorem 2.3.4 *The functional equation*

$$f = x^2 t + x \int_y \left(y \partial_{x,y} f\right), \tag{2.3.22}$$

where t is known to be the enufunction of plane trees and $\partial_{x,y}$ is the $\langle x, y\rangle$-difference shown by (1.6.8), is well defined among all the power series with coefficients being nonnegative integers and the solution is $f = u$, the enufunction of unicyclic maps.

Proof The first term of (2.3.22) is the contribution of $\mathcal{U}_{\rm I}$ to u as discussed above. Then, we are going to show that the second term is the contribution of $\mathcal{U}_{\rm II}$.

In fact, Lemma 2.3.6 leads to

$$\begin{aligned} u_{\mathcal{U}_{\rm II}} &= \sum_{U\in\mathcal{U}} \sum_{i=2}^{m(U)} x^i y_{m(U)+2-i} \underline{y}^{\underline{n}} \\ &= \sum_{U\in\mathcal{U}} \int_y \Big(\sum_{i=2}^{m(U)} x^i y^{m(U)+2-i}\Big) \underline{y}^{\underline{n}} \\ &= \int_y \Big(\sum_{U\in\mathcal{U}} \Big(xy\frac{yx^{m(U)} - xy^{m(U)}}{x-y}\Big)\underline{y}^{n}\Big) \\ &= x \int_y \Big(y\partial_{x,y}u(z)\Big). \end{aligned}$$

Hence, u is a solution of (2.3.22).

The uniqueness can be directly derived from the recursion obtained by equating the corresponding coefficients on the two sides of Eq.(2.3.22). □

From what was mentioned above we are also allowed to enumerate other unicyclic maps such as those having the root-edge not on the circuit and those with the circuit not forming the boundary of the inner face.

§2.4 General outerplanar maps

An *outerplanar map* is a rooted planar map such that the boundary of the root-face contains all the vertices. Of course, it is easily checked that plane trees and unicyclic maps are all special kinds of outerplanar maps. Here, we discuss outerplanar maps in general.

First, we investigate *strict* outerplanar maps, *i.e.*, those are no loops appearing in the inner domain of the root-loop, if any, and those from which no one of the resultant maps of contracting the root-edges consecutively has a loop in the inner domain of a root-loop. Let $\hat{\mathcal{O}}$ be the set of all the strict outerplanar maps. For convenience, the vertex map ϑ is assumed to be strict.

The set $\hat{\mathcal{O}}$ is divided into three parts: $\hat{\mathcal{O}}_0$, $\hat{\mathcal{O}}_{\mathrm{I}}$ and $\hat{\mathcal{O}}_{\mathrm{II}}$ where $\hat{\mathcal{O}}_0$ consists of only one map, the vertex map ϑ and $\hat{\mathcal{O}}_{\mathrm{I}}$ is the set of those with the root-edge being a loop. For any $M \in \hat{\mathcal{O}}$, let a be the root-edge.

Lemma 2.4.1 *Let* $\hat{\mathcal{O}}_{\langle \mathrm{I}\rangle} = \{M - a | \forall M \in \hat{\mathcal{O}}_{\mathrm{I}}\}$. *Then we have*

$$\hat{\mathcal{O}}_{\langle \mathrm{I}\rangle} = \hat{\mathcal{O}} \tag{2.4.1}$$

and

$$\left|\hat{\mathcal{O}}_{\mathrm{I}}\right| = \left|\hat{\mathcal{O}}\right|. \tag{2.4.2}$$

Proof For any $M \in \hat{\mathcal{O}}_{\langle \mathrm{I}\rangle}$, there is a unique map $M' \in \hat{\mathcal{O}}_{\mathrm{I}}$ obtained by adding a loop which is chosen to be its root-edge predecessive to a on the boundary of the root- face. From the strictness of M', M is strict as well. Therefore, $M \in \hat{\mathcal{O}}$.

Conversely, for any $M \in \hat{\mathcal{O}}$, because the map M' obtained by adding a loop as described above from M is a member of $\hat{\mathcal{O}}_{\mathrm{I}}$, we have $M \in \hat{\mathcal{O}}_{\langle \mathrm{I}\rangle}$.

Further, since the above procedure provides a bijection between $\hat{\mathcal{O}}_I$ and $\hat{\mathcal{O}}$, the lemma follows. □

Let $g_{\hat{\mathcal{O}}}$ be the enufunction of strict outerplanar maps with vertex partition. Then from Lemma 2.4.1, we have

$$g_{\hat{\mathcal{O}}_{\mathrm{I}}} = x^2 g_{\hat{\mathcal{O}}} \tag{2.4.3}$$

where $g_{\hat{\mathcal{O}}_{\mathrm{I}}}$ is the contribution of $\hat{\mathcal{O}}_{\mathrm{I}}$ to $g_{\hat{\mathcal{O}}}$.

Lemma 2.4.2 *Let* $\hat{\mathcal{O}}_{(\mathrm{II})} = \{M \bullet a | \forall M \in \hat{\mathcal{O}}_{\mathrm{II}}\}$, *where a is the root-edge of M. Then we have*

$$\hat{\mathcal{O}}_{(\mathrm{II})} = \hat{\mathcal{O}}. \tag{2.4.4}$$

Proof Because the root-edge a of a map $M \in \hat{\mathcal{O}}_{\mathrm{II}}$ is not a loop, it is a link on the root-face boundary. The contraction of a remains the strictness and the outerplanarity. This leads that $\hat{\mathcal{O}}_{(\mathrm{II})}$ is a subset of $\hat{\mathcal{O}}$.

On the other hand, for a map $M \in \hat{\mathcal{O}}$, we may always find a map M' by splitting the root-vertex of M into the root-edge a' of M'. It is easily seen that M' is strict outerplanar as well. Since a' is not a loop, we have that $M' \in \hat{\mathcal{O}}_{\mathrm{II}}$. Hence, $M \in \hat{\mathcal{O}}_{(\mathrm{II})}$. □

Unfortunately, the lemma does not provide a bijection between $\hat{\mathcal{O}}_{\mathrm{II}}$ and $\hat{\mathcal{O}}$. A refinement of the lemma is presented in the following form.

Lemma 2.4.3 *For $\hat{\mathcal{O}}_{\mathrm{II}}$, we have the following relation:*

$$\left|\hat{\mathcal{O}}_{\mathrm{II}}\right| = \sum_{M\in\hat{\mathcal{O}}} \left|\left\{M_{(i,m(M)-i+2)} \mid 1 \le i \le m(M)+1\right\}\right| \tag{2.4.5}$$

where $M_{(i,m(M)-i+2)}$ is the map obtained by splitting the root-vertex of M such that its root-vertex has valency i, hence the valency of the nonrooted end of its root-edge is $m(M)-i+2$, and $m(M)$ is the valency of the root-vertex of M.

Proof Because for any $M \in \hat{\mathcal{O}}$, we have

$$M_{(1,m(M)+1)},\ M_{(2,m(M))},\ M_{(3,m(M)-1)},\ \cdots,\ M_{(m(M)+1,1)}$$

in correspondence with it. Moreover, it can be verified that the correspondence determines a bijection between the two sets on the two sides of (2.4.5). From Lemma 2.4.2, the lemma follows. □

According to Lemma 2.4.3, we may see that

$$\begin{aligned}
g_{\hat{\mathcal{O}}_{\mathrm{II}}} &= \sum_{M\in\hat{\mathcal{O}}} \Big(\sum_{i=1}^{m(M)+1} x^i y_{m(M)-i+2} \Big) \underline{y}^{\underline{n}} \\
&= \sum_{M\in\hat{\mathcal{O}}} \int_y \Big(\sum_{i=1}^{m(M)+1} x^i y^{m(M)-i+2} \Big) \underline{y}^{\underline{n}} \\
&= x \int_y \sum_{M\in\hat{\mathcal{O}}} \left(y \frac{x^{m(M)+1} - y^{m(M)+1}}{x-y} \underline{y}^{\underline{n}} \right)
\end{aligned}$$

$$= x \int_y \left(y\delta_{x,y}(zg_{\hat{\mathcal{O}}}(z)) \right) \tag{2.4.6}$$

where $\delta_{x,y}$ is the (x,y)-difference as defined in §1.6.

Theorem 2.4.1 *The following functional equation*

$$(1-x^2)f = 1 + x \int_y \left(y\delta_{x,y}\left(zf(z)\right) \right) \tag{2.4.7}$$

is well defined among all the power series with coefficients nonnegative integers and the solution is

$$f = g_{\hat{\mathcal{O}}} = \sum_{M\in\hat{\mathcal{O}}} \sum_{\underline{n}\geq 0} x^{m(M)} \underline{y}^{\underline{n}(M)}. \tag{2.4.8}$$

Proof Since

$$g_{\hat{\mathcal{O}}} = g_{\hat{\mathcal{O}}_0} + g_{\hat{\mathcal{O}}_{\mathrm{I}}} + g_{\hat{\mathcal{O}}_{\mathrm{II}}},$$

from (2.4.3), (2.4.6) and that $g_{\hat{\mathcal{O}}_0} = 1$ the theorem is derived. □

Let $\hat{\underline{g}} = (\hat{g}_1, \hat{g}_2, \cdots)$ be the vector with

$$\hat{g}_i = \hat{g}_i(\underline{y}),\ i \geq 1,$$

as the coefficient of x^i in $g_{\hat{\mathcal{O}}}$ as a power series of x. Then, we are allowed to discuss $\hat{\underline{g}}$ instead of $g_{\hat{\mathcal{O}}}$.

Lemma 2.4.4 *The vector $\hat{\underline{g}}$ satisfies the following linear equation system:*

$$(I - Y_{\hat{\mathcal{O}}})\hat{\underline{g}}^{\mathrm{T}} = y_1 e_1^{\mathrm{T}} + e_2^{\mathrm{T}} \tag{2.4.9}$$

where e_i, $i = 1,2$, is the vector with all the components 0 but only the i-th 1, I is the unit matrix and

$$Y_{\hat{\mathcal{O}}} = \begin{pmatrix} y_2 & y_3 & y_4 & y_5 & y_6 & \cdots \\ y_1 & y_2 & y_3 & y_4 & y_5 & \cdots \\ 1 & y_1 & y_2 & y_3 & y_4 & \cdots \\ 0 & 1 & y_1 & y_2 & y_3 & \cdots \\ 0 & 0 & 1 & y_1 & y_2 & \cdots \\ 0 & 0 & 0 & 1 & y_1 & \cdots \\ \cdot & \cdot & \cdot & \cdot & \cdot & \cdots \end{pmatrix} \tag{2.4.10}$$

Proof From Theorem 2.4.1, by expanding the two sides of (2.4.7) into power series of x and equating the correspondent coefficients on the two sides, Eq.(2.4.9) is soon derived. □

From the lemma we may find

Theorem 2.4.2 *The enufunction has the following explicit form as*

$$g_{\hat{O}} = 1 + \underline{x} \sum_{i \geq 0} Y_{\hat{O}}^i (y_1 e_1 + e_2)^{\mathrm{T}} \tag{2.4.11}$$

where $\underline{x} = (x, x^2, x^3, \cdots)$.

Proof From Lemma 2.4.4, by employing

$$(I - Y_{\hat{O}})^{-1} = \sum_{i \geq 0} Y_{\hat{O}}^i, \tag{2.4.12}$$

the formula (2.4.11) is obtained. □

Now, we are allowed to investigate general outerplanar maps on the basis of strict ones. A planar map which consists of only one vertex with all the edges being loops is called a *loop bundled* map. For convenience, the vertex map is treated as a degenerate loop bundled map.

Let $\mathcal{L}$ be the set of all loop bundled maps. For a map M, the *inner addition* with the loop map O, denoted by $O \hat{+} M$, is defined to be the map obtained by identifying their root-vertices and putting M in the inner domain of O such that its root-edge is the same as O's and the edge which corresponds to the root-edge of M is the successor of the root-edge in the rotation at its root- vertex.

For a set $\mathcal{M}$ of maps, $O \hat{\odot} \mathcal{M}$ is defined to be the *inner product* where O is considered to be the set which consists of the loop map itself.

Lemma 2.4.5 *For* $\mathcal{L}$, *we have*

$$\mathcal{L} = \sum_{i \geq 0} \left(O \widehat{\odot} \mathcal{L} \right)^{\odot i} \tag{2.4.13}$$

where $\odot$ *is the* 1-*production defined in* §2.1.

Proof Since $(O\hat{\odot}\mathcal{L})^{\odot 0} = \vartheta$ is considered, the vertex map ϑ appears in both sides of (2.4.13).

In general, for any $L \in \mathcal{L}$, assume the valency of the root- face of L is i, $i \geq 1$. Then we can see that L is a member of $(O\hat{\odot}\mathcal{L})^{\odot i}$ on the right hand side of (2.4.13).

Conversely, it is easily seen that for any i, a map $L \in (O\hat{\odot}\mathcal{L})^{\odot i}$, L is of course a member of $\mathcal{L}$. □

Let φ be the enufunction of $\mathcal{L}$ with the valency of the root-vertex as a parameter. From this lemma, we have that φ satisfies the following relation

$$\varphi = \sum_{i\geq 0} (x^2\varphi)^i. \tag{2.4.14}$$

Since φ is a power series of x with all coefficients nonnegative, Eq.(2.4.14) has the solution

$$\varphi = \frac{1}{2x^2}\left(1 - \sqrt{1-4x^2}\right). \tag{2.4.15}$$

By expanding $\sqrt{1-4x^2}$ into a power series of x, we find

$$\varphi = \sum_{n\geq 0} \frac{(2n)!}{(n+1)!n!} x^{2n}. \tag{2.4.16}$$

Let $\mathcal{O}$ be the set of all general outerplanar maps. Moreover, $\mathcal{O}$ can be classified into three classes as well as above, *i.e.*,

$$\mathcal{O} = \mathcal{O}_0 + \mathcal{O}_{\mathrm{I}} + \mathcal{O}_{\mathrm{II}}.$$

Of course, $\mathcal{O}_0 = \hat{\mathcal{O}}_0$.

Lemma 2.4.6 *For* $\mathcal{O}_{\mathrm{I}}$, *we have*

$$\mathcal{O}_{\mathrm{I}} = \left(O\widehat{\odot}\mathcal{L}\right)\odot\mathcal{O} \tag{2.4.17}$$

where $\odot$ *is the inner production.*

Proof For any $O \in \mathcal{O}_{\mathrm{I}}$, because the root-edge is a loop, from the outerplanarity the inner domain only covers loops. It has to be a member of the set on the right hand side of (2.4.17).

Conversely, it is easy to see that any

$$O \in \left(O\widehat{\odot}\mathcal{L}\right)\odot\mathcal{O}$$

is also a member of the set on the left hand side of (2.4.17). □

By this lemma we are allowed to find the contribution of $\mathcal{O}_{\mathrm{I}}$ to the enufunction $g_{\mathcal{O}}$, *i.e.*,

$$g_{\mathcal{O}_{\mathrm{I}}} = x^2\varphi g_{\mathcal{O}}. \tag{2.4.18}$$

Further, from (2.4.16), we have

$$g_{\mathcal{O}_{\mathrm{I}}} = \sum_{n\geq 1}\frac{(2n-2)!x^{2n}}{n!(n-1)!}g_{\mathcal{O}}. \tag{2.4.19}$$

Lemma 2.4.7 *For $\mathcal{O}_{\mathrm{II}}$ we have*

$$\mathcal{O}_{\mathrm{II}} = \sum_{O\in\mathcal{O}}\left\{O_{(i,m(O)-i+2)}\middle|\ 1\leq i\leq m(O)+1\right\} \tag{2.4.20}$$

where $O_{(i,m(O)-i+2)}$ is the resultant map of splitting the root- vertex of O such that the valency of its root-vertex is i and $m(O)$ is the valency of the root-vertex of O.

Proof Because for any $O\in\mathcal{O}_{\mathrm{II}}$, the map obtained by contracting the root-edge of O is still an outerplanar map, the set on the left hand side is a subset of that on the right side of (2.4.20).

Conversely, for an $O\in\mathcal{O}$ and an i, $1\leq i\leq m(O)+1$, $O_{(i,m(O)-i+2)}$ is the unique one obtained from O by splitting the root-vertex such that the valency of its root-vertex is i. Since $O_{(i,m(O)-i+2)}$ is planar and its root-edge is not a loop, we see that

$$O_{(i,m(O)-i+2)}\in\mathcal{O}_{\mathrm{II}}.$$

Or in other words, the set on the right side is a subset of that on the left side of (2.4.20). □

According to this lemma we can find the contribution of $\mathcal{O}_{\mathrm{II}}$ to $g_{\mathcal{O}_{\mathrm{II}}}$ as

$$g_{\mathcal{O}_{\mathrm{II}}} = x\int_y\left(y\delta_{x,y}(zg_{\mathcal{O}}(z))\right) \tag{2.4.21}$$

in a way similar to the procedure of obtaining (2.4.6) in the strict case.

Theorem 2.4.3 *The following functional equation:*

$$(1 - x^2\varphi)\psi = 1 + x\int_y \left(y\delta_{x,y}(z\psi(z))\right) \tag{2.4.22}$$

is well defined and the solution is $\psi = g_O$.

Proof Since

$$g_O = g_{O_0} + g_{O_{\rm I}} + g_{O_{\rm II}},$$

from (2.4.18) and (2.4.21) with the consideration of $g_{O_0} = 1$, it is checked that g_O satisfies the equation (2.4.22).

The well definedness of (2.4.22) can be verified by showing that of the recursion determined by (2.4.22). □

Let $\underline{g} = (g_1, g_2, \cdots)$ be the vector determined by g_O such that $g_i = g_i(\underline{y})$, $i \geq 1$, is the coefficient of x^i in the power series of g_O.

Lemma 2.4.8 *The vector* $\underline{g}$ *satisfies the following linear equation system as*

$$(I - Y_O)\underline{g}^{\rm T} = y_1 e_1^{\rm T} + \sum_{n\geq 1} \frac{(2n-2)!}{n!(n-1)!} e_{2n}^{\rm T} \tag{2.4.23}$$

where e_i, $i = 1, 2n$,*is the vector of all entries* 0 *but the* i*-th* 1 *and*

$$Y_O = \begin{pmatrix} y_2 & y_3 & y_4 & y_5 & y_6 & \cdots \\ y_1 & y_2 & y_3 & y_4 & y_5 & \cdots \\ & y_1 & y_2 & y_3 & y_4 & \cdots \\ & & y_1 & y_2 & y_3 & \cdots \\ & & & y_1 & y_2 & \cdots \\ * & & & & & \ddots \end{pmatrix} \tag{2.4.24}$$

in which '' stands for*

$$y_{i,j} = \begin{cases} \dfrac{(2n-2)!}{n!(n-1)!}, & \text{if } i - j = 2n; \\[2ex] 0, & \text{if } i - j = 2n+1 \end{cases}$$

for $i - j \geq 2$ *and* $n = 1, 2, 3, \cdots$.

Proof It can be obtained in the usual way as used before by comparing the coefficients of the equation (2.4.22). □

Theorem 2.4.4 *The enufunction of general outerplanar maps has the explicit expression as*

$$g_O = 1 + \underline{x} \sum_{i \geq 0} Y_O^i \Big(y_1 e_1^{\mathrm{T}} + \sum_{n \geq 1} \frac{(2n-2)!}{n!(n-1)!} e_{2n}^{\mathrm{T}} \Big) \tag{2.4.25}$$

where $\underline{x} = (x, x^2, x^3, \cdots)$.

Proof Obtained directly from the equation (2.4.23). □

However, it seems no simpler relationship among the components of $\underline{g}$ as trees, wintersweets and unicyclic maps do in previous sections.

§2.5 Notes

2.5.1 Planted trees were introduced by Tutte in the sixties [Tut8]. In that paper the enumeration of this kind of colored trees with vertex partitions was also discussed to find a solution. However, in §2.1 a much simpler way is used. By this method it is also allowed to solve the colored problems. The enumeration of plane trees has been considered by many authors [HaT1, HPT1]. In [HaT1] and [HPT1], the nonrooted cases for uncolored and colored plane trees were respectively investigated. It seems that the matrix and the functional forms were firstly appeared in [Liu21].

2.5.2 Wintersweets were introduced to generalize plane trees to find an explicit expression of the enufunction with vertex partition in [Liu21]. However, the final formula (2.2.19) had not yet been extracted at that time. The unrooted and/or the colored cases can also be investigated for probably finding simpler explicit expressions.

2.5.3 The enumeration of unicyclic maps in [Liu21] provided the formula (2.3.9). The functional equation (2.3.22) can be seen in [Liu53] and [Liu54]. In those articles only the main ideas were described.

However, in §2.3 there is the simpler explicit expression (2.3.19) which is its first occurrence in the literature.

2.5.4 Eqs.(2.4.11) and (2.4.25) respectively appeared in [Liu25] and [Liu30]. Although a simpler explicit formula was obtained for the enufunction of nonseparable outerplanar maps with face partition [Liu36], nothing has been done for the case of vertex partition up to now. This means that the equations (2.4.11) and (2.4.25) have not been found to have a simpler explicit solution yet.

2.5.5 Naturally, one might ask how to count outerplanar maps on orientable or nonorientable surfaces. It causes to have some ways similar to those provided in §2.4. Of course, we have not known whether there are some results in publications right now.

2.5.6 Except for trees some unrooted cases have been done, all others including wintersweets, unicyclic maps and outerplanar maps have not been found to have simpler explicit formulae. From what we have obtained here, it looks to have some way, which can be used especially for wintersweets and unicyclic maps.

2.5.7 Outerplanar maps may also be generalized to those on surfaces such that there is a face whose boundary covers all the vertices, and further for those having a face boundary which is a circuit (without common vertex) covering all the vertices.

2.5.8 Although the enufunction for outerplanar maps with vertex partition has not yet been found to have a simpler formula, a number of simpler formulae have been obtained for the number of vertices, the number of faces and the number of edges as parameters [DoY1, Liu30, Liu32, Liu36, *etc.*].

Chapter 3

Triangulations

§3.1 Outerplanar triangulations

A *triangulation* is such a planar map whose faces are all triangles, *i.e.*, the boundary of faces are all 3-gons. If a map has all but one of its faces triangles, then it is called a *near-triangulation*. Planar near-triangulations are treated as triangulations on the disk.

Because only one triangulation, the triangle itself, is outerplanar, an outerplanar near-triangulation is allowed to be called an *outerplanar triangulation* without confusion. Of course, an outerplanar triangulation is a triangulation on the disk such that all the vertices are on the boundary of the disk.

For convenience we employ the description of maps in §1.1. Let $\mathcal{T}_{\text{ot}}$ be the set of all outerplanar triangulations. For $M = (\mathcal{X}_{\alpha,\beta}(X), \mathcal{P}) \in \mathcal{T}_{\text{ot}}$, $r \in X$ is chosen to be the root. Then, $v_r(M), e_r(M)$ and $f_r(M)$ stand for the root-vertex, root-edge and root-face of M respectively. Write

$$\mathcal{T}_{\text{ot}} = \mathcal{T}_{\text{ot}}^{(0)} + \mathcal{T}_{\text{ot}}^{(\text{I})} + \mathcal{T}_{\text{ot}}^{(\text{II})}, \tag{3.1.1}$$

i.e., $\mathcal{T}_{\text{ot}}$ is classified into $\mathcal{T}_{\text{ot}}^{(i)}, i = 0$, I, II, such that

$$\begin{cases} \mathcal{T}_{\text{ot}}^{(0)} = \{\vartheta\}, \ \vartheta \text{ is the vertex map;} \\ \mathcal{T}_{\text{ot}}^{(\text{I})} = \{M | e_r(M) \text{ is a cut-edge}\}; \\ \mathcal{T}_{\text{ot}}^{(\text{II})} = \{M | e_r(M) \text{ is on a circuit}\}. \end{cases}$$

Here we have to make the convention that the link map is in $\mathcal{T}_{\text{ot}}^{(\text{I})}$.

Lemma 3.1.1 *Let $\mathcal{T}^*_{\rm ot} = \{M|v_r$ is cut vertex $\}$ and $\mathcal{T}^{\triangle}_{\rm ot} = \mathcal{T}_{\rm ot} - \mathcal{T}^{(0)}_{\rm ot} - \mathcal{T}^*_{\rm ot}$. Then,*

$$\mathcal{T}^*_{\rm ot} = \sum_{k\geq 2} \mathcal{T}^{\triangle \odot k}_{\rm ot}. \tag{3.1.2}$$

Proof For any $M \in \mathcal{T}^*_{\rm ot}$ there exists a $k \geq 2$ such that $M = M_1 \odot M_2 \odot \cdots \odot M_k$, $M_i \in \mathcal{T}^{\triangle}_{\rm ot}$, $1 \leq i \leq k$. Hence, $M \in \mathcal{T}^{\triangle \odot k}_{\rm ot}$.

Conversely, for any member M of the set on the right hand side of (3.1.2), or in other words, there is $k \geq 2$ such that $M \in \mathcal{T}^{\triangle \odot k}_{\rm ot}$. Since $M = M_1 \odot M_2 \odot \cdots \odot M_k, M_i \in \mathcal{T}^{\triangle}_{\rm ot}, 1 \leq i \leq k$, the root-vertex of M is a cut vertex. Thus, $M \in \mathcal{T}^*_{\rm ot}$. □

Because all vertices are on the boundary of the root-face in an outerplanar triangulation, we are allowed to enumerate them with the valency of root-face as the parameter. Let

$$g_{\mathcal{T}_{\rm ot}} = \sum_{M\in\mathcal{T}_{\rm ot}} x^{m(M)}$$

$$= \sum_{m\geq 0} c_{\mathcal{T}_{\rm ot}}(m)x^m \tag{3.1.3}$$

where $m(M)$ is the valency of the root-face of M. Then from Lemma 3.1.1, the contribution of $\mathcal{T}^*_{\rm ot}$ to $g_{\mathcal{T}_{\rm ot}}$ is given by

$$g_{\mathcal{T}^*_{\rm ot}} = \sum_{k\geq 2} (g_{\mathcal{T}_{\rm ot}} - 1 - g^*_{\mathcal{T}_{\rm ot}})^k$$

$$= \frac{(g_{\mathcal{T}_{\rm ot}} - 1 - g_{\mathcal{T}^*_{\rm ot}})^2}{2 - g_{\mathcal{T}_{\rm ot}} + g_{\mathcal{T}^*_{\rm ot}}}. \tag{3.1.4}$$

By simplification we obtain that

$$g_{\mathcal{T}^*_{\rm ot}} = \frac{(g_{\mathcal{T}_{\rm ot}} - 1)^2}{g_{\mathcal{T}_{\rm ot}}}. \tag{3.1.5}$$

Lemma 3.1.2 *Let $\mathcal{T}^{(\rm I)}_{\langle \rm ot\rangle} = \{M - a|\forall M \in \mathcal{T}^{(\rm I)}_{\rm ot}\}$, where $a = e_r(M)$ is the root-edge of M. Then, we have*

$$\mathcal{T}^{(\rm I)}_{\langle \rm ot\rangle} = \mathcal{T}_{\rm ot} \times \mathcal{T}_{\rm ot}. \tag{3.1.6}$$

Proof Because any $M \in \mathcal{T}_{\rm ot}^{(\rm I)}$ the root-edge a of M is a cut edge, $M - a = M_1 + M_2, M_1, M_2 \in \mathcal{T}_{\rm ot}$. That implies $M - a \in \mathcal{T}_{\rm ot} \times \mathcal{T}_{\rm ot}$. Hence $\mathcal{T}_{\langle \rm ot\rangle}^{(\rm I)} \subseteq \mathcal{T}_{\rm ot} \times \mathcal{T}_{\rm ot}$.

Conversely, for $M \in \mathcal{T}_{\rm ot} \times \mathcal{T}_{\rm ot}$ we have $M = M_1 + M_2, M_1, M_2 \in \mathcal{T}_{\rm ot}$. The map M' obtained by adding an edge a connecting the two root-vertices $v_r(M_1)$ and $v_r(M_2)$ is a member of $\mathcal{T}_{\rm ot}^{(\rm I)}$. Thus $M = M' - a \in \mathcal{T}_{\langle \rm ot\rangle}^{(\rm I)}$. This means that $\mathcal{T}_{\rm ot} \times \mathcal{T}_{\rm ot} \subseteq \mathcal{T}_{\langle \rm ot\rangle}^{(\rm I)}$. □

According to this lemma we are allowed to evaluate the contribution of $\mathcal{T}_{\rm ot}^{(\rm I)}$ to $g_{\mathcal{T}_{\rm ot}}$. That is

$$g_{\mathcal{T}_{\rm ot}^{(\rm I)}} = x^2 g_{\mathcal{T}_{\rm ot}}^2 \tag{3.1.7}$$

in which x^2 represents the contribution of the root-edge to the root-face boundary of a map in $\mathcal{T}_{\rm ot}^{(\rm I)}$.

For a map $M \in \mathcal{T}_{\rm ot}^{(\rm II)}$, $a = Kr$ is the root-edge of M, let $M - \hat{a}$ be the map obtained by deleting the edge a such that the root of $M - \hat{a}$ is $\mathcal{P}\alpha\beta\mathcal{P}r$ where $(\mathcal{P}r, \mathcal{P}\alpha\beta\mathcal{P}r, (\mathcal{P}\alpha\beta)^2\mathcal{P}r)$ is the triangle incident to a.

Lemma 3.1.3 *Let* $\mathcal{T}_{\langle \rm ot\rangle}^{(\rm II)} - \{M - \hat{a} | \forall M \in \mathcal{T}_{\rm ot}^{(\rm II)}\}$. *Then we have*

$$\mathcal{T}_{\langle \rm ot\rangle}^{(\rm II)} = \mathcal{T}_{\rm ot}^{*} \tag{3.1.8}$$

where $\mathcal{T}_{\rm ot}^{*}$ *is defined as in Lemma* 3.1.1.

Proof From the outerplanarity, any map $M \in \mathcal{T}_{\rm ot}^{(\rm II)}$ has $v_{\mathcal{P}\alpha\beta\mathcal{P}r}$ as a cut vertex of $M - \hat{a}$. By the definition, $M - \hat{a}$ is a member of $\mathcal{T}_{\rm ot}^{*}$.

Conversely, for any $M^* \in \mathcal{T}_{\rm ot}^{*}$, because the root-vertex is a cut-vertex, we may join an edge $a^* = \langle v_{(\mathcal{P}\alpha\beta)^m r}, v_{\beta r}\rangle$ where the root-face of M^* is assumed to be $(r, \mathcal{P}\alpha\beta r, \cdots, (\mathcal{P}\alpha\beta)^m r)$ to obtain a map $M^* + \hat{a}^*$. Because a^* is an edge on a circuit in $M^* + \hat{a}^*$, it is easily checked that $M^* + \hat{a}^* \in \mathcal{T}_{\rm ot}^{(\rm II)}$. Hence, $M^* \in \mathcal{T}_{\langle \rm ot\rangle}^{(\rm II)}$. □

According to Lemma 3.1.3 the contribution of $\mathcal{T}_{\rm ot}^{(\rm II)}$ to $g_{\mathcal{T}_{\rm ot}}$ is evaluated to be

$$g_{\mathcal{T}_{\rm ot}^{(\rm II)}} = x^{-1} g_{\mathcal{T}_{\rm ot}^{*}} \tag{3.1.9}$$

by considering that the valency of the root-face of a map in $\mathcal{T}_{\rm ot}^{(\rm II)}$ is one less than that of the corresponding map in $\mathcal{T}_{\rm ot}^{*}$. Further from (3.1.5), we have

$$g_{\mathcal{T}_{\rm ot}^{(\rm II)}} = \frac{1}{x}\frac{(g_{\mathcal{T}_{\rm ot}} - 1)^2}{g_{\mathcal{T}_{\rm ot}}}. \tag{3.1.10}$$

Theorem 3.1.1 *The enufunction $g_{\mathcal{T}_{\rm ot}}$ satisfies the equation as follows:*

$$x^3 g_{\mathcal{T}_{\rm ot}}^3 + (1-x)g_{\mathcal{T}_{\rm ot}}^2 + (x-2)g_{\mathcal{T}_{\rm ot}} + 1 = 0. \tag{3.1.11}$$

Proof Since $g_{\mathcal{T}_{\rm ot}} = g_{\mathcal{T}_{\rm ot}^{(0)}} + g_{\mathcal{T}_{\rm ot}^{(\rm I)}} + g_{\mathcal{T}_{\rm ot}^{(\rm II)}}$, from (3.1.7) and (3.1.10) taking into consideration that $g_{\mathcal{T}_{\rm ot}^{(0)}} = 1$, we have

$$g_{\mathcal{T}_{\rm ot}} = 1 + x^2 g_{\mathcal{T}_{\rm ot}}^2 + \frac{1}{x}\frac{(g_{\mathcal{T}_{\rm ot}} - 1)^2}{g_{\mathcal{T}_{\rm ot}}}.$$

Multiplying by $x g_{\mathcal{T}_{\rm ot}}$ the two sides, after grouping we may find the equation (3.1.11). □

Although from (3.1.11) it is allowed to find $g_{\mathcal{T}_{\rm ot}}$ directly by using Lagrangian inversion, because of the cubicness of the equation the result is rather complicated for usage.

If we introduce the substitution

$$f = x g_{\mathcal{T}_{\rm ot}}, \tag{3.1.12}$$

then (3.1.11) becomes

$$f^3 + \frac{1-x}{x^2}f^2 + \frac{x-2}{x}f + 1 = 0 \tag{3.1.13}$$

which is in fact the form as

$$x^2(f^3 + f + 1) - x(f^2 + 2f) + f^2 = 0. \tag{3.1.14}$$

From (3.1.14), we find that

$$\frac{1}{x} = \frac{f + 2 \pm f\sqrt{1-4f}}{2f}. \tag{3.1.15}$$

According to $f(0) = 0$ and $f'(0) = 1$, we have

$$x = \frac{2f}{f + 2 - f\sqrt{1 - 4f}}. \tag{3.1.16}$$

Let $\sqrt{1 - 4f} = 1 - 2\theta$. Then,

$$\begin{cases} f = \theta(1 - \theta); \\ x = \dfrac{\theta(1 - \theta)}{1 + \theta^2(1 - \theta)}. \end{cases} \tag{3.1.17}$$

Theorem 3.1.2 *The enufunction $g_{\mathcal{T}_{\rm ot}}$ defined by (3.1.3) has the following explicit expression:*

$$g_{\mathcal{T}_{\rm ot}} = 1 + \sum_{\substack{m \geq 2 \\ 1 \leq k \leq \lfloor \frac{m}{2} \rfloor}} \frac{m!}{(k-1)!(m-k)!} \times \frac{(2m - 3k - 1)!}{(m - k + 1)!(m - 2k)!} x^m. \tag{3.1.18}$$

Proof By introducing a function $g = g(x, z)$ with the following parametric expressions:

$$\begin{cases} x = \dfrac{\theta(1 - \theta)}{1 + z\theta^2(1 - \theta)}; \\ g = \theta(1 - \theta), \end{cases} \tag{3.1.19}$$

from (3.1.17) we see that

$$g(x, 1) = f(x). \tag{3.1.20}$$

However, by employing Corollary 1.5.1 (Lagrangian inversion), we may find that

$$\partial_{(x,z)}^{(m,k)} g = \frac{(m-1)!}{(k-1)!(m-k-1)!} \times \frac{(2m - 3k - 3)!}{(m - k)!(m - 2k - 1)!} \tag{3.1.21}$$

for $m \geq 3$, $1 \leq k \leq \lfloor \frac{m-1}{2} \rfloor$, where ∂ is the coefficient operator defined by (1.5.11).

From (3.1.20), we obtain that

$$\begin{aligned} c_{\mathcal{T}_{\mathrm{ot}}}(m) &= \partial_x^m g_{\mathcal{T}_{\mathrm{ot}}} \\ &= \sum_{1 \leq k \leq \lfloor \frac{m}{2} \rfloor} \frac{m!}{(k-1)!(m-k)!} \\ &\quad \times \frac{(2m-3k-1)!}{(m-k+1)!(m-2k)!} \end{aligned} \tag{3.1.22}$$

for $m \geq 2$.

Moreover, it is easily checked that

$$c_{\mathcal{T}_{\mathrm{ot}}}(0) = 1 \text{ and } c_{\mathcal{T}_{\mathrm{ot}}}(1) = 0.$$

Therefore the theorem holds. □

The method used for deriving the equations satisfied by $g_{\mathcal{T}_{\mathrm{ot}}}$ can be generalized to enumerate this type of maps for considering two parameters, *e.g.*, the *size* (the edge number) and the valency of the root-face or the *order* (the vertex number) and the valency of the root-face.

Further, it can also be generalized for enumerating general near-triangulations on the plane, or in other words on the sphere, that will be discussed later.

§3.2 Planar triangulations

The simplest planar triangulation is of course the triangle with only two faces both of which are 3-gons.

Lemma 3.2.1 *If loops are not considered, then no planar triangulation has a cut-vertex.*

Proof By contradiction. Suppose M is a triangulation without loop and M has a cut-vertex v. Then, M has the form as

$$M = M_1 \dot{+} M_2,$$

i.e., $M = M_1 \cup M_2$ provided $M_1 \cap M_2 = v$. Because M is a triangulation, the root-face $f_r(M)$ is a 3-gon. Since the boundary of $f_r(M)$ is a disjoint union of those of $f_r(M_1)$ and $f_r(M_2)$, at least one of $f_r(M_1)$ and $f_r(M_2)$ is a 1-gon. This implies that there is a loop on M which contradicts the assumption. □

A loopless planar triangulation without multi-edge which contains the root-edge is called a *fair triangulation*. Let $\mathcal{T}_{\rm ft}$ ($\mathcal{T}_{\rm nl}$) be the set of all fair (loopless) triangulations.

If a loopless planar near-triangulation has no multi-edge which contains the root-edge, then it is said to be *fair* as well. If a map has that the root-face is a 2-gon, then it is said to be 2-*boundary*. A 2-boundary near-triangulation is allowed to call a 2-*boundary triangulation* for short some where.

Lemma 3.2.2 *A 2-boundary near-triangulation without inner multi-edge with the root-edge is fair iff it is without loop.*

Proof From the definition and the same reason as that of Lemma 3.2.1, the necessity is obvious. For the sufficiency, the only thing we have to do is to show that for a 2-boundary near-triangulation under consideration, without loop implies without cut vertex. Because if otherwise, the cut vertex is not on the boundary of the root-face which is a 2-gon. In fact, this is impossible by considering the triangularity of inner faces. □

The set of all fair (or loopless) 2-boundary near-triangulations is denoted by $\mathcal{T}_{\rm 2ft}$ (or $\mathcal{T}_{\rm 2nl}$).

Lemma 3.2.3 *Let* $\mathcal{T}_{\langle \rm 2x\rangle} = \{M - a | \forall M \in \mathcal{T}_{\rm 2x} - L\}$ *where* $a = e_r(M)$, *the root-edge of* $M = (\mathcal{X}, \mathcal{J})$ *and* L, *the link map. Then we have*

$$\mathcal{T}_{\langle \rm 2x\rangle} = \mathcal{T}_{\rm x} \tag{3.2.1}$$

where x=nl (*or* ft).

Proof For any $M \in \mathcal{T}_{\langle \rm 2nl\rangle}$, since there is a map $M' \in \mathcal{T}_{\rm 2nl}$, $M \neq L$, such that $M = M' - a$ where a is the root-edge of M', from Lemma 3.2.1 we know that $M \in \mathcal{T}_{\rm nl}$.

Conversely, for any $M \in \mathcal{T}_{\rm nl}$, we may construct a map M' by adding an edge which forms the root-face boundary with the root-edge of M. Of course, it is not L. From Lemma 3.2.2, we find that $M' \in \mathcal{T}_{\rm 2nl}$ and hence $M = M' - a \in \mathcal{T}_{\langle \rm 2nl\rangle}$. □

A 2-boundary map which has no inner edge whose two ends are the same as those of the root-edge is said to be *reduced*(fair as well when it is a loopless near-triangulation from Lemma 3.2.2). For convenience, we may always make the convention that the link map is not reduced. The composition of two maps $M_1 = (\mathcal{X}_1, \mathcal{P}_1)$ and $M_2 = (\mathcal{X}_2, \mathcal{P}_2)$ into another map M by identifying the root-edge Kr_2 of M_2 with the nonroot-edge $K\mathcal{P}_1^{-1}r_1$ adjacent to the root-edge of M_1 is called 1*e-addition* which is denoted by $M = M_1 \uplus M_2$. The root of M is chosen to be the same as that of M_1. In contrast, the 1-addition mentioned in §2.1 is also called 1*v-addition.*

For two sets of maps $\mathcal{M}_1$ and $\mathcal{M}_2$, let $\mathcal{M}_1 \oslash \mathcal{M}_2 = \{M_1 \uplus M_2 | \forall M_1 \in \mathcal{M}_1, \forall M_2 \in \mathcal{M}_2\}$ which is called 1*e-production.* And likewise, the 1-product in §2.1 is also called 1*v- product.* For a sequence of sets of maps $\mathcal{M}_1$, $\mathcal{M}_2$, $\cdots$, $\mathcal{M}_k$, we define that

$$\bigoslash_{i=1}^{k} \mathcal{M}_i = \left(\bigoslash_{i=1}^{k-1} \mathcal{M}_i \right) \oslash \mathcal{M}_k. \tag{3.2.2}$$

In particular, if $\mathcal{M}_i = \mathcal{M},\ i = 1, 2, \cdots, k$, we may write

$$\bigoslash_{i=1}^{k} \mathcal{M}_i = \mathcal{M}^{\oslash k}.$$

Let $\mathcal{T}_{\rm 2nl}^{\rm (R)}$ be the set of all reduced 2-boundary near-triangulations. It is easily verified that

$$\mathcal{T}_{\rm 2nl} = L + \sum_{i \geq 1} \mathcal{T}_{\rm 2nl}^{{\rm (R)}^{\oslash i}}. \tag{3.2.3}$$

Lemma 3.2.4 *We have the relation as*

$$\mathcal{T}_{\rm nl} = \mathcal{T}_{\rm ft} + (\mathcal{T}_{\rm 2nl} - L) \oslash \mathcal{T}_{\rm 2ft} \tag{3.2.4}$$

where L is the link map.

Proof A direct result derived from the definitions. □

Let $M = (\mathcal{X}, \mathcal{J}) \in \mathcal{T}_{2\text{nl}}$, $M \neq L$, with $v_r = (r, R_0, s, T_0, l)$ and $v_{\beta r} = (\delta r, \delta l, T_1, \delta s, R_1)$. Write M_1 and M_2 be the extended submaps of M such that $v_{r_1} = (r, R_0, s)$ and $v_{r_2} = (\delta l, T_1)$ with $v_{\beta r_1} = (\delta r, \delta s, R_1)$ and $v_{\beta r_2} = (l, T_0)$ respectively. Then, M is called the $2v$-*addition* of M_1 and M_2, denoted by $M = M_1 +: M_2$. Meanwhile, for two sets $\mathcal{M}_1$ and $\mathcal{M}_2$ of maps,

$$\mathcal{M}_1 \times: \mathcal{M}_2 = \{M_1 +: M_2 | \forall M_1 \in \mathcal{M}_1, \forall M_2 \in \mathcal{M}_2\}$$

is called the $2v$-*production* of $\mathcal{M}_1$ and $\mathcal{M}_2$.

Lemma 3.2.5 *We have*

$$\mathcal{T}_{2\text{nl}} = L + \mathcal{T}_{2\text{nl}} \times: \mathcal{T}_{\text{ft}} \tag{3.2.5}$$

where $\times:$ is the $2v$- product defined as above.

Proof From what is just mentioned, the conclusion is soon obtained. □

For a near-triangulation M another near-triangulation M' can be obtained by contracting the root-edge into a vertex while the two edges on the boundary of the inner triangle incident with the root-edge are identified. This operation is called *shrinking* the root-edge. Or write $M' = M\hat{\bullet}a$, $a = e_r(M)$.

For a map $M = (\mathcal{X}, \mathcal{J})$ with the root-vertex of valency $m(M)$, write $M_{[i]} = (\mathcal{X}_{[i]}, \mathcal{I})$ where $\mathcal{X}_{[i]} = \mathcal{X} - Kl_i + Ks + Kt + Kr'$, $l_i = \mathcal{J}^{m(M)-i+1}r$, $Ks = \{l_i, \alpha s, \beta l_i, \delta s\}$, $Kt = \{t, \alpha l_i, \beta t, \delta l_i\}$ and $Kr' = \{r', \alpha r', \beta r', \delta r'\}$ such that $\mathcal{I}r' = t$, $\mathcal{I}s = \delta r'$ and $\mathcal{I}\delta t = \delta s$ for $i = 2, 3, \cdots, m(M)+1$. Easy to check that $m(M_{[i]}) = i$.

The procedure of finding $M_{[i]}$, $i = 2, 3, \cdots, m(M)+1$ from M mentioned above is called *splitting* the edge into a triangle.

Lemma 3.2.6 *For a map M, let*

$$\mathcal{M}_{[m(M)+1]} = \{M_{[m(M)+1]}, M_{[m(M)]}, \cdots, M_{[2]}\} \tag{3.2.6}$$

where $M_{[i]}$ is the map obtained by splitting the edge into a triangle such that the valency of its root-vertex is i from M for $i = 2, 3, \cdots, m(M)+1$. Then, we have

$$\mathcal{T}_{\text{ft}} = \sum_{M \in \mathcal{T}_{\text{2nl}}} \mathcal{M}_{[m(M)+1]}. \tag{3.2.7}$$

Proof For $M \in \mathcal{T}_{\text{ft}}$, we may obtain M' by shrinking the root-edge. Because $M' \in \mathcal{T}_{\text{2nl}}$ from the fairness, it is soon found that M is a member of the set on the right hand side of (3.2.7).

Conversely, for any M in the set on the right hand side of (3.2.7), since M is obtained by splitting an edge into a triangle on a map M' in $\mathcal{T}_{\text{2nl}}$, from Lemma 3.2.1 we have $M \in \mathcal{T}_{\text{ft}}$. □

For convenience let us write

$$f_{\mathcal{T}}(x, y) = \sum_{M \in \mathcal{T}} x^{m(M)} y^{s(M)} \tag{3.2.8}$$

and

$$h_{\mathcal{T}}(y) = f_{\mathcal{T}}(1, y) \tag{3.2.9}$$

where $m(M)$ and $s(M)$ are respectively the valency of the root-vertex and the size of M.

Theorem 3.2.1 *The function $f = f_{\mathcal{T}_{\text{ft}}}(x, y)$ defined as in (3.2.8) satisfies the following equation:*

$$(1-h)f^2 - \Big(\frac{x^2y^3}{1-x} + 1 - h\Big)f + x^2y^3\Big(1 + \frac{xh}{1-x}\Big) = 0 \tag{3.2.10}$$

where $h = f_{x=1} = f_{\mathcal{T}_{\text{ft}}}(1, y)$ as shown in (3.2.9).

Proof According to Lemma 3.2.6 we have

$$\begin{aligned} f &= \sum_{M \in \mathcal{T}_{\text{2nl}}} \sum_{i=2}^{m(M)+1} x^i y^{s(M)+2} \\ &= x^2y^2 \sum_{M \in \mathcal{T}_{\text{2nl}}} \frac{1 - x^{m(M)}}{1-x} y^{s(M)} \\ &= x^2y^2 \frac{h_2 - f_2}{1-x} \end{aligned}$$

where $h_2 = f_{\mathcal{T}_{2\mathrm{nl}}}(1, y)$ and $f_2 = f_{\mathcal{T}_{2\mathrm{nl}}}(x, y)$ as defined above.

From Lemma 3.2.5, we find that

$$f_2 = xy + f_2 f.$$

Therefore, by combining them (3.2.10) can be soon found within rearrangement. □

Let $F = f - 1$ and $H = h - 1$. Then, (3.2.10) becomes

$$(1-x)HF^2 + (x^2Y + (1-x)H)F - x^3YH = 0 \tag{3.2.11}$$

where $Y = y^3$.

The discriminant denoted by $\Delta = \Delta(Y, H; x)$ of (3.2.11) is

$$\Delta = (x^2Y + (1-x)H)^2 + 4(1-x)x^3H^2. \tag{3.2.12}$$

If $x = \xi$ is considered to be a power series of Y such that Δ has a double root, then H and Y should satisfy the following equations:

$$\Delta|_{x=\xi} = 0; \quad \left.\frac{\partial \Delta}{\partial x}\right|_{x=\xi} = 0. \tag{3.2.13}$$

In fact, they are solved with the parametric expressions as

$$\xi = 1 + \frac{\xi^3}{(\xi-2)^2}Y; \ H = \frac{(\xi-2)^2}{\xi(2\xi-3)}. \tag{3.2.14}$$

Let $\eta = \xi - 1$. Then, (3.2.14) becomes

$$\eta = \frac{(\eta+1)^3}{(\eta-1)^2}Y, \ H = \frac{(\eta-1)^2}{(\eta+1)(2\eta-1)}. \tag{3.2.15}$$

This enables to employ Corollary 1.5.1 in the way as

$$H = -1 + \sum_{n\geq 1} \frac{Y^n}{n!} \frac{\mathrm{d}^{n-1}}{\mathrm{d}x^{n-1}} \left(\frac{(\eta+1)^{3n-2}(5\eta-1)}{(\eta-1)^{2n-1}(2\eta-1)^2} \right)\Bigg|_{\eta=0}$$

$$= -1 + \sum_{n\geq 1} \frac{A_{n-1}^{(n)} - 5A_{n-2}^{(n)}}{n} Y^n \tag{3.2.16}$$

where

$$A_s^{(n)} = \sum_{t=0}^{\min\{3n-2,s\}} \sum_{q=0}^{s-t} \binom{2n+s-t-q-2}{2n-2} \times \binom{3n-2}{t} (1+q)2^q \tag{3.2.17}$$

for $A_0^{(1)} = 1$, $A_{-1}^{(1)} = 0$, $s = n-1$ or $n-2$, and $n \geq 1$.

Corollary 3.2.1 *The number of reduced 2-boundary near- triangulations with $3m+1$ edges is just the number of fair triangulations of the size $3m$ for $m \geq 1$.*

Proof Since a 2-boundary near-triangulation is reduced iff it is fair, from above enumeration it can be checked that the number of fair triangulations of size $3m$ is the same as the number of 2-boundary ones of size $3m+1$ for $m \geq 1$. And, the 1-to-1 correspondence between them is given in Lemma 3.2.3 in this case. □

Theorem 3.2.2 *The function $g = f_{\mathcal{T}_{\rm nl}}(x, y)$ satisfies the following equation:*

$$(1-x)g = x^2y^3(1+g)(1+d-x(1+g)) \tag{3.2.18}$$

where $d = g|_{x=1} = f_{\mathcal{T}_{\rm nl}}(1, y)$.

Proof From Lemma 3.2.4, we have

$$g = f + \frac{(g_2 - xy)f}{xy}$$

where $f = f_{\mathcal{T}_{\rm ft}}(x, y)$ and $g_2 = f_{\mathcal{T}_{\rm 2nl}}(x, y)$ as defined in (3.2.8–9).

In virtue of Lemma 3.2.3, we have

$$g_2 = xy + xyg.$$

By combining them we may then find that

$$f = \frac{g}{1+g}, \quad h = \frac{d}{1+d}.$$

By substituting it into (3.2.10), Eq.(3.2.18) can be soon obtained within simpification. □

Let $G = g + 1$ while $D = d + 1$ and let $Y = y^3$. Then, (3.2.18) becomes

$$x^3 Y G^2 + (1 - x - x^2 Y D)G + x - 1 = 0. \tag{3.2.19}$$

In a manner similar to (3.2.10), by extracting the expressions of D and Y with ξ as a parameter from the restriction that the discriminant has a double root at $x = \xi$, an explicit form of D can be found as

$$D = 1 + \sum_{n \geq 1} \frac{B_n}{n} Y^n \tag{3.2.20}$$

where

$$B_n = \binom{3n}{n-1} + \sum_{t=0}^{n-2} \binom{3n}{t} \times \left[\binom{3n-t+1}{2n+2} - 5 \binom{3n-t}{2n+2} \right] \tag{3.2.21}$$

for $B_1 = 1$ and $n \geq 2$.

Let $T_1 = (\mathcal{X}_1, \mathcal{J}_1)$ and $T_2 = (\mathcal{X}_2, \mathcal{J}_2)$ be two triangulations. Then, we may construct another triangulation $T = (\mathcal{X}, \mathcal{J}_1)$ with $\mathcal{X} = \mathcal{X}_1 + \mathcal{X}_2$ such that r_1 is the root of T,

$$\mathcal{J} r_1 = \alpha \mathcal{J}_2^{m(T_2)-1} r_2, \ \mathcal{J} \alpha r_2 = \mathcal{J}_1 r_1, \ \mathcal{J} \delta r_2 = \alpha \mathcal{J}_1 \delta r_1$$

and $\mathcal{J} \beta r_1 = \mathcal{J}_2 \delta r_2$. This operation is called the *inner 2v-addition*. And, write

$$T = T_2 +\!\!:\!\rangle\, T_1.$$

For two sets of triangulations $\mathcal{T}_1$ and $\mathcal{T}_2$, let

$$\mathcal{T}_1 \times\!\!:\!\rangle\, \mathcal{T}_2 = \{T_1 +\!\!:\!\rangle T_2 | \forall T_1 \in \mathcal{T}_1, \forall T_2 \in \mathcal{T}_2\}$$

which is called the *inner 2v-product* of $\mathcal{T}_1$ and $\mathcal{T}_2$. Generally, write

$$\mathcal{T}_1 \times\!\!:\!\rangle\, \mathcal{T}_2 \times\!\!:\!\rangle \cdots \times\!\!:\!\rangle\, \mathcal{T}_k = \mathcal{T}^{\times:\rangle k}$$

when $\mathcal{T}_i = \mathcal{T}$ for $i = 1, 2, \cdots, k$, $k \geq 1$. From what has been obtained above, the following relations are found as

$$\mathcal{T}_{\mathrm{nl}} = \sum_{i \geq 1} \mathcal{T}_{\mathrm{ft}}^{\times:\rangle i} \tag{3.2.22}$$

and

$$\mathcal{T}_{\rm nl} = \mathcal{T}_{\rm ft} + \mathcal{T}_{\rm nl} \times\!\!\rangle\, \mathcal{T}_{\rm ft}. \qquad (3.2.23)$$

Corollary 3.2.2 *The number of loopless triangulations of size 3m with at least one multi-edge which contains the root-edge is*

$$\frac{B_m - A_{m-1}^{(m)} + 5A_{m-1}^{(m)}}{m}. \qquad (3.2.24)$$

Proof A direct result of Theorems 3.2.1–2. □

§3.3 Triangulations on the disc

As mentioned in §3.1, a planar near-triangulation is, in fact, a triangulation on the disc. A planar near-triangulation is said to be *nonseparable* if its underlying graph has no cut-vertex; *separable*, otherwise. Because a nonseparable map allows to have multi-edges from the definition, those of nonseparable near-triangulation without multiedge are said to be *strict*. In a strict near-triangulation, it is allowed to have a circuit which consists of three vertices with the property that there is a vertex in each of the inner and the outer domains. Such a circuit is called a *separating* triangle. A strict near-triangulation which has no separating triangle is said to be *simple*. Of course, a simple near-triangulation does not imply to be 3-connected.

Let $\mathcal{T}_{\rm ns}$, $\mathcal{T}_{\rm st}$, and $\mathcal{T}_{\rm sm}$ be the sets of all nonseparable, strict, and simple planar near-triangulations.

For $\mathcal{T}_{\rm ns}$, we may partition it into three parts: $\mathcal{T}_1, \mathcal{T}_2$ and $\mathcal{T}_3$ where $\mathcal{T}_1$ consists of only one map, the link map L, or write $\mathcal{T}_1 = L$ for simplicity without confusion, and

$$\mathcal{T}_2 = \{T | T - a \text{ is separable}\}.$$

As usual, a represents the root-edge $e_r(T)$ of T. Of course,

$$\mathcal{T}_3 = \mathcal{T}_{\rm ns} - \mathcal{T}_1 - \mathcal{T}_2 = \{T | T - a \text{ is nonseparable}\}.$$

Lemma 3.3.1 *Let* $\mathcal{T}_{\langle 2\rangle} = \{T - a | \forall T \in \mathcal{T}_2\}$. *Then we have*

$$\mathcal{T}_{\langle 2\rangle} = \mathcal{T}_{\rm ns} \odot \mathcal{T}_{\rm ns} \qquad (3.3.1)$$

where $\odot$ *is the* 1*-production defined in* §2.1.

Proof Because for any $M \in \mathcal{T}_2$, $M' = M - a$ has a cut-vertex, $M - a = M_1 \dot{+} M_2$ such that $M_1, M_2 \in \mathcal{T}_{\rm ns}$. Hence,

$$M' \in \mathcal{T}_{\rm ns} \bigodot \mathcal{T}_{\rm ns}.$$

Conversely, for any $M \in \mathcal{T}_{\rm ns} \odot \mathcal{T}_{\rm ns}$, we may add a new edge as the root-edge a in the inner domain of root-face to form a triangle face in the resultant map M'. Because $M' - a = M \in \mathcal{T}_{\rm ns} \odot \mathcal{T}_{\rm ns}$, we have $M' \in \mathcal{T}_2$ and hence $M \in \mathcal{T}_{\langle 2\rangle}$. This implies that $\mathcal{T}_{\rm ns} \odot \mathcal{T}_{\rm ns}$ is a subset of $\mathcal{T}_{\langle 2\rangle}$. □

Lemma 3.3.2 *Let* $\mathcal{T}_{\langle 3\rangle} = \{M - a | \forall M \in \mathcal{T}_3\}$. *Then we have*

$$\mathcal{T}_{\langle 3\rangle} = \mathcal{T}_{\rm ns} - \mathcal{T}_{\rm ns2} \tag{3.3.2}$$

where $\mathcal{T}_{\rm ns_2}$ *is the set of all such maps in* $\mathcal{T}_{\rm ns}$ *that the valencies of the root-faces are* 2.

Proof Because for any $M \in \mathcal{T}_3$, $M - a$ is nonseparable and of course is still planar near-triangulation, by considering that there is without 2-gon as a nonroot-face boundary we have $M - a \in \mathcal{T}_{\rm ns} - \mathcal{T}_{\rm ns_2}$.

Conversely, for any $M \in \mathcal{T}_{\rm ns} - \mathcal{T}_{\rm ns_2}$, we may always construct the map M' by adding an edge a which connects the root-vertex $v_r(M)$ and the vertex $v_{\beta\mathcal{P}\alpha r}(M)$ as the root-edge of M' in the inner domain of the root-face $f_r(M)$. It is easily seen that M' is still a nonseparable planar near-triangulation and that $M' - a$ is without cut-vertex. Hence, $M = M' - a \in \mathcal{T}_{\langle 3\rangle}$. □

Now, let us write

$$f = f_{\mathcal{T}_{\rm ns}} = \sum_{M \in \mathcal{T}_{\rm ns}} x^{n(M)} y^{s(M)}, \tag{3.3.3}$$

where $n(M)$ and $s(M)$ are the valency of the root-face and the size of M. And, write

$$f = \sum_{n \geq 2} F_n x^n \tag{3.3.4}$$

where F_n is a function of y. Of course, we have

$$F_2 = \frac{1}{x^2} f_{\mathcal{T}_{\rm ns_2}}. \tag{3.3.5}$$

Theorem 3.3.1 *The function f with F_2 satisfies the following equation:*

$$yf^2 + (y - x)f + x^2y(x - F_2) = 0, \tag{3.3.6}$$

where F_2 is defined in (3.3.4–5).

Proof First, we may see that

$$f_{\mathcal{T}_1} = x^2y. \tag{3.3.7}$$

Then for $\mathcal{T}_2$, in view of (3.3.1) we have

$$f_{\mathcal{T}_2} = \frac{y}{x}f^2. \tag{3.3.8}$$

Further for $\mathcal{T}_3$, from lemma 3.3.2 we have

$$f_{\mathcal{T}_3} = \frac{y}{x}\left(f - x^2F_2\right). \tag{3.3.9}$$

In consequence, from (3.3.7–9) we find that

$$f = x^2y + \frac{y}{x}f^2 + \frac{y}{x}\left(f - x^2F_2\right)$$

according to which, (3.3.6) is soon derived by multiplying the two sides by x and then grouping the terms. □

Let D be the discriminant of equation (3.3.6), that is

$$D = y^2 - 2yx + (1 + 4y^2F_2)x^2 - 4y^2x^3. \tag{3.3.10}$$

From Theorem 1.4.1, we have the following form:

$$\begin{aligned} D &= (y + ux)^2(1 - vx) \\ &= y^2 + (2yu - y^2v)x \\ &\quad +(u^2 - 2yuv)x^2 - u^2vx^3 \end{aligned} \tag{3.3.11}$$

where u and v are functions of y.

By identifying the coefficients of $x^i, i = 0, 1, 2, 3$, in (3.3.10) and (3.3.11), we find the parametric expressions:

$$2yu - y^2v = -2y,\ u^2v = 4y^2,$$

$$u^2 - 2yuv = 1 + 4y^2 F_2. \tag{3.3.12}$$

From (3.3.12), we have

$$y^3 = \frac{u^2(u+1)}{2}, \; v = \frac{2(u+1)}{y} \tag{3.3.13}$$

with u as a parameter.

According to Theorem 3.31, from (3.3.11) in expanding $\sqrt{(1-vx)}$ into a power series we obtain

$$f = \frac{1}{2y}\Big\{ x - y + (y + xu)$$

$$\times\Big(1 - \sum_{m\geq 1} \frac{(2m-2)!v^m x^m}{2^{2m-1}m!(m-1)!}\Big)\Big\}. \tag{3.3.14}$$

Let $\theta = u + 1$. Then, (3.3.13) becomes

$$\theta = y^3 \frac{2}{(1-v)^2}, \; v = \frac{2\theta}{y}. \tag{3.3.15}$$

Substituting (3.3.15) into (3.3.14), we have

$$f = \sum_{m\geq 2} \Big(\frac{(2m-4)!(1-\theta)v^{m-1}}{2^{2m-3}(m-1)!(m-2)!}$$

$$- \frac{(2n-2)!yv^m}{2^{2m-1}m!(m-1)!}\Big) \frac{x^m}{2y}. \tag{3.3.16}$$

By Corollary 1.5.1, we obtain that

$$v^m = \sum_{k\geq m} \frac{2^{m+k+1}m(3k-m-1)!}{(2k)!(k-m)!} y^{3k-m};$$

$$(1-\theta)v^{m-1} = \sum_{k\geq m-1} \frac{2^{m+k-1}(2m-3)}{(2k-1)!}$$

$$\times \frac{(3k-m-1)!}{(k-m+1)!} y^{3k-m+1}. \tag{3.3.17}$$

Therefore, by substituting (3.3.17) into (3.3.16), we find

$$f = \sum_{\substack{m\geq 2\\ k\geq m-1}} \frac{2^{k-m+2}(2m-3)!}{(2k)!(k-m+1)!(m-2)!} \times \frac{(3k-m-1)!}{(m-2)!} x^m y^{3k-m}. \quad (3.3.18)$$

For $T \in \mathcal{T}_{\rm st}$, or write $T = T(e_1, e_2, \cdots, e_\epsilon)$ where $e_i, i = 1, 2, \cdots, \epsilon$, are all the edges of T. For any $M_1, M_2, \cdots, M_\epsilon \in \mathcal{T}_{\rm ns}$, let $T(M_1, M_2, \cdots, M_\epsilon)$ be the map obtained by replacing e_i by M_i, $i = 1, 2, \cdots, \epsilon$, on T.

Lemma 3.3.3 *Suppose* $\mathcal{T}_{\rm st}(\mathcal{T}_{\rm ns_2}) = \{T(M_1, M_2, M_3, \cdots, M_\epsilon) | \forall M_1, \cdots, M_\epsilon \in \mathcal{T}_{\rm ns_2}\}$. *Then, we have*

$$\mathcal{T}_{\rm ns} = \mathcal{T}_{\rm st}(\mathcal{T}_{\rm ns_2}) + \mathcal{T}_{\rm ns_2} \quad (3.3.19)$$

where + is the union without common element.

Proof For any $M \in \mathcal{T}_{\rm ns} - \mathcal{T}_{\rm ns_2}$ if the outer domain of a pair of edges which has the same ends does not have any pair whose inner domain covers the inner domain of that pair, then it is said to be *maximal*. It is easily verified that the resultant map of shrinking each maximal pair of edges into an edge in M is strict, and hence $M \in \mathcal{T}_{\rm st}(\mathcal{T}_{\rm ns_2})$. That implies the set on the left hand side of (3.3.19) is a subset of that on the right.

Conversely, it is obvious that any map in the set on the right hand side of (3.3.19) is also a member of that on the left from the near-triangularity and the nonseparability. □

Let us write

$$f_{\rm st} = f_{\rm st}(x, y) = f_{\mathcal{T}_{\rm st}}(x, y)$$

as the enumerating function of $\mathcal{T}_{\rm st}$ with the valency of root-face and the edge number as parameters. Then, we soon from Lemma 3.3.3 see that

$$f = f_{\rm st}(x, F_2) + x^2 F_2. \quad (3.3.20)$$

Theorem 3.3.2 *The function $f_{\rm st}$ has the following expression:*

$$f_{\rm st} = \sum_{\substack{m\geq 3\\ t\geq m-1}} \frac{2(4t-2m-1)!}{(3t-m)!(t-m+1)!} \times \frac{(2m-3)!}{(m-1)!(m-3)!} x^m z^{3t-m} \tag{3.3.21}$$

where F_2 is defined in (3.3.4–5).

Proof Write $z = F_2$. Then, from (3.3.12) we have

$$z^3 = \frac{-(u+1)(1+3u)^3}{16u^4}, \quad y = \frac{-2u^2}{1+3u} z \tag{3.3.22}$$

with only u as the parameter. Now, we may take y as a function of z.

From (3.3.20) and (3.3.16), we find

$$f_{\rm st} + x^2 z = -\sum_{m\geq 2} \left(\frac{(2m-2)! v^m y}{2^{2m} m!(m-1)!} + \frac{(2m-4)! u v^{m-1}}{2^{2m-2}(m-1)!(m-2)!} \right) \frac{x^m}{y} \tag{3.3.23}$$

where

$$v = \frac{2(1+u)}{y}.$$

Hence,

$$v^m z^m = \frac{(-1)^m (1+u)^m (1+3u)^m}{u^{2m}};$$

$$\frac{uv^{m-1}}{y} z^m = \frac{(-1)^m (1+u)^{m-1} (1+3u)^m}{2u^{2m-1}}. \tag{3.3.24}$$

Let

$$w = -\frac{1+u}{u}.$$

Then we find

$$w = \frac{16}{(2-w)^3} z^3, \quad v^m z^m = w^m (2-w)^m,$$

$$\frac{uv^{m-1}}{y} z^m = -\frac{(2-w)^m w^{m-1}}{2}. \tag{3.3.25}$$

By employing Corollary 1.5.1, we have

$$v^m = \sum_{t\geq m} \frac{2^{2m+1}m(4t-2m-1)!}{(3t-m)!(t-m)!} z^{3t-m};$$

$$\frac{uv^{m-1}}{y} = \sum_{t\geq m-1} \frac{2^{2m-2}(3-2m)(4t-2m)!}{(3t-m)!(t-m+1)!} z^{3t-m}. \qquad (3.3.26)$$

And hence from (3.3.23), we may soon obtain (3.3.21). □

Let $\mathcal{T}_{st_3}$ be the set of all strict planar triangulations. Then, we write

$$F_3 = \partial_x^3 f_{st} = \sum_{M\in\mathcal{T}_{st_3}} y^{s(M)}, \qquad (3.3.27)$$

the coefficient of x^3 in f_{st}.

For a map $M \in \mathcal{T}_{sm}$, let $M\langle\mathcal{T}_{st_3}\rangle$ be the set of those produced by replacing some inner faces of M with maps in $\mathcal{T}_{st_3}$. And, write

$$\mathcal{T}_{sm}\langle\mathcal{T}_{st_3}\rangle = \bigcup_{M\in\mathcal{T}_{sm}} M\langle\mathcal{T}_{st_3}\rangle. \qquad (3.3.28)$$

Lemma 3.3.4 *Let L and T be the link map and the triangle map, or the respective sets of themselves. Then, we have*

$$\mathcal{T}_{st} - \{L, T\} = \mathcal{T}_{sm}\langle\mathcal{T}_{st_3}\rangle - \mathcal{T}_{st_3} \qquad (3.3.29)$$

where $\mathcal{T}_{sm}\langle\mathcal{T}_{st_3}\rangle$ is defined in (3.3.28).

Proof For any $M \in \mathcal{T}_{st}, M \neq L$ or T, let $T_1, T_2, \cdots, T_k$ be all maximal triangles each of which is not a face boundary in M. Since any two triangles only have two possibilities: they are without common inner point or one overlaps the other, we are allowed to treat all $T_i, i = 1, \cdots, k$, as triangular faces to find a map $M' \in \mathcal{T}_{sm}$. Of course, $M' \neq T$ and hence $M \notin \mathcal{T}_{st_3}$. That implies $M \in \mathcal{T}_{sm}\langle\mathcal{T}_{st_3}\rangle - \mathcal{T}_{st_3}$.

Conversely, for any $M \in \mathcal{T}_{sm}\langle\mathcal{T}_{st_3}\rangle - \mathcal{T}_{st_3}$, it is easily seen that $M \neq L$, or T. Moreover, it is also easily seen that M does not have multiedge from the strictness of maps in $\mathcal{T}_{st_3}$ and the simplicity of maps in $\mathcal{T}_{sm}$. That is $M \in \mathcal{T}_{st} - \{L, T\}$. □

For any $M \in \mathcal{T}_{\rm sm}$, it is known that the relation:

$$3t(M) + n(M) = 2s(M) \tag{3.3.30}$$

is satisfied where $t(M), n(M)$ and $s(M)$ are the number of nonrooted faces (triangles), the valency of the root-face and the size of M respectively.

Let us write

$$g_{\rm sm} = \sum_{M \in \mathcal{T}_{\rm sm}} x^{n(M)} z^{t(M)}.$$

Then, we by (3.3.30) have

$$\begin{aligned} f_{\rm sm} &= \sum_{M \in \mathcal{T}_{\rm sm}} x^{n(M)} y^{s(M)} \\ &= \sum_{M \in \mathcal{T}_{\rm sm}} x^{n(M)} y^{\frac{3t(M)+n(M)}{2}} \\ &= \sum_{M \in \mathcal{T}_{\rm sm}} (x\sqrt{y})^{n(M)} (\sqrt{y^3})^{t(M)} \\ &= g_{\rm sm}(x\sqrt{y}, \sqrt{y^3}). \end{aligned}$$

That implies

$$g_{\rm sm} = f_{\rm sm}\left(\frac{x}{\sqrt[3]{z}}, \sqrt[3]{z^2}\right). \tag{3.3.31}$$

Theorem 3.3.3 *The relation*

$$f_{\rm st} = x^2 y + x^3 y^3 + f_{\rm sm}\left(\frac{xy}{F_3^{\frac{1}{3}}}, \frac{1}{y} F_3^{\frac{2}{3}}\right) - x^3 F_3 \tag{3.3.32}$$

is satisfied.

Proof From (3.3.9) and (3.3.31), by considering that each edge on the root-face boundary has 1 contribution to $s(M)$ and that each triangle from $\mathcal{T}_{\rm st_3}$ has three residual contributions to $s(M)$, we have

$$f_{\rm st} = f_L + f_T + f_{\rm sm}\left(\frac{xy}{F_3^{\frac{1}{3}}}, \frac{1}{y} F_3^{\frac{2}{3}}\right) - x^3 F_3 \tag{3.3.33}$$

where f_T and f_L are the enufunctions of the sets which consist of a single triangle map and the link map respectively. Since $f_L = x^2 y$ and $f_T = x^3 y^3$, the relation (3.3.32) is soon found. □

From (3.3.22),(3.3.23) and (3.3.25), we find that

$$z^3 = \frac{w(2-w)^3}{16}, \ F_3 = \frac{w(1-w)}{2}$$

and

$$f_{\rm st} - x^2 z = \sum_{m\geq 3}\left(\frac{(2m-4)!w^{m-1}(2-w)^m}{2^{2m-1}(m-1)!(m-2)!} - \frac{(2m-2)!w^m(2-w)^m}{2^{2m}m!(m-1)!}\right)\frac{x^m}{z^m}. \tag{3.3.34}$$

Let

$$\tilde{x} = \frac{xy}{F_3^{\frac{1}{3}}}, \ \tilde{y} = \frac{1}{y}F_3^{\frac{2}{3}}. \tag{3.3.35}$$

Then, we find that

$$\tilde{y}^3 = \frac{4w(1-w)^2}{(2-w)^3} \tag{3.3.36}$$

and

$$f_{\rm sm} = -\frac{w(1-w)}{2}\tilde{x}^3 + \frac{4w(1-w)^2}{(2-w)^3}\tilde{x}^3 + \sum_{m\geq 3}\frac{2^{m+1}(2m-4)!w^{m-1}(1-w)^m}{m!(m-2)!} \times \frac{(m-2mw+3w)}{(2-w)^{2m}}\frac{\tilde{x}^m}{\tilde{y}^m}. \tag{3.3.37}$$

Take $\theta = \frac{w}{2-w}$. Then we obtain

$$\tilde{y}^3 = \theta(1-\theta)^2 \tag{3.3.38}$$

and

$$f_{\rm sm} = -\frac{\theta(1-\theta)}{(1+\theta)^2}\tilde{x}^3 + \theta(1-\theta)^2\tilde{x}^3 + \sum_{m\geq 3}\frac{(2m-4)!(m-3m\theta+6\theta)}{(m-2)!} \times \frac{\theta^{m-1}(1-\theta)^m}{m!}\frac{\tilde{x}^m}{\tilde{y}^m}. \tag{3.3.39}$$

In consequence, by employing Corollary 1.5.1 we may find the following expression

$$f_{\rm sm} = x^3y^3 + \sum_{j\geq 2}\left(\frac{(3j-5)!}{(2j-3)!(j-1)!}\right.$$
$$\left. - \sum_{1\leq i\leq \frac{j}{2}} \frac{(i+1)(4i-1)(3j-2i-5)!}{(2j-3)!(j-2i-1)!}\right)x^3y^{3j}$$
$$+ \sum_{\substack{m\geq 4\\ j\geq m-1}} \frac{(3j-2m-1)!}{(m-1)!(m-4)!(j-m+1)!}$$
$$\times\frac{(2m-4)!}{(2j-m)!}x^my^{3j-m}. \tag{3.3.40}$$

§3.4 Triangulations on the projective plane

First, we observe what the smallest sized triangulations on the projective plane look like. Since the size of a triangulation has to be divisible by 3, only two possibilities may happen for graphs of size 3. They are all of order 2. One consists of a loop and two links which are with the same pair of ends. Let Kx_2 be the loop and Kx_1, Kx_3 be the links. Then the triangulation is determined by the permutation as

$$\mathcal{P}_1 = (x_1, x_2, x_3, \beta x_2)(\beta x_1, \alpha\beta x_3).$$

The other is conformed by two loops and a link. Let Kx_1 be the link and Kx_2, Kx_3 be the loops. Then the triangulation on the projective plane is determined by the permutation as

$$\mathcal{P}_2 = (\beta x_1)(x_1, x_2, x_3, \beta x_3, \alpha\beta x_2).$$

For near-triangulations on the projective plane, those having a single face are of course simplest. Among them, the smallest one is the loop map $L_2 = (Kx, (x, \beta x))$ which was mentioned in §1.1. Further, we may also see that the underlying graph of any map on the projective plane is not without cycle. From Theorem 1.1.3 (or called the Euler formula), a map on the projective plane has a single face iff its underlying graph is unicyclic, because any unicyclic graph has its order equal to its size. If only maps whose underlying graphs have no cut-edge are considered, then the underlying graphs of single face maps on the projective plane are all conformed by a single circuit itself.

All maps considered here are for convenience assumed to be without a submap which is isomorphic to

$$L_1 = (Kx, (x, \alpha\beta x))$$

as mentioned in §1.1.

Lemma 3.4.1 *Let $H(x)$ be the enufunction of single face maps whose underlying graphs are without cut-edge on the projective plane. Then, we have*

$$H(x) = \frac{x}{1-x} \tag{3.4.1}$$

where x corresponds to the parameter of size.

Proof From the uniqueness of embedding a circuit on the projective plane, the lemma follows. □

For a map M with a single face on the projective plane, we may write

$$M = \Big(\bigcup_{i=1}^{k} Kx_i, (x_1, \beta x_k)\prod_{i=2}^{k}(x_i, \alpha\beta x_{i-1})\Big). \tag{3.4.2}$$

The underlying graph of M is a circuit of length k in view of Lemma 3.4.1. And the face of M can be checked to be

$$f(M) = (x_1, x_2, \cdots, x_k, \alpha x_1, \alpha x_2, \cdots, \alpha x_k) \tag{3.4.3}$$

which is a $2k$-gon.

A triangulation which has a Hamiltonian circuit on the projective plane is called a $\tilde{1}$H-*triangulation.* In general, a map which has a Hamiltonian circuit on the projective plane is called a $\tilde{1}$H-*map.*

Let M be a $\tilde{1}$H-map. Because M is on the projective plane, if all the edges not on the Hamiltonian circuit are deleted, then what remains is just a single face map. Thus, for a nonseparable outerplanar map, it is allowed to transform the Hamiltonian circuit into the boundary as shown in (3.4.3) and then to recover all the edges deleted. The resultant map is called the *boundary map* of M.

Lemma 3.4.2 *A map is a* $\tilde{1}$H*-map iff the underlying graph of its boundary map is nonseparable and outerplanar.*

Proof The necessity is obvious from what was discussed above. Conversely, for a map whose boundary map is nonseparable and outerplanar, it is easily checked that the Euler characteristic of the boundary map is 1, which is equal to that of the map itself. Therefore, the map is only allowed to be on the projective plane. Further, from the nonseparability and the outerplanarity which imply the Hamiltonicity, it is a $\tilde{1}$H-map. □

Let $\mathcal{T}_{\tilde{1}\mathrm{H}}$ be the set of all $\tilde{1}$H-triangulations. From Lemma 3.4.2, we may find that there is a bijection (or in other words, a 1-to-1 correspondence) between $\mathcal{T}_{\tilde{1}\mathrm{H}}$ and $\mathcal{T}_{\mathrm{NOT}}^{\mathrm{ev}}$. Here,

$$\mathcal{T}_{\mathrm{NOT}}^{\mathrm{ev}} = \sum_{i\geq 2} \mathcal{T}_{\mathrm{NOT}_{2i}}. \tag{3.4.4}$$

In this formula, $\mathcal{T}_{\mathrm{NOT}_{2i}}$ is the set of all the nonseparable outerplanar triangulations with the valency of the root-face being $2i$.

Lemma 3.4.3 *Let* $t_{\tilde{1}\mathrm{H}}(y)$ *be the enufunction of* $\tilde{1}$H*-triangulations with the parameter as the size corresponding to* y*. And, let* $f_{\mathrm{NOT}}(x,y)$ *be the enufunction of nonseparable outerplanar triangulations with the root-face valency and the number of edges not on the root-face boundary as two parameters which are correspond to* x *and* y *respectively. Then, we have*

$$t_{\tilde{1}\mathrm{H}}(y) = f_{\mathrm{NOT}}(\sqrt{y}, y) - yF_2(y) \tag{3.4.5}$$

where $F_2(y)$ *is the coefficient of the term* x^2 *and all the terms with noninteger powers of* x *in the power series of* $f_{\mathrm{NOT}}(x,y)$ *are all defined to be* 0.

Proof Since an outerplanar near-triangulation with the root-face, which is a $2i$-gon for $i \geq 2$, uniquely determines a $\tilde{1}$H-triangulation on the projective plane from what was discussed about (3.4.4) above, the formula (3.4.5) is soon found by considering the assumptions on $t_{\tilde{1}\mathrm{H}}(y)$ and $f_{\mathrm{NOT}}(x,y)$. □

Let $g(x,y)$ be the enufunction of outerplanar triangulations with the valency of the root-face and the number of edges not on the root-face boundary as parameters corresponding to x and y, respectively.

Lemma 3.4.4 *For $f_{\mathrm{NOT}}(x,y)$, we have the following relation:*

$$g(x,y) = x^2\{1+g(x,y)\}^2 + f_{\mathrm{NOT}}\Big(x\{1+g(x,y)\},y\Big) \tag{3.4.6}$$

with $g(x,y)$ as mentioned above.

Proof Because any outerplanar triangulation with its root-edge not a cut-edge can always be expressed as a 1-addition of a nonseparable outerplanar triangulation with the same number of general outerplanar triangulations as that of vertices on its outerface boundary of the nonseparable one, the lemma is soon found. □

Let M be a nonseparable outerplanar triangulation. Then from the nonseparability, we always have that

$$m(M) = n(M) + 3 \tag{3.4.7}$$

where $m(M)$ and $n(M)$ are the valency of the root-face and the number of edges not on the root-face boundary. It allows us only to consider $f_{\mathrm{NOT}}(x)$ instead of $f_{\mathrm{NOT}}(x,y)$ as described above.

Moreover, it can be shown that $f = f_{\mathrm{NOT}}(x)$ satisfies the equation as

$$f = x^2 + \frac{1}{x}f^2 \tag{3.4.8}$$

where the link map is treated as the degenerate case. Then, we have

$$f = \frac{x}{2}(1-\sqrt{1-4x}). \tag{3.4.9}$$

By the Taylor enpansion of the square root,

$$f = \sum_{m\geq 2}\frac{(2m-4)!}{(m-1)!(m-2)!}x^m. \tag{3.4.10}$$

In other words, we find

$$\partial_x^n f_{\mathrm{NOT}}(x) = \frac{(2n-4)!}{(n-1)!(n-2)!} \tag{3.4.11}$$

for $n \geq 2$.

Theorem 3.4.1 *The number of $\tilde{1}$H-triangulations of size n with $n = 0 (\mathrm{mod}\ 3)$ is*

$$T_{\tilde{1}\mathrm{H}}(n) = \frac{(\frac{4}{3}n)!}{(\frac{2}{3}n+1)!(\frac{2}{3}n)!} \tag{3.4.12}$$

for $n \geq 3$.

Proof Suppose m is the number of edges on the Hamiltonian circuit of a $\tilde{1}$H-triangulation. In view of (3.4.7), we have that $n = m + 2m - 3 = 3m - 3$. From (3.4.5) in Lemma 3.4.3, we have

$$T_{\tilde{1}\mathrm{H}}(n) = \partial_z^{2m} f_{\mathrm{NOT}}(z),$$

and hence the theorem by (3.4.11). □

For a map M not on the plane, according to the discussion on the discrete forms of Jordan curve theorem in [Liu58,§4.2], there exists a circuit C such that $M - C$ is still a connected component. Such a circuit is said to be *essential.*

If a map can be expressed in the form $M = M_1 \dot{+} M_2$, then M is said to be *splitable.* The vertex $v = M_1 \cap M_2$ is called a *split point* of M. For a planar map, it is splitable iff it is separable. However, for a nonplanar map it is not always true. A map which is without split point is said to be *nonsplitable.* In general, a nonsplitable map is allowed to have loops and/or multi-edges.

Let $C = (Kx_1, Kx_2, \cdots, Kx_l)$ be an essential circuit of a map M on the projective plane. Then, the dual of the resultant map $\mu(C)$ of deleting all the edges not on C from M is with a single vertex as

$$\mu(C^*) = \Big(\bigcup_{i=1}^{l} Kx_i, (x_1, x_2, \cdots, x_l, \alpha x_1, \alpha x_2, \cdots, \alpha x_l)\Big). \tag{3.4.13}$$

The underlying graph of $\mu(C)$, the dual of $\mu(C^*)$, is C itself.

Now, we may introduce a map M', called a $\tilde{1}$B-*map* of M, whose face set is the union of the face set $F(M)$ of M and the face set $F(\mu C)$ of $\mu(C)$ with the substitution of $\hat{x}_i$ and $\tilde{x}_i$ for x_i such that

$$K\hat{x}_i = \{x_i, \alpha\hat{x}_i, \beta x_i, \alpha\beta\hat{x}_i\}$$

and

$$K\tilde{x}_i = \{\tilde{x}_i, \alpha x_i, \beta\tilde{x}_i, \alpha\beta x_i\}$$

for $i = 1, 2, \cdots, l$. In fact, $F(\mu C)$ consists of the single face

$$f_0 = (\alpha\hat{x}_1, \alpha\hat{x}_2, \cdots, \alpha\hat{x}_l, \tilde{x}_1, \tilde{x}_2, \cdots, \tilde{x}_l) \tag{3.4.14}$$

which is called the *boundary* of M'.

Lemma 3.4.5 *For a map on the projective plane, all of its* $\tilde{1}$B-*maps are planar.*

Proof Let M be a map on the projective plane. For any $\tilde{1}$B-map M' of M, we see that the difference between the order and the size of M' is the same as that of M and that the face number of M' is one less than that of M. This implies that the Euler characteristic of M' is 2. Therefore, M' is a map on the sphere. The lemma is proved. □

Let T be a triangulation on the projective plane. From Lemma 3.4.5, we know that all the $\tilde{1}$B-maps are planar near-triangulations. Moreover, all the near-triangulations have the valency of root-face as an even number.

For a $\tilde{1}$B-map M' of a map M on the projective plane, we may always orient the faces such that each edge not on the boundary has the two directions occupied, while each edge on the boundary has only one of the two from the planarity of M'.

Accordingly, an orientation on M is determined such that each of the edges on the essential circuit has exactly one direction occupied twice to form a dicircuit, while each of others have both directions occupied once each by what is on M'.

A rooted map with such an orientation of faces is said to be *face-oriented*. Two face-oriented maps are isomorphic iff there is an isomorphism between them as rooted maps such that the essential circuits of them are in correspondence.

Lemma 3.4.6 *Let* $\mathcal{T}_{\tilde{1}\text{fot}}$ *be the set of all the* $\tilde{1}$B-*maps for nonsplitable face-oriented triangulations on the projective plane. And, let* $\mathcal{T}_{\text{nt}_{2i}}$ *be the set of all the rooted nonseparable planar near-triangulations* *with the valency of root-face being* $2i$. *Then, we have*

$$\mathcal{T}_{\tilde{1}\text{fot}} = \sum_{i \geq 2} \mathcal{T}_{\text{nt}_{2i}}. \tag{3.4.15}$$

Proof It is obvious that the set on the left hand side of (3.4.15) is a subset of that on the right, from what was mentioned above. Conversely, since any rooted nonseparable planar near-triangulation with the valency $2i$, $i \geq 2$, of the root-face can be transformed into a face-oriented triangulation which should be nonsplitable by marking the edges on the root-face boundary as

$$(x_1, x_2, \cdots, x_i, x_1, x_2, \cdots, x_i)$$

in cyclic order and then identifying the same edges in pairs to form the essential circuit $(x_1, x_2, \cdots, x_i)$, the set on the right hand side of (3.4.15) is a subset of that on the left. The lemma is obtained. □

Let $f = f_{\mathcal{T}_{\tilde{1}\text{fot}}}$ be the enufunction of nonsplitting face-oriented triangulations on the projective plane. And, write

$$f = \sum_{n \geq 3} F_{\tilde{1}\text{fot}}(n) y^n \tag{3.4.16}$$

where $F_{\tilde{1}\text{fot}}(n)$ is the number of nonsplitting face-oriented triangulations on the projective plane with n edges.

Theorem 3.4.2 *For* $n \geq 3$, *we have*

$$F_{\tilde{1}\text{fot}}(n) = \sum_{i=2}^{\frac{n}{3}-1} \frac{2^{k-2i+2}(4i-3)!(n-2i-1)!}{(2k)!(k-2i+1)!((2i-2)!)^2} \tag{3.4.17}$$

where $n = 3k$, $k \geq 1$.

Proof From the uniqueness for a nonsplitable face-oriented triangulation on the projective plane in correspondence with a rooted nonseparable planar near-triangulation of even valency of the root-face at least 4 and vice versa, in view of Lemma 3.4.6 we are allowed to employ (3.3.18). By considering the relationship (3.4.7) between the boundary valency and the number of edges not on the boundary for an outerplanar triangulation and the availability of the range of values in a term of (3.3.18), the theorem is soon found. □

§3.5 Triangulations on the torus

First, we see what the smallest map on the torus looks like. Because maps on the torus should have two edges on different circuits from the Euler formula, the smallest one denoted by M_0 is of order 1 with two loops Kx_1 and Kx_2. The dual of M_0 is

$$M_0^* = \Big(Kx_1 \cup Kx_2, (x_1, \alpha\beta x_2, \alpha\beta x_1, x_2)\Big).$$

If a new edge Kx_3 is added on M_0^* to obtain M_1 and M_2 as

$$M_1 = \Big(Kx_1 \cup Kx_2 \cup Kx_3, (x_1, x_3, x_2, \alpha\beta x_1, \alpha\beta x_3, \alpha\beta x_2)\Big)$$

and

$$M_2 = \Big(Kx_1 \cup Kx_2 \cup Kx_3, (x_1, x_2, x_3, \alpha\beta x_1, \alpha\beta x_2, \alpha\beta x_3)\Big),$$

then it can easily be shown that both M_1 and M_2 are triangulations which are actually the same in the rooted case on the torus.

For a map M on the torus, from the basic theory of maps described in §1.1 we may find that the resultant map M' of deleting an edge which is on the common boundary of two faces, if any, has only one face and that the dual M'^* of M' has the following form within topological equivalence as

$$M'^* = \Big(Kx_0 \cup Kx_1 \cup Kx_2, (x_0, x_1, x_2, \alpha\beta x_0, \alpha\beta x_1, \alpha\beta x_2)\Big) \tag{3.5.1}$$

where x_0, x_1 and x_2 are all on paths in general and one of them is allowed to be at a vertex. Such a map M' is called a 1-*basic submap* of M.

In addition, one might pay attention to that all the maps just mentioned above are self-dual.

If a map has three disjoint paths with the same pair of ends such that its vertices are all on the paths, then it is said to be *thetalic*.

For a map M on the torus, let $B = B(M)$ be a 1-basic submap of M. Then, the map obtained by cutting along the edges of B on M is called 1B-*map* of M. It is easily shown that any 1B-map is planar with a boundary of even lengths.

As to triangulations on the torus, unlike $\tilde{1}$H-maps on the projective plane, a thetalic triangulation has its 1-basic submap not unique in general. If a rooted triangulation has a 1-basic submap marked, then it is said to be *basic oriented.* Two basic oriented maps on the torus are treated to be the same if there is an isomorphism with the marked basic submaps in correspondence.

What should be mentioned here is to assume all maps considered in this section are without a submap isomorphic to $L_1 = (Kx, (x, \alpha\beta x))$. That is no loss of generality anyway.

Lemma 3.5.1 *A triangulation T on the torus is thetalic iff there is a* 1*-basic submap of T such that the* 1B*-map of T is a nonseparable outerplanar triangulation of even order at least* 4.

Proof The necessity is obvious from what was discussed above. Conversely, from the nonseparability, we are allowed to assume that the boundary of the 1B–map of T is a circuit of length $2k$, we may write the $2k$ edges in the cyclic order as

$$(x_1, x_2, \cdots, x_k, \alpha\beta x_1, \alpha\beta x_2, \cdots, \alpha\beta x_k) \tag{3.5.2}$$

where x_1 is the root. By identifying the pairs of edges as shown in (3.5.2), we find that T is a triangulation on the torus from the triangularity of the 1B-map and the Euler formula. Thus, in virtue of the outerplanarity and (3.5.1) for $k \geq 2$, the sufficiency is obtained. □

For any outerplanar triangulation of even order at least 4, the face boundary is assumed to be as shown in (3.5.2), one might ask how many ways can be used to form basic triangulations on the torus.

There are two possibilities which have to be considered. One is to find the ways to separate x_1, x_2, $\cdots$, x_k into two consecutive parts, in other words two segments in linear order. We may see that the number of ways in this case is $k - 1$. The other is to determine the ways to separate x_1, x_2, $\cdots$, x_k into three segments. That is

$$\binom{k-1}{2}.$$

Lemma 3.5.2 *For a nonseparable outerplanar triangulation T_{2k} of order $2k$, $k \geq 2$, we have*

$$\binom{k}{2}$$

distinct basic oriented thetalic triangulations of order k on the torus, which has T_{2k} as a 1B-map of them.

Proof In view of the nonseparability, the outerface of the outerplanar triangulation can be written as in the form (3.5.2). According to (3.5.1), any partition of x_1, x_2, $\cdots$, x_k into three segments can form a basic oriented thetalic triangulation on the torus, whose 1B-map is T_{2k}. However, one of the three segments is allowed to be null (the degenerate case of the segment being a vertex). From what was mentioned on the number of ways for such partitions, we see that

$$(k-1) + \binom{k-1}{2} = \binom{k}{2}$$

basic oriented thetalic triangulations are found on the torus from T_{2k}, $k \geq 2$. □

Let $t_{1\mathrm{T}}(x)$ be the enufunction of basic oriented thetalic triangulations on the torus. Then, we may also write

$$t_{1\mathrm{T}}(x) = \sum_{m \geq 3} T_{1\mathrm{T}}(m) x^m \tag{3.5.3}$$

where $T_{1\mathrm{T}}(m)$ is the number of such maps of size $\epsilon = m$, $m \geq 3$.

Because 1B-maps of a thetalic triangulation T of size $m(T)$ on the torus are all with $2k(T)$ edges on the outerface boundary such that

$$m(T) = k(T) + (2k(T) - 3) = 3k(T) - 3,$$

and hence

$$k(T) = \frac{m(T)+3}{3}, \tag{3.5.4}$$

we always have $m(T) = 0 \pmod 2$ and that the order of T is

$$\nu = k(T) - 1 = \frac{m(T)}{3}. \tag{3.5.5}$$

Theorem 3.5.1 *The number of basic oriented thetalic triangulations with size m on the torus is*

$$T_{1\mathrm{T}}(m) = \binom{\frac{m}{3}+1}{2} \frac{(\frac{4}{3}m)!}{(\frac{2}{3}m+1)!(\frac{2}{3}m)!} \tag{3.5.6}$$

for $m \geq 3$.

Proof Because the outerface boundary of 1B-maps of a thetalic triangulations with size m on the torus has

$$2k = 2\Big(\frac{m}{3}+1\Big) = \frac{2m}{3} + 2$$

edges from (3.5.4), the theorem is soon obtained by Lemma 3.5.2 and (3.4.11). □

Let T be a triangulation on the torus. If T is splitable, then we have $T = T_1 + T_2$ such that the fact: T_1 is not in the inner domain of a face f of T_2 implies the fact: T_2 is in the inner domain of a face f of T_1. From the triangularity of F, it is only possible that if f is not a 1-gon then f is a 2-gon. This leads that if T_1 does not have a submap L_1, then T_2 has a submap L_1. Therefore, there is no difference between nonsplitability and without submap L_1 for a triangulation on the torus.

In what follows, we adopt $\mathcal{T}_{\mathrm{nt}}(2i)$ to stand for the set of all the rooted nonseparable planar near-triangulations with outerface boundary of length $2i$.

Lemma 3.5.3 *Let $\mathcal{T}_{1\mathrm{B}}$ be the set of all the* 1B*-maps of nonsplitable basic oriented triangulations on the torus. Then, we have*

$$\mathcal{T}_{1\mathrm{B}} = \bigcup_{i\geq 2} \mathcal{T}_{\mathrm{nt}}(2i) \tag{3.5.7}$$

in any way.

Proof First, it is easy to check that the case of $i = 1$ does not make sense because no 1-basic submap is of size 1.

For any $T \in \mathcal{T}_{1B}$, all the 1B-maps of T for basic submaps of size i are planar near-triangulations by the Euler formula and the nonseparability is derived from the nonsplitability. Thus, the set on the left hand side of (3.5.7) is a subset of the set on the right.

Conversely, for each $T \in \mathcal{T}_{\rm nt}(2i)$, the outerface boundary of T is assumed to be in the cyclic order as

$$(x_1, x_2, \cdots, x_i, \alpha\beta x_1, \alpha\beta x_2, \cdots, \alpha\beta x_i). \tag{3.5.8}$$

By partitioning $x_1 x_2, \cdots, x_i$ into two or three segments and then identifying with the corresponding segments in $\alpha\beta x_1, \alpha\beta x_2, \cdots, \alpha\beta x_i$, we may find a 1-basic oriented triangulation T' for what is just obtained in identifying as the basic submap. Of course, T' is splitable from the nonseparability of T and T is a 1B-map of T'. Hence, the set on the right hand side of (3.5.7) is a subset of the left. □

Lemma 3.5.4 *Let $\mathcal{T}_{1\rm tr}(i)$ be the set of all the basic oriented triangulations (without L_1) with a 1-basic submap of size i. Then, we have*

$$\left|\mathcal{T}_{1\rm tr}(i)\right| = \binom{i}{2} \left|\mathcal{T}_{\rm nt}(2i)\right| \tag{3.5.9}$$

for $i \geq 2$.

Proof First for $T \in \mathcal{T}_{1\rm tr}(i)$, by cutting along the edges in the given 1-basic submap of size i a unique rooted planar near- triangulation T' with the outerface boundary of length $2i$ is obtained. From the nonsplitability of T, T' has no submap L_1 and hence no cut vertex. Therefore, T' is nonseparable.

Conversely for any $T' \in \mathcal{T}_{\rm nt}(2i)$, assume that its outerface boundary is in the cyclic order as shown in (3.5.8). From Lemma 3.5.2, we may get $\binom{i}{2}$ distinct basic oriented triangulations on the torus.

In consequence, a 1-to-$\binom{i}{2}$ correspondence from $\mathcal{T}_{\rm nt}(2i)$ onto $\mathcal{T}_{1\rm tr}(i)$ is found. That implies the lemma. □

Let $f = f_{\mathcal{T}_{1\text{tr}}}(y)$ be the enufunction of nonsplitable basic oriented triangulations on the torus with size as the parameter. Or we write

$$f = \sum_{n\geq 3} F_{1\text{tr}}(n) y^n \tag{3.5.10}$$

where $F_{1\text{tr}}(n)$ is the number of such triangulations of size n on the torus.

Theorem 3.5.2 *For $F_{1\text{tr}}(n)$ defined in (3.5.10), we have*

$$F_{1\text{tr}}(n) = \sum_{i=2}^{\frac{n}{3}+1} \frac{2^{k-2i+1} i!(4i-3)!(n-2i-1)!}{(2k)!(k-2i+1)!((2i-2)!)^2(i-2)!} \tag{3.5.11}$$

where $n = 3k$, $k \geq 1$.

Proof From Lemma 3.5.4, by employing (3.3.18) with considering the available upper bound of i from (3..4.7), the theorem may soon be obtained. □

§3.6 Notes

3.6.1 The enumeration of triangulations was first introduced by Tutte in the beginning of sixties [Tut3, Tut11, Tut21–24, Tut27]. In fact, the topics he worked on at that time were much concentrated on planar triangulations for the sake of attaching the four color problem. Many elegant results related to nonseparable planar triangulations were found. Soon after him, Brown [Bro1] and Mullin [Mul4, Mul6] developed his methods to count triangulations on the disk and some more general cases of planar triangulations respectively with simplification. More than a dozen of years later, Dong and Liu worked on the most general case to obtain a number of simpler results [DoL1].

3.6.2 Outerplanar triangulations were investigated in [Liu35] as a special case without inner vertices of a near-triangulation on the sphere. However, the enumerating formula in that paper was rather sophisticated.

3.6.3 The method used in §3.2 looks new in that the fair triangulations are introduced such that the independent equations for enumerating themselves and loopless triangulations have been found as (3.2.10) and (3.2.18). However, the functions f and g of two variables are needed to extract an explicit expression in a simpler form for each. Meanwhile, the explicit formulae of h and d are still needed to simplify for those of summation free or a summation with only positive terms.

3.6.4 Although simple near-triangulations are not 3- connected in general, it can be shown that simple planar triangulations are all 3-connected and further it can also be shown that a planar triangulation is 4-connected iff it is simple. The first recursion formula for enumerating simple planar triangulations occurred in [Tut3]. However, a more simpler recursion and an explicit formula were presented in [Liu5].

3.6.5 For $f_{\text{NOT}}(z)$, if Lemma 3.4.4 is considered, since $g = g_{\mathcal{T}_{\text{ot}}}$ as shown in (3.1.11), by (3.1.17) it can be seen that

$$z = x(1+g) = \frac{\theta(1-\theta)(2+\theta^2(1-\theta))}{1+\theta^2(1-\theta)}.$$

By empoying Corollary 1.5.1, we have

$$\partial_z^n f_{\text{NOT}}(z) = \frac{1}{n}\frac{\mathrm{d}^{n-1}}{\mathrm{d}\theta^{n-1}}\left\{\frac{(1+\theta^2(1-\theta))^n\theta(2-3\theta)}{(1-\theta)^n(2+\theta^2(1-\theta))^n}\right\}\Bigg|_{\theta=0}$$

which looks rather complicated in form.

3.6.6 From face-oriented triangulations which are enumerated in §3.4 on the projective plane, one might look for how many face-oriented triangulations which correspond to a rooted triangulation in enumeration.

3.6.7 There also are problems of determining how many 1B-maps which correspond to a triangulation on the torus in §3.5.

Chapter 4

Cubic Maps

§4.1 Planar cubic maps

A map with all of its vertices 3-valent is said to be *cubic*. A *near-cubic map* is a map such that all of its vertices but one is 3-valent. The exceptional vertex is always marked as the root-vertex.

Let $\mathcal{M}_{\rm nc}$ be the set of all rooted near-cubic planar maps. For convenience, the vertex map ϑ is assumed to be a member of $\mathcal{M}_{\rm nc}$.

For $M \in \mathcal{M}_{\rm nc}$, let $m(M)$ and $n(M)$ be the valency of the root-vertex and the number of nonrooted vertices respectively. Of course, nonrooted vertices all have valency 3. Then we have

$$m(M) + 3n(M) = 2\epsilon(M) \tag{4.1.1}$$

where $\epsilon(M)$ is the size of M. From (4.1.1) it is also known that

$$m(M) = n(M)(\text{mod } 2). \tag{4.1.2}$$

Write $\mathcal{M}_{\rm nc_i} = \{M|\forall M \in \mathcal{M}_{\rm nc}, m(M) = i\}$ for $i \geq 1$. Then we have

$$\mathcal{M}_{\rm nc} = \vartheta + \sum_{i\geq 1} \mathcal{M}_{\rm nc_i} \tag{4.1.3}$$

where ϑ stands for the set of the single element ϑ itself as well.

Further, let

$$\mathcal{M}_{\rm nc}^{\rm I} = \{M|\forall M \in \mathcal{M}_{\rm nc}, a \text{ is a loop}\},$$

$$\mathcal{M}_{\rm nc}^{\rm II} = \{M|\forall M \in \mathcal{M}_{\rm nc}, a \text{ is a link}\}$$

where a is the root-edge of M. Then we also have

$$\mathcal{M}_{\rm nc} = \vartheta + \mathcal{M}^{\rm I}_{\rm nc} + \mathcal{M}^{\rm II}_{\rm nc}. \tag{4.1.4}$$

Lemma 4.1.1 *Let $\mathcal{M}^{\rm I}_{\langle \rm nc\rangle} = \{M - a|\forall M \in \mathcal{M}^{\rm I}_{\rm nc}\}$. Then, we have*

$$\mathcal{M}^{\rm I}_{\langle \rm nc\rangle} = \mathcal{M}_{\rm nc} \odot \mathcal{M}_{\rm nc}, \tag{4.1.5}$$

where $\odot$ is the 1v-production.

Proof For any $M \in \mathcal{M}^{\rm I}_{\langle \rm nc\rangle}$, since there is a map $M' \in \mathcal{M}^{\rm I}_{\rm nc}$ such that $M = M' - a$, $a = e_r(M')$ is a loop, we have $M = M_1 \dot{+} M_2$, where M_1 and M_2 are the inner and outer parts of the loop a respectively. Because both M_1 and M_2 are near-cubic planar maps, we see that M is a member of the set on the right hand side of (4.1.5).

Conversely, for any $M \in \mathcal{M}_{\rm nc} \odot \mathcal{M}_{\rm nc}$, since $M = M_1 \dot{+} M_2$ we may construct a map M' by adding a loop a as its root-edge such that M_1 and M_2 are in the inner and outer domains of a on M, respectively. By virtue of the near-cubicness of M_1 and M_2, we see that $M' \in \mathcal{M}^{\rm I}_{\rm nc}$. From $M = M' - a$, M is a member of the set on the left of (4.1.5). □

Lemma 4.1.2 *Let $\mathcal{M}^{\rm II}_{(\rm nc)} = \{M \bullet a|\forall M \in \mathcal{M}^{\rm II}_{\rm nc}\}$. Then, we have*

$$\mathcal{M}^{\rm II}_{(\rm nc)}(\rm N) = \mathcal{M}_{\rm nc} - \mathcal{M}^{\rm II}_{(\rm nc)}(\rm S) - \mathcal{M}_{\rm nc_1} - \vartheta,$$

$$\mathcal{M}^{\rm II}_{(\rm nc)}(\rm S) = \mathcal{M}_{\rm nc} \odot \mathcal{M}_{\rm nc_1} \tag{4.1.6}$$

where $\mathcal{M}^{\rm II}_{(\rm nc)}(\rm S)$ and $\mathcal{M}^{\rm II}_{(\rm nc)}(\rm N)$ are the sets of those with the root-edge separable and not respectively.

Proof Only the first is shown. For any $M \in \mathcal{M}^{\rm II}_{(\rm nc)}(\rm N)$, since there is a map $M' \in \mathcal{M}^{\rm II}_{\rm nc}(\rm N)$ such that $M = M' \bullet a$, $a = e_r(M')$ is a link, and the nonrooted end of a has to be 3-valent, we have that M is allowed to be neither the vertex map ϑ nor a map in $\mathcal{M}_{\rm nc_1}$. Because of the nonseparability of a, $M \notin \mathcal{M}^{\rm II}_{(\rm nc)}(\rm S)$ as well. In view of the near-cubicness which is easily checked, we see that M is a member of the set on the right hand side of (4.1.6).

Conversely, for any M in the set on the right hand side of (4.1.6), because M is neither ϑ nor those with the valency of the root-vertex

$m(M) = 1$ or 2 and hence $m(M) \geq 3$, we are allowed to construct a map M' by splitting the root-vertex into two vertices o_1 and o_2 such that o_1 is the root-vertex and that the nonrooted end o_2 is of valency 3. Since $M \notin \mathcal{M}^{\mathrm{II}}_{(\mathrm{nc})}(\mathrm{S})$, it is easily seen that $M' \in \mathcal{M}^{\mathrm{II}}_{\mathrm{nc}}(\mathrm{N})$. Therefore, from

$$M = M' \bullet a,\ a = e_r(M') = (o_1, o_2),$$

we have that M is a member of the set on the left of (4.1.6). □

Let $f(x, y, z)$ be the enufunction of rooted near-cubic planar maps and write

$$f(x, y, z) = \sum_{M \in \mathcal{M}_{\mathrm{nc}}} x^{m(M)} y^{l(M)} z^{n(M)} \tag{4.1.7}$$

where $m(M)$, $l(M)$ and $n(M)$ are the valency of root-vertex, the valency of root-face and the number of nonrooted vertices of M respectively.

Theorem 4.1.1 *The enufunction $f = f(x, y, z)$ as shown in (4.1.7) satisfies the equation as*

$$\left(1 - x^2yu - \frac{yz}{x}\right)f = 1 - \frac{yz}{x}\left(1 + xF_1\right) \tag{4.1.8}$$

where $u = f_{y=1}$ and F_1 is the coefficient of the term involving x in f.

Proof Let f_ϑ, $f_{\mathcal{M}^{\mathrm{I}}_{\mathrm{nc}}}$ and $f_{\mathcal{M}^{\mathrm{II}}_{\mathrm{nc}}}$ be the contributions of ϑ, $\mathcal{M}^{\mathrm{I}}_{\mathrm{nc}}$ and $\mathcal{M}^{\mathrm{II}}_{\mathrm{nc}}$ to $f(x, y, z)$ respectively. Then, from (4.1.4) we have

$$f = f_\vartheta + f_{\mathcal{M}^{\mathrm{I}}_{\mathrm{nc}}} + f_{\mathcal{M}^{\mathrm{II}}_{\mathrm{nc}}}. \tag{4.1.9}$$

First, since the vertex map ϑ is with $m(\vartheta)$, $l(\vartheta)$ and $n(\vartheta)$ all zero, we may see that

$$f_\vartheta = 1. \tag{4.1.10}$$

For any $M \in \mathcal{M}^{\mathrm{I}}_{\mathrm{nc}}$, let $M - a = M_1 \dot{+} M_2$ from Lemma 4.1.1 where M_2 is assumed to be in the inner domain of a. Then,

$$m(M) = m(M_1) + m(M_2) + 2,$$

$$l(M) = l(M_1) + 1,$$
$$n(M) = n(M_1) + n(M_2). \tag{4.1.11}$$

By Lemma 4.1.1 and (4.1.11) we have

$$f_{\mathcal{M}^{\mathrm{I}}_{\mathrm{nc}}} = x^2yuf. \tag{4.1.12}$$

For $M \in \mathcal{M}^{\mathrm{II}}_{\mathrm{nc}}$, since

$$\begin{aligned} m(M) &= m(M \bullet a) - 1, \\ l(M) &= l(M \bullet a) + 1, \\ n(M) &= n(M \bullet a) + 1, \end{aligned} \tag{4.1.13}$$

by Lemma 4.1.2 we find that

$$f_{\mathcal{M}^{\mathrm{II}}_{\mathrm{nc}}} = \frac{yz}{x}\Big(f - xF_1 - 1\Big). \tag{4.1.14}$$

According to (4.1.9), (4.1.10), (4.1.12) and (4.1.14), we obtain

$$f = 1 + x^2yuf + \frac{yz}{x}\Big(f - xF_1 - 1\Big)$$

from which Eq.(4.1.8) is derived within grouping. □

If $y = 1$ is fixed, then from (4.1.8) we see that the function $u = u(y, z)$ with $U_1 = F_1|_{y=1}$ satisfies the following quadratic equation:

$$x^2u^2 - \Big(1 - \frac{z}{x}\Big)u - \frac{z}{x}(xU_1 + 1) + 1 = 0. \tag{4.1.15}$$

The discriminant of the Eq.(4.1.15) in u is

$$D = \Big(1 - \frac{z}{x}\Big)^2 - 4x^2\Big(1 - \frac{z}{x}(xU_1 + 1)\Big),$$

and hence

$$x^2D = 4x^4(zU_1 - 1) + 4zx^3 + x^2 - 2xz + z^2 \tag{4.1.16}$$

which is not a perfect square anyway.

According to the greatest power of x and the last term in (4.1.16), it can be verified that

$$x^2D = (z - ax)^2(1 - 2bx + cx^2), \tag{4.1.17}$$

where

$$a = 1 - 2\theta,\ b = \frac{2\theta}{x},\ c = \frac{12\theta^2 - 4\theta}{z^2},$$

and further

$$\theta = \frac{z^2}{(1-2\theta)(1-4\theta)},$$

$$zU_1 = \frac{\theta(1-6\theta)}{1-4\theta}. \tag{4.1.18}$$

Lemma 4.1.3 *Let $\mathcal{M}_{(\mathrm{nc}_1)} = \{M \bullet a | \forall M \in \mathcal{M}_{\mathrm{nc}_1}\}$ for $a = e_r(M)$. Then we have*

$$\mathcal{M}_{(\mathrm{nc}_1)} = \mathcal{M}_{\mathrm{nc}_2} \tag{4.1.19}$$

where $\mathcal{M}_{\mathrm{nc}_2}$ is defined in (4.1.3).

Proof For any $M \in \mathcal{M}_{(\mathrm{nc}_1)}$, there is a map $M' \in \mathcal{M}_{\mathrm{nc}_1}$ such that $M = M' \bullet a$ with $a = e_r(M')$. Since the valency of the nonrooted end of a in M' is 3, the valency of the root-vertex of M is 2. From the cubicness of M' we see that $M \in \mathcal{M}_{\mathrm{nc}_2}$.

Conversely, for any $M \in \mathcal{M}_{\mathrm{nc}_2}$ we may always find a map M' by adding a new vertex o with an edge $(o, v_r(M))$ from M such that $v_r(M') = o$ and $e_r(M') = (o, v_r(M)) = a'$. Of course, the valency of o is 1 and the valency of $v_r(M)$ in M' is 3. By the cubicness of M, we see that $M = M' \bullet a' \in \mathcal{M}_{(\mathrm{nc}_1)}$.

As consequence the lemma is obtained. □

Lemma 4.1.4 *Let $\mathcal{M}_{(\mathrm{nc}_2)} = \{M \bullet a | \forall M \in \mathcal{M}_{\mathrm{nc}_2}\}$ for $a = e_r(M)$. Then we have*

$$\mathcal{M}_{(\mathrm{nc}_2)} - \vartheta = \mathcal{M}_{\mathrm{nc}_3} \tag{4.1.20}$$

where $\mathcal{M}_{\mathrm{nc}_3}$ is defined in (4.1.3).

Proof For $M \in \mathcal{M}_{(\mathrm{nc}_2)} - \theta$ let $M' \in \mathcal{M}_{\mathrm{nc}_2}$ be the map that $M = M' \bullet a$, $a = e(M')$. Since the rooted end of a is of valency 2 and M is not the loop map, the valency of root-vertex of M is 3. From the cubicness of M', we see that $M \in \mathcal{M}_{\mathrm{nc}_3}$.

On the other hand, for a map $M \in \mathcal{M}_{\mathrm{nc}_3}$, we may always construct a map M' by splitting its root-vertex into two vertices connected with an edge a as the root-edge of M' such that the rooted end is of valency 2 and the other, of valency 3. From the cubicness of M, we see that

$M' \in \mathcal{M}_{\text{nc}_2}$. By considering $M = M' \bullet a$, $a = e_r(M')$, we have $M \in \mathcal{M}_{(\text{nc}_2)}$.

Therefore, the lemma is true. □

Let t_{nc_2} and t_{nc_3} be the enufunctions of rooted near-cubic planar maps of the respective root-vertex valency 2 and 3 with the number of nonrooted vertices as the parameter. Since U_1 is the enufunction of rooted near-cubic planar maps of the root-vertex valency 1 with the number of nonrooted vertices as the parameter, from Lemma 4.1.3 we have

$$U_1 = zt_{\text{nc}_2}. \tag{4.1.21}$$

Further, from Lemma 4.1.4 we have

$$t_{\text{nc}_2} - 1 = zt_{\text{nc}_3}. \tag{4.1.22}$$

Combining (4.1.21) and (4.1.22), we soon find

$$U_1 = z + z^2 t_{\text{nc}_3}. \tag{4.1.23}$$

Theorem 4.1.2 *The number of rooted planar cubic maps with $n(\geq 1)$ nonrooted vertices is*

$$T_{\text{nc}_3}(n) = \begin{cases} \displaystyle\sum_{i=0}^{k-1} \frac{2^{2k-i+1}(k+i-1)!(2k-i+1)!}{(k+2)!(k+1)!(k-i-1)!i!}, \\ \qquad \textit{if } n = 2k-1; \\ 0, \quad \textit{otherwise} \end{cases} \tag{4.1.24}$$

for $k \geq 2$.

Proof Based on (4.1.18) it can be calculate that

$$H(z) = \frac{\mathrm{d}zU_1}{\mathrm{d}z^2} = \frac{1}{(1-4\theta)^2}. \tag{4.1.25}$$

By employing Corollary 1.5.1 we may find that

$$\partial_z^{2k} H(x) = \frac{1}{k!} \frac{\mathrm{d}^{k-1}}{\mathrm{d}\theta^{k-1}} \left(\frac{1}{(1-2\theta)^k (1-4\theta)^{k+3}} \right) \Bigg|_{\theta=0}$$

$$= \sum_{i=0}^{k-1} \frac{2^{2k-i+1}(k+i-1)!(2k+l-i)!}{k!(k+2)!(k-i-1)!i!}$$

for $k \geq 1$.

Since $\partial_z^{2k-1} U_1 = \partial_z^{2k} z U_1$, by (4.1.23) and (4.1.25) we may obtain (4.1.24). This is the theorem. □

§4.2 Bipartite cubic maps

A map M whose vertex set V can be divided into two classes: V_1 and V_2 such that no edge has its two ends in the same class is said to be *bipartite*. A *bipartite cubic map* is such a map which is both bipartite and cubic in its own right. In this section, only planar maps are dealt with.

Now we investigate bipartite near-cubic planar maps.

Lemma 4.2.1 *Any bipartite near-cubic map (not necessary to be planar) has no cut-edge.*

Proof By contradiction, suppose M is a near-cubic map and has a cut-edge (u, v). We may write $M = M_1 \cup (u, v) \cup M_2$ such that u and v are incident with M_1 and M_2 respectively. Of course, $M_1 \cap M_2 = \emptyset$.

Because at least one of u and v is not the root-vertex of M, If M_1 is fixed to be with the root-vertex, then M_2 is a bipartite near-cubic map with the root-vertex v of valency 2. By the bipartiteness, let ν_1 and ν_2 be the cardinalities of the two classes of the vertex set $V(M_2)$ of M_2. Assume that the root-vertex is in the part of ν_1 vertices without loss of generality since, otherwise, the interchange of the subscripts between M_1 an M_2 has to be made. By the cubicness we have that $3\nu_1 - 1 = 3\nu_2$ and hence $3(\nu_1 - \nu_2) = 1$. This is a contradiction. □

For a bipartite near-cubic map M (not necessary to be planar), let $m(M)$ and $n(M)$ be the valency of the root-vertex and the number of nonrooted vertices respectively. From a similar discussion in the proof of Lemma 4.2.1, we may find that

$$m(M) = 0(\mathrm{mod}\ 3). \tag{4.2.1}$$

Further, since $3(\nu(M) - 1) + m(M) = 2\epsilon(M)$ and $n(M) = \nu(M) - 1$, we may also find that

$$n(M) = m(M)(\mathrm{mod}\ 2),\ \epsilon(M) = 0(\mathrm{mod}\ 3). \tag{4.2.2}$$

Let $\mathcal{M}_{\mathrm{bc}}$ be the set of all bipartite near-cubic planar maps. The vertex map ϑ is assumed to be in $\mathcal{M}_{\mathrm{bc}}$. And, let

$$\mathcal{M}_{\mathrm{bc}_i} = \{M | M \in \mathcal{M}_{\mathrm{bc}}, (v_r, v_{\beta r}) \text{ is an } i\text{-multiedge}\}$$

for $i = 1, 2$, and 3. From the cubicness, we have

$$\mathcal{M}_{\mathrm{bc}} = \sum_{i=0}^{3} \mathcal{M}_{\mathrm{bc}_i} \tag{4.2.3}$$

where $\mathcal{M}_{\mathrm{bc}_0} = \vartheta$.

For $M = (\mathcal{X}_{\alpha,\beta}, \mathcal{P})$ with r being the root, suppose the two edges incident with $v_{\beta r}$ are e_t and e_l, $t, l \in \mathcal{X}_{\alpha,\beta}$, where $l = \mathcal{P}\alpha\beta r$. Let $M \bullet v_{\beta r}$ be the map obtained by contracting $v_{\beta r}$ and its three incident edges into a vertex with the root being $\mathcal{P}\alpha\beta l$, and let

$$\mathcal{M}_{\mathrm{bc}_1}^{(v_{\beta r})} = \{M \bullet v_{\beta r} | \forall M \in \mathcal{M}_{\mathrm{bc}_1}\}. \tag{4.2.4}$$

Then, we have

$$\mathcal{M}_{\mathrm{bc}_1}^{(v_{\beta r})} = \mathcal{M}_{\mathrm{bc}_1}^{(v_{\beta r})_1} + \mathcal{M}_{\mathrm{bc}_1}^{(v_{\beta r})_2} \tag{4.2.5}$$

where

$$\mathcal{M}_{\mathrm{bc}_1}^{(v_{\beta r})_1} = \{M | \forall M \in \mathcal{M}_{\mathrm{bc}_1}^{(v_{\beta r})}, v_{\beta t} = v_{\beta l}\};$$

$$\mathcal{M}_{\mathrm{bc}_1}^{(v_{\beta r})_2} = \{M | \forall M \in \mathcal{M}_{\mathrm{bc}_1}^{(v_{\beta r})}, v_{\beta t} \neq v_{\beta l}\}. \tag{4.2.6}$$

Lemma 4.2.2 *For $\mathcal{M}_{\mathrm{bc}_1}$, we have*

$$\mathcal{M}_{\mathrm{bc}_1}^{(v_{\beta r})_1} = \mathcal{M}_{\mathrm{bc}} - \vartheta;$$

$$\mathcal{M}_{\mathrm{bc}_1}^{(v_{\beta r})_2} = \mathcal{M}_{\mathrm{bc}} - \mathcal{M}_{\mathrm{bc}}^{(3)} - \vartheta \tag{4.2.7}$$

where

$$\mathcal{M}_{\rm bc}^{(3)} = \{M | \forall M \in \mathcal{M}_{\rm bc}, m(M) = 3\} \tag{4.2.8}$$

and ϑ is the vertex map.

Proof First, let us prove the first statement.

For any $M \in \mathcal{M}_{\rm bc_1}^{(v_{\beta r})_1}$, there is a map $M' \in \mathcal{M}_{\rm bc}$, $v_{\beta t} = v_{\beta l}$, such that $M = M' \bullet v_{\beta r}$. It can be seen that $M \neq \vartheta$. From the bipartiteness and the cubicness, we have $M \in \mathcal{M}_{\rm bc} - \vartheta$.

Conversely, for any $M \in \mathcal{M}_{\rm bc}, M \neq \vartheta$, we may construct a map M' obtained by splitting the root-vortex of M into two vertices v and u such that u is of valency 1 and incident with the root of M, and by adding a new vertex o with three edges:(o, v) and two of (o, u). It is easily checked that $M' \in \mathcal{M}_{\rm bc_1}$. Because $v_t = v_l = u$ in this case, we know that $M = M' \bullet v_{\beta r} \in \mathcal{M}_{\rm bc_1}^{(v_{\beta r})_1}$.

Now let us prove the second statement of the lemma.

For $M \in \mathcal{M}_{\rm bc_1}^{(v_{\beta r})_2}$, there is a map $M' \in \mathcal{M}_{\rm bc_1}$ such that $M = M' \bullet v_{\beta r}$. Because $v_{\beta l} \neq v_{\beta t}$, $m(M) \geq 6$ and hence M is a member of $\mathcal{M}_{\rm bc} - \mathcal{M}_{\rm bc}^{(3)} - \vartheta$.

Conversely, for any $M \in \mathcal{M}_{\rm bc}$, $m(M) \geq 6$, we may construct the map M' obtained by splitting the root-vertex of M into three vertices, named o, u and v, and by adding a new vertex w with three edges: $(o.w)$, (w, u) and (w, v) such that $w = v_{\beta r}(M')$, $o = v_r(M')$ and both u and v are incident exactly with two edges each. It is easily seen that $M' \in \mathcal{M}_{\rm bc_1}$ and $M = M' \bullet v_{\beta r}$. Because $u \neq v$, *i.e.*, $v_{\beta l} \neq v_{\beta t}$, we have $M \in \mathcal{M}_{\rm bc_1}^{(v_{\beta r})_2}$. □

Let M be a map in $\mathcal{M}_{\rm bc_2}$. Then, we may see that $M = M_1 \dot{+} M_2$ where M_2 is the part of M in the inner domain of the circuit C which consists of the 2-edge $(v_r, v_{\beta r})$ when C is not on the root-face boundary; or the outer domain of C, otherwise. Write

$$\begin{cases} \mathcal{J} = \{M_1 | \forall M \subset \mathcal{M}_{\rm bc_2}, C \text{ is not} \\ \qquad\qquad \text{on the root-face boundary}\}; \\ \hat{\mathcal{J}} = \{M_1 | \forall M \in \mathcal{M}_{\rm bc_2}, C \text{ is} \\ \qquad\qquad \text{on the root-face boundary}\}. \end{cases} \tag{4.2.9}$$

Lemma 4.2.3 *For $\mathcal{M}_{\mathrm{bc}_2}$, we have*

$$\mathcal{M}_{\mathrm{bc}_2} = (\mathcal{J} + \hat{\mathcal{J}})\bigodot\mathcal{M}_{\mathrm{bc}} \tag{4.2.10}$$

where $\odot$ is the $1v$-production.

Proof For $M \in \mathcal{M}_{\mathrm{bc}_2}$, there are two possibilities. One is the circuit formed by the two edges joining the two vertices v_r and $v_{\beta r}$ on the root-face boundary. Let M_2 be the part in the outer domain of the circuit having only one vertex, the root-vertex in common with the other part named by M_1. That implies $M = M_1 \dot{+} M_2$. Because $M_2 \in \mathcal{M}_{\mathrm{bc}}$ and $M_1 \in \hat{\mathcal{J}}$ we have $M \in \hat{\mathcal{J}} \odot \mathcal{M}_{\mathrm{bc}}$. The other leads that $M_2 \in \mathcal{M}_{\mathrm{bc}}$ and $M_1 \in \mathcal{J}$. That implies $M = M_1 \dot{+} M_2 \in \mathcal{J} \odot \mathcal{M}_{\mathrm{bc}}$.

Conversely, for any map M taken from the set on the right hand side of (4.2.10), because $M = M_1 \dot{+} M_2$ with two edges joining the two vertices v_r and $v_{\beta r}$, from the bipartiteness and the cubicness of M_1 and M_2 we soon see that M is a member of the set on the left hand side of (4.2.10). □

Lemma 4.2.4 *Let $\mathcal{J}^{(v_{\beta r})} = \{J \bullet v_{\beta r} | \forall J \in \mathcal{J}\}$ and $\hat{\mathcal{J}}^{(v_{\beta r})} = \{J \bullet v_{\beta r} | \forall J \in \hat{\mathcal{J}}\}$, we have*

$$\mathcal{J}^{(v_{\beta r})} = \hat{\mathcal{J}}^{(v_{\beta r})} = \mathcal{M}_{\mathrm{bc}} - \vartheta \tag{4.2.11}$$

where ϑ is the vertex map.

Proof First, from the definition of $\mathcal{J}$ and $\hat{\mathcal{J}}$, it is easily checked that $\mathcal{J}^{(v_{\beta r})} = \hat{\mathcal{J}}^{(v_{\beta r})}$.

Then, we show the second equality.

For $M \in \hat{\mathcal{J}}^{(v_{\beta r})}$, there is a map $M' \in \hat{\mathcal{J}}$ such that $M = M' \bullet v_{\beta r}$. Because the valency of the root-vertex of M is not less than 3, from the bipartiteness and the cubicness of M by (4.2.1) we see that M is a member of the set on the right hand side of (4.2.11).

Conversely, for a map $M \in \mathcal{M}_{\mathrm{bc}} - \vartheta$, we can uniquely construct a map M' by splitting the root-vertex of M into two vertices o and u such that u is of valency 1 and then by adding a new vertex v with one edge (o, v) and two edges (v, u) which form the outerface boundary. It is easily seen that $M' \in \hat{\mathcal{J}}$. Because $M = M' \bullet v$, $v = v_{\beta r}(M')$, we find that M is a member of the set on the left of (4.2.11). □

Let us write the map

$$L^3 = \Big(\bigcup_{i=1}^{3} Kx_i, (x_1, x_2, x_3)(\alpha\beta x_3, \alpha\beta x_2, \alpha\beta x_1)\Big).$$

Lemma 4.2.5 *For $\mathcal{M}_{\mathrm{bc}_3}$, we have*

$$\mathcal{M}_{\mathrm{bc}_3} = L^3 \odot \mathcal{M}_{\mathrm{bc}}^{\odot 3} \tag{4.2.12}$$

where $\odot$ is the 1v-production.

Proof Because for any $M \in \mathcal{M}_{\mathrm{bc}_3}$, there is an L^3 as a submap of M. Let the three parts of M in the inner domains of the three faces of L^3 be M_1, M_2 and M_3. Then, we have $M = L^3 \dot{+} M_1 \dot{+} M_2 \dot{+} M_3$. Because $M_i \in \mathcal{M}_{\mathrm{bc}}$, $i = 1, 2, 3$, we know that M is a member of the set on the right hand side of (4.2.12).

Conversely, for any $M = L^3 \dot{+} M_1 \dot{+} M_2 \dot{+} M_3$, $M_i \in \mathcal{M}_{\mathrm{bc}}$, $i = 1, 2, 3$, it is easy to check that the map $M \in L^3 \odot \mathcal{M}_{\mathrm{bc}}^{\odot 3}$ is of course a member of the set on the left of (4.2.12) by considering the construction of M with (4.2.3). □

Let $q = q_{\mathcal{M}_{\mathrm{bc}}}$ be the enufunction of bipartite near-cubic planar maps with the valency of root-vertex and the number of nonrooted vertices of the part the root-vertex is not in as the two parameters, *i.e.*,

$$q = \sum_{M \in \mathcal{M}_{\mathrm{bc}}} x^{m(M)} y^{s(M)} \tag{4.2.13}$$

where $m(M)$ and $s(M)$ are the valency of root-vertex and the number of nonrooted vertices in the part not containing the root-vertex respectively. From (4.2.1), we may write

$$m(M) = 3k(M), \tag{4.2.14}$$

and hence we have

$$q = \sum_{M \in \mathcal{M}_{\mathrm{bc}}} z^{k(M)} y^{n(M)} \tag{4.2.15}$$

where $z = x^3$. On the other hand, if q is written in the form as

$$q = \sum_{k \geq 0} \sum_{s \geq 0} Q_{\mathcal{M}_{\mathrm{bc}}}(k, s) z^k y^s, \tag{4.2.16}$$

then we have that

$$q_1 = \sum_{s\geq 0} Q_{\mathcal{M}_{\mathrm{bc}}}(1,s)y^s \tag{4.2.17}$$

is the enufunction of bipartite cubic planar maps with the number of nonrooted vertices in the nonrooted part as the parameter.

Theorem 4.2.1 *For the function q with q_1, we have the following equation:*

$$\begin{aligned} &yzq^3 + 2yq^2 + (yz^{-1} - y - 1)q \\ &+(1 - y - yz^{-1} - yz^{-1}q_1) = 0 \end{aligned} \tag{4.2.18}$$

where q_1 is shown in (4.2.17).

Proof From (4.2.3) we know that

$$q = \sum_{i=0}^{3} q_{\mathcal{M}_{\mathrm{bc}_i}} \tag{4.2.19}$$

where $q_{\mathcal{M}_{\mathrm{bc}_i}}$ are the contributions of $\mathcal{M}_{\mathrm{bc}_i}$ to q for $i = 0, 1, 2, 3$.

First, since $\mathcal{M}_{\mathrm{bc}_0}$ consists of only one map, the vertex map, we soon find that

$$q_{\mathcal{M}_{\mathrm{bc}_0}} = 1. \tag{4.2.20}$$

Secondly for $\mathcal{M}_{\mathrm{bc}_1}$, from (4.2.5) we have

$$q_{\mathcal{M}_{\mathrm{bc}_1}} = y\Bigg(\sum_{M\in\mathcal{M}_{\mathrm{bc}_1}^{(v_{\beta r})_1}} z^{m(M)}y^{s(M)} + z^{-1} \sum_{M\in\mathcal{M}_{\mathrm{bc}_1}^{(v_{\beta r})_2}} z^{m(M)}y^{s(M)} \Bigg)$$

by Lemma 4.2.2,

$$\begin{aligned} &= y\big(q - 1 + z^{-1}(q - q_1 - 1)\big) \\ &= y\big((1 + z^{-1})q - z^{-1}q_1 - (1 + z^{-1})\big). \end{aligned} \tag{4.2.21}$$

Thirdly for $\mathcal{M}_{\mathrm{bc}_2}$, from Lemma 4.2.5 we have

$$q_{\mathcal{M}_{\mathrm{bc}_2}} = \Bigg(\sum_{M_1\in\mathcal{J}} z^{m(M_1)}y^{s(M_1)} + \sum_{M_1\in\hat{\mathcal{J}}} z^{m(M_1)}y^{s(M_1)} \Bigg)$$

$$\times \sum_{M_2 \in \mathcal{M}_{\rm bc}} z^{m(M_2)} y^{s(M_2)}$$

by Lemma 4.2.4,

$$= \Big(y(q-1+q-1)\Big)q$$

$$= 2yq^2 - 2yq. \qquad (4.2.22)$$

Finally for $\mathcal{M}_{\rm bc_3}$, from Lemma 4.2.5 we have

$$q_{\mathcal{M}_{\rm bc_3}} = zyq^3. \qquad (4.2.23)$$

In consequence, by combining (4.2.19– 23), Eq.(4.2.18) is soon obtained after grouping. □

Although Eq.(4.2.18) is cubic about q, it can be checked that the following parametric expressions are satisfied for this equation:

$$\xi = 1 + 2y\xi^2; \ \eta = 1 + yz\xi^2\eta^3;$$

$$q = \xi\Big(1 - \tfrac{1}{2}(\eta - 1)(\xi - 1)\Big). \qquad (4.2.24)$$

Theorem 4.2.2 *The enufunction q has the following explicit expression:*

$$q(z,y) = 1 + \sum_{k\geq 1}\sum_{s\geq 1} \frac{2^{s-k}(2s-1)!}{(s-k)!(s+k+1)!}$$

$$\times \frac{(3k)!}{k!(2k-1)!} z^k y^s. \qquad (4.2.25)$$

Proof From (4.2.24), let

$$\alpha = \xi - 1; \ \beta = \eta - 1,$$

then we have

$$\alpha = 2y(\alpha+1)^2; \ \beta = zy(\alpha+1)^2(\beta+1)^3;$$

$$q = (\alpha + 1)\Big(1 - \tfrac{1}{2}\alpha\beta\Big). \qquad (4.2.26)$$

According to (4.2.26), we are allowed to employ Theorem 1.5.2 for obtaining (4.2.25) and hence obtain the theorem. □

Lemma 4.2.6 *In any bipartite near-cubic map (not necessary to be planar), no 3-valent vertex is a cut-vertex.*

Proof By contradiction. If there is a 3-valent vertex v which is a cut-vertex in a bipartite near-cubic map M, then M can be obtained by 1-addition of M_1 and M_2 at v such that one, assumed to be M_1 without loss of generality, of M_1 and M_2 has to be with valency 1 of v. Because M_1 is also a bipartite near-cubic map, this is a contradiction to (4.2.1). □

This lemma allows us to determine the enufunction q_{np} of rooted bipartite nonseparable near-cubic planar maps by q with the same parameters. That is

$$q_{\mathrm{np}} = \frac{q-1}{q}. \tag{4.2.27}$$

Theorem 4.2.3 *The enufunction $q_{\mathrm{np1}} = q_1$ with the number of nonrooted vertices in the nonrooted part as the parameter. Further, we have*

$$q_1 = \sum_{s\geq 1} \frac{3 \cdot 2^s (2s-1)!}{(s-1)!(s+2)!} y^s. \tag{4.2.28}$$

Proof From Lemma 4.2.6, the first statement is true. Further, by (4.2.25) with $k = 1$, (4.2.28) is soon found. □

§4.3 Cubic Hamiltonian maps

A map is said to be *Hamiltonian* if all the vertices of the map are on one circuit. We may choose the root to be incident with an edge (the root-edge) on the circuit. Such a circuit is also said to be Hamiltonian.

If the Hamiltonian circuit is the boundary of the outerface, then the map is called outerplanar as mentioned in Chapter 2. On the other hand, if it is the boundary of an inner face, then the map is said to be *innerplanar*. Of course, any outerplanar map is also topologically equivalent to an innerplanar map when one of its inner face is topologically transformed into the outerface. A Hamiltonian cubic planar map is in fact the resultant one of identifying the boundaries which correspond to the Hamiltonian circuits of an outerplanar cubic map and an inner planar cubic map such that there is no vertex in common.

First, we enumerate rooted innerplanar cubic maps, as well as rooted outerplanar cubic maps.

Let $\mathcal{I}$ be the map of all rooted innerplanar cubic maps. For $I \in \mathcal{I}$, the root is usually chosen to be incident with an edge on the inner face boundary of I. Let $I_{(\mathrm{H})}$ be the map obtained by contracting all the edges on the Hamiltonian circuit with the root $r(I_{(\mathrm{H})}) = \mathcal{P}\alpha\beta r(I)$. Then, it is easily verified that $I_{(\mathrm{H})}$ is a planar map with only one vertex and its planar dual is a tree in correspondence with their roots.

Lemma 4.3.1 *Let $\mathcal{I}(n)$ and $\mathcal{T}(n)$ be the sets of all rooted innerplanar cubic maps with n edges not on the Hamiltonian circuit and rooted plane trees of size n respectively. Then we have*

$$|\mathcal{I}(n)| = |\mathcal{T}(n)| \tag{4.3.1}$$

for $n \geq 1$.

Proof For any $I \in \mathcal{I}(n)$, we may uniquely find $I_{(\mathrm{H})}$ whose planar dual $I^*_{(\mathrm{H})}$ is a rooted plane tree of size n, *i.e.*, $I^*_{(\mathrm{H})} \in \mathcal{T}(n)$, for $n \geq 1$. This shows $|\mathcal{I}(n)| \leq |\mathcal{T}(n)|$.

On the other hand, for any $T \in \mathcal{T}(n)$, T^* is the planar dual of T. Since T^* is a planar map with a single vertex, we may uniquely find a map I_{T^*} by replacing the vertex by a circle with $2n$ points marked as vertices such that they appear in the cyclic order as determined by the rotation of the vertex of T^*. It is easily seen that I_{T^*} is an inner planar cubic map with n edges not on the Hamiltonian circuit. Its root is

$$r(I_{T^*}) = \alpha\beta\mathcal{P}^{-1}r(t^*).$$

This implies $I_{T^*} \in \mathcal{I}(n)$. Thus, $|\mathcal{T}(n)| \leq |\mathcal{I}(n)|$. □

The map which is obtained by replacing each vertex of a map M with a circle as shown in the proof of Lemma 4.3.1 is said to be the *vertex circled* map of M. Of course, vertex circled maps are all cubic maps.

Let $t(x)$ be the enufunction of rooted plane trees with the size as the parameter. Of course, the vertex map is defined to be a tree for convenience. Because for any plane tree T except the vertex map, $T - a$ where a is the root-edge of T has two connected components which are all treated to be rooted plane trees including the vertex map, we see that $t(x)$ satisfies the following equation

$$t(x) = 1 + xt^2(x). \tag{4.3.2}$$

Since $t(x)$ has to be a power series of x with all coefficients being positive integers, we find that

$$\begin{aligned} t(x) &= \frac{1 - \sqrt{1 - 4x}}{2x} \\ &= \sum_{n \geq 0} \frac{(2n)!}{n!(n+1)!} x^n. \end{aligned} \tag{4.3.3}$$

The coefficient of x^n in (4.3.3) is also called the *Catalan number.*

Lemma 4.3.2 *Let* In(m) (*or* Ou(m)) *be the number of rooted innerplanar* (*or outerplanar*) *cubic maps of size* m. *Then we have*

$$\mathrm{In}(m)\Big(or\ \mathrm{Ou}(m)\Big) = \frac{(2k)!}{k!(k+1)!} \tag{4.3.4}$$

where $k = m/3$ *for* $m \geq 1$.

Proof Because for any cubic map of order n and size m we always have

$$3n = 2m \tag{4.3.5}$$

which implies that

$$\begin{cases} m = 0(\text{mod } 3); \\ \\ n = 0(\text{mod } 2). \end{cases} \tag{4.3.6}$$

The number k is always an integer. Further from (4.3.5), we see that the number of edges not on the Hamiltonian circuit is

$$m - \frac{2}{3}m = \frac{m}{3} = k.$$

From Lemma 4.3.1 the lemma is obtained. □

Let M_1 and M_2 be two rooted maps. A map M obtained from M_1 and M_2 by identifying their Hamiltonian circuits such that no vertex is in common is said to be *H-merged*, or precisely *vertex disjoint H-merged*. Of course, an H-merged map M is cubic iff M_1 and M_2 are cubic and it can also be treated as a rooted one by choosing its root as the same as that of M_1.

Lemma 4.3.3 *Assume M_1 and M_2 are an innerplanar cubic map of order $2j$ and an outerplanar cubic map of order $2i$, $i + j = n$, respectively. Then, we have*

$$\binom{2n-1}{2i} \tag{4.3.7}$$

distinct H-merged cubic Hamiltonian maps.

Proof Because on the Hamiltonian circuit of an H-merged map M, there are $2n$ positions with 1 position for the rooted vertex which belongs to M_1. Then, there is a 1-to-1 correspondence between the H-merged maps and the possibilities of choosing $2i$ from $2n-1$, which implies the lemma. □

According to Lemmas 4.3.1–3, we may soon find the following

Theorem 4.3.1 *The number of rooted planar cubic Hamiltonian maps with $2n$ vertices is*

$$H_{\text{Cub}}(n) = \frac{2(2n-1)!(2n+1)!}{(n+1)!n!(n-1)!(n+2)!} \tag{4.3.8}$$

for $n \geq 1$.

Proof Because the number of innerplanar maps equals the number of outerplanar maps of the same size, by (4.3.4) and (4.3.7) we have

$$H_{\text{Cub}}(n) = \sum_{i=1}^{n-1} \binom{2n-1}{2i} \text{In}(i)\text{Ou}(n-i)$$

$$= \sum_{i=0}^{n-1} \binom{2n-1}{2i} \frac{(2i)!}{i!(i+1)!} \frac{(2n-2i)!}{(n-i)!(n-i+1)!}$$

$$= \sum_{i=0}^{n-1} \frac{2(2n-1)!}{i!(i+1)!(n-i-1)!(n-i+1)!}$$

$$= \frac{2(2n-1)!}{n!(n+1)!} \sum_{i=0}^{n-1} \binom{n+1}{i} \binom{n}{i+1}.$$

Since $\binom{n+1}{i}$ is the coefficient of x^i in $(1+x)^{n+1}$ and $\binom{n}{i+1}$ is the coefficient of x^{n-1-i} in $(1+x)^n$, then

$$\sum_{i=0}^{n-1} \binom{n+1}{i} \binom{n}{i+1}$$

is the coefficient of x^{n-1} in $(1+x)^{2n+1}$. That is

$$\binom{2n+1}{n-1}.$$

Hence, (4.3.8) is true. □

Although there is a simple formula for counting Hamiltonian cubic map as shown in Theorem 4.3.1, it seems to have more complicated cases for general Hamiltonian planar maps, even for 3-connected Hamiltonian cubic maps. However, it looks without much difficulty for Hamiltonian 4-regular maps in a way similar to what we have shown in this section.

§4.4 Cubic maps on surfaces

Let $M = (\mathcal{X}_{\alpha,\beta}, \mathcal{P})$ be a map on a surface. Write $V = V(M)$, $E = E(M)$ and $F = F(M)$ to be the order, the size and the dual order of M, respectively.

For a subset $C^* = \{e_1, e_2, \cdots, e_s\}$ of E, if there are faces f_1, f_2, $\cdots$, f_s such that $e_i \in f_i \cap f_{i+1}$, $i = 1, 2, \cdots, s$, $f_{s+1} = f_1$, then C^* is called an *S-cocycle*.

According to the theory described in [Liu58], it is seen that a map M is on the sphere iff all of its S-cocycles are cocycles. This implies that if M is not on the sphere, then there is at least one S-cocycle which is not a cocycle. Such an S-cocycle is said to be *essential.*

If an S-cocycle has no proper subset which is still an S-cocycle, then it is called an *S-cocircuit.* Because any S-cocycle has a subset which is an S-cocircuit, we are allowed only, unless otherwise stated, to discuss S-cocircuits here.

Let C^* be an S-cocycle of M. We may introduce a vertex v_i on the edge e_i of C^*, $i = 1, 2, \cdots, s$, and then joining the edges (v_i, v_{i+1}), $i = 1, 2, \cdots, s$, $v_{s+1} = v_1$, in the inner domain of corresponding faces. Such a map obtained in this way from M is called a *C^*-subdivision* of M and denoted by $H_{C^*}(M)$. A *C^*-transversal*, denoted by $T_{C^*}(M)$, is defined to be the map obtained by cutting along the edges (v_i, v_{i+1}), $i = 1, 2, \cdots, s$, to form a new face in cyclic orders on $H_{C^*}(M)$. From what was mentioned in §1.1, it is easily checked that $H_{C^*}(M)$ is a map on the same surface as M. However, $T_{C^*}(M)$ is on the surface whose genus is one less than that of M.

Now maps on the projective plane are considered.

Theorem 4.4.1 *A map M is on the projective plane iff there is an essential S-cocircuit C^* of M such that C^*-transversal of M is on the sphere.*

Proof The first statement of the necessity is obvious because if there is no essential S-cocircuit in M then M has to be on the sphere, from what has been mentioned above. This contradicts the condition on M. Then, since the genus of the surface $T_{C^*}(M)$ is on is one less than that of the projective plane M is on, the only possibility is that $T_{C^*}(M)$ is on the sphere.

Conversely, since $T_{C^*}(M)$ is on the sphere and $T_{C^*}(M)$ has exactly one face different from M, all the edges corresponding to (v_i, v_{i+1}), $i = 1, 2, \cdots, s$, as mentioned above form a face boundary. Further, by Euler formula we see that the C^*-subdivision of M is on the projective plane and hence M is on the projective plane as well. The sufficiency is true. □

Let us see what is the difference between the maps of a graph H which consists of three edges: one link and two loops with two vertices such that each vertex is 3-valent on the sphere and on the projective plane. Write Kx_i, $i = 1, 2, 3$, to be the three edges. Now, we may find three distinct rooted planar maps. All of them are cubic. That is, from

$$\Big(\bigcup_{i=1}^{3} Kx_i, (x_1, \alpha\beta x_1, x_2)(\alpha\beta x_2, x_3, \alpha\beta x_3)\Big), \tag{4.4.1}$$

with x_1, x_2 and αx_1 as the roots we obtain H_1, H_2 and H_3, respectively.

On the other hand, from the map

$$\Big(\bigcup_{i=1}^{3} Kx_i, (x_1, x_2, \beta x_1)(x_3, \alpha\beta x_3, \alpha\beta x_2)\Big) \tag{4.4.2}$$

on the projective plane, we may obtain six distinct rooted cubic maps $\tilde{H}_i$, $i = 1, 2, \cdots, 6$, by choosing x_1, αx_1, x_2, βx_2, x_3 and αx_3 as the roots respectively.

This looks as though it suggests to us, by considering what is mentioned in §3.4, that we should investigate nonsplitable cubic maps to avoid some complication which is not necessary.

For a rooted map M on the projective plane, if an essential S-cocircuit C^* is marked, then it is said to be C^**-oriented.* Two C^*-oriented maps are treated as being the same if there is an isomorphism between them such that their marked essential S-cocircuits are in correspondence.

A nonseparable planar map which has no edge with both of its ends on the root-face boundary is said to be *out-chordless.*

Lemma 4.4.1 *Let $\mathcal{M}^{l}_{\mathrm{nlt}_{\tilde{1}}}$ be the set of all C^*-oriented nonsplitable cubic maps with the essential S-circuit of l edges on the projective plane and let $\mathcal{M}^{2l}_{\mathrm{npr}_0}$ be the set of all rooted out-chordless cubic maps with the root-face of valency $2l$. Then we have*

$$\left|\mathcal{M}^{l}_{\mathrm{nlt}_{\tilde{1}}}\right| = \left|\mathcal{M}^{2l}_{\mathrm{npr}_0}\right|. \tag{4.4.3}$$

Proof For any $M \in \mathcal{M}^{l}_{\mathrm{nlt}_{\tilde{1}}}$ from Theorem 4.4.1 we see that there is a unique planar cubic map with the root- face of valency $2l$. That is the C^*- transversal $T_{C^*}(M)$ of M. Because of the nonsplitability of M, it can be shown that $T_{C^*}(M)$ is nonseparable by the reason similar to that described in §3.4. Of course, it is out-chordless since any inner edge is not allowed to have both of its ends on the outer face boundary. Hence, $M \in \mathcal{M}^{2l}_{\mathrm{npr}_0}$.

Conversely, for any $M \in \mathcal{M}^{2l}_{\mathrm{npr}_0}$ let the root-face of M be

$$(x_1, x_2, \cdots, x_l, x_{l+1}, x_{l+2}, \cdots, x_{2l}).$$

Then, we may find a map $\tilde{M}$ by identifying γx_i with γx_{l+i}, $i = 1, 2, \cdots, l$, where $\gamma = 1$, α, β and $\alpha\beta$ with all 2-valent vertices being ignored. It is easily seen that $\tilde{M}$ has its face number one less than that of M. From the near-cubicness of M each vertex incident with an x_i (or x_{l+i} as well) is of valency 4. Thus, we may uniquely obtain a map M' by deleting all the l edges which are incident with x_i, $i = 1, 2, \cdots, l$, on $\tilde{M}$. It is easily checked that $\tilde{M}$ is C^*-oriented with the S-cocircuit of l edges and that $M = T_{C^*}(M')$. From Theorem 4.4.1 and the out-chordlessness we have that $M' \in \mathcal{M}^{l}_{\mathrm{nlt}_{\bar{1}}}$. □

In what follows we discuss how to determine the enufunction of out-chordless maps from what has been obtained in §4.1.

First, nonseparable cubic planar maps should be considered. Let $\mathcal{H}$ be the set of rooted maps of this kind. Our purpose is to determine the function

$$h = h_{\mathcal{H}} = \sum_{H \in \mathcal{H}} y^{s(M)} z^{l(M)} \tag{4.4.4}$$

where $s(H)$ and $l(H)$ are the size and the valency of the root-face of H respectively. The set $\mathcal{H}$ is divided into two classes as $\mathcal{H}_0$ and $\mathcal{H}_1$ such that $\mathcal{H}_0$ is consists of only one map which is of order 2 with size 3 of three links. Easy to see that the contribution of $\mathcal{H}_0$ to h is

$$h_{\mathcal{H}_0} = y^3 z^2. \tag{4.4.5}$$

For $H \in \mathcal{H}_1$, write v_1, v_2, $\cdots$, v_k as the vertices on the root-face boundary which appear in the order along the direction from the non-rooted end v_1 of the root-edge a to the root-vertex v_k such that v_i, $2 \leq i \leq k-1$, are all the cut-vertices of $H - a$. From the cubicness of H, k is an even number, v_{2j-1}, and v_{2j} are incident with a nonseparable component which is cubic map H_j when the two vertices v_{2j-1} and v_{2j} are ignored, and (v_{2j}, v_{2j+1}) is an edge of $H - a$ for any possible j. Because each H_j is allowed to be any map in $\mathcal{H}$, we may find the contribution of $\mathcal{H}_1$ to h is

$$h_{\mathcal{H}_1} = \sum_{j\geq 1} y^{3j} z^j \Big(\sum_{H\in\mathcal{H}} \frac{z - z^{l(H)}}{1-z} y^{s(H)}\Big)^j$$

$$= \frac{y^3 z(z\tilde{h} - h)}{1 - z - y^3 z(z\tilde{h} - h)} \tag{4.4.6}$$

where $\tilde{h} = h(y, 1)$, the enufunction of such maps considered here with the size as the parameter.

Lemma 4.4.2 *The enufunction h satisfies the equation as*

$$xz^2\Big(1 - z + (1 - xz^2)\tilde{h}\Big) + \Big(x^2 z^3$$

$$-xz - (1-z) + xz^2\tilde{h}\Big)h - xzh^2 = 0 \tag{4.4.7}$$

where $x = y^3$ and $\tilde{h} = h(x^{\frac{1}{3}}, 1)$.

Proof Since $h = h_{\mathcal{H}_0} + h_{\mathcal{H}_1}$, from (4.4.5–6) we have

$$h = xz^2 + \frac{xz(z\tilde{h} - h)}{1 - z - xz(z\tilde{h} - h)}.$$

By rearrangement, Eq.(4.4.7) is soon obtained. □

The discriminant of the equation (4.4.7) is

$$D = \Big(x^2z^3 - xz - (1-z) + xz^2\tilde{h}\Big)^2$$

$$-4x^2xz^3\Big(1 - z + (1 - xz^2)\tilde{h}\Big)$$

$$= z^6x^4 - 2z^4(1 + z\tilde{h})x^3 + z^2\Big(2z(1-z)$$

$$+(1 + z\tilde{h})^2\Big)x^2 + 2z(1-z)(1+z\tilde{h})x$$

$$+(1-z)^2 \tag{4.4.8}$$

from which we are allowed to find the parametric expressions of x and $\tilde{h}$ with $\xi = z(x)$ as the parameter. Thus, $\tilde{h}$ and h can be determined.

Then out-chordless maps are investigated. Let $\mathcal{Q}$ be the set of all out-chordless maps. For a map $Q \in \mathcal{Q}$, assume that $Q - a$, where a is the root-edge of Q, has $k - 1$ *out-chords* with the occurrences in the order of

$$e_1,\ e_2,\ \cdots,\ e_{k-1}$$

when one walks from the root-vertex to the nonrooted end of the root-edge along the outerface boundary of $Q - a$. Write

$$M_1,\ M_2,\ \cdots,\ M_k$$

as the out-chordless parts of $Q - a$ such that M_i is with e_{i-1} and e_i for $i = 1,2, \cdots, k$ while e_0 and e_k are the edges adjacent to a on the boundary of the inner face incident with a. From the cubicness and the out-chordlessness of Q, we can see that all M_i, $i = 1, 2, \cdots, k$, are out-chordless cubic maps when 2-valent vertices are ignored except only for polygons themselves.

Let $q = q_{\mathcal{Q}}(y, z)$ be the enufunction of $\mathcal{Q}$ with the size and the valency of root-face as two parameters, *i.e.*,

$$q = \sum_{Q \in \mathcal{Q}} y^{s(Q)} z^{l(Q)}. \tag{4.4.9}$$

Lemma 4.4.3 *The function q in (4.4.9) satisfies the following quadratic equation:*

$$y^3q^2 - \Big((1-z)(1-y^3) + y^3z(1+z\tilde{q})\Big)q$$
$$+y^3z\Big(1 - z(1+z\tilde{q})\Big) = 0 \tag{4.4.10}$$

with $\tilde{q} = q(y, 1)$.

Proof In what is mentioned above, by considering that the procedure of constructing M_i, $i = 1, 2, \cdots, k$, from maps in $\mathcal{Q}$ causes to produce $4k - 2$ edges and then of constructing $Q - a$ to reduce in number of $k - 1$ on account of repetition, we see that there are

$$4k - 2 - (k - 1) = 3k - 1$$

additional edges in the process to $Q - a$ from maps in $\mathcal{Q}$ all together. This allows us to find

$$q = yz \sum_{M = Q - a,\ Q \in \mathcal{Q}} y^{s(M)} z^{l(M)}$$

$$= z\sum_{k\geq 1} y^{3k}\left(1+\sum_{Q\in\mathcal{Q}} y^{s(Q)}\sum_{i=2}^{l(Q)-1} z^i\right)^k$$

$$= z\sum_{k\geq 1} y^{3k}\left(1+\frac{z\tilde{q}-q}{1-z}\right)^k$$

$$= z\frac{y^3\left(1+\dfrac{z^2\tilde{q}-q}{1-z}\right)}{1-y^3\left(1+\dfrac{z^2\tilde{q}-q}{1-z}\right)}.$$

Thus, from rearrangement of what has just been obtained, the equation (4.4.10) can be soon derived. □

§4.5 Notes

4.5.1 Because U_1 is known from Theorem 4.1.2 we can solve the quadratic equation (4.1.15) to obtain

$$u = \frac{1}{2x}\left\{x - z + (z - ax)\sqrt{1 - 2bx + cx^2}\right\} \tag{4.5.1}$$

where a, b and c appear in the parametric expression (4.1.18) by considering that u has to be a power series of x and z with all the coefficients nonnegative integers on the basis of Theorem 1.4.2.

Although an explicit expression of u can be found from (4.5.1) by employing the Corollary 1.5.1, or from

$$u = \left(1 - \frac{xz}{x-z}U_1\right) + \left(\frac{x^3}{x-z}\right)u^2, \tag{4.5.2}$$

to get

$$u = \sum_{n\geq 0}\frac{(2n)!}{n!(n+1)!}\left(\frac{x^3}{x-z}\right)^n\left(1-\frac{xz}{x-z}U_1\right)^{n+1} \tag{4.5.3}$$

by employing the Corollary 1.5.2 as shown in [Liu35], or [MNS1] respectively, no simpler explicit formula has been derived till now.

4.5.2 The main results in §4.2 are from the manuscripts which appeared at the end of sixties by Tutte. However, the parameter should be the number of edges not on the root-face boundary instead of the number of nonrooted vertices there.

4.5.3 How to enumerate the number of C^*-oriented bipartite splitable cubic maps on the projective plane is still an open problem based on what has been discussed in §4.2.

4.5.4 Hamiltonian cubic planar maps used to be very important for the four color problem. People tried to solve the problem by showing that every 3-connected cubic planar map is Hamiltonian. However, Tutte firstly denied the statement by a counterexample which is famous for graph theorists. On the other hand, cubic maps have also paid an important role in solving the Gaussian crossing problem[Liu58]. The topic on enumerating cubic Hamiltonian planar maps is firstly studied by Tutte as well [Tut4]. There still are the similar problems for general Hamiltonian maps on the plane, or the sphere and the other surfaces.

4.5.5 Following Eqs.(4.4.7–8) determine $\tilde{h}$ and then h. It is also allowed to determine them from general cubic maps discussed in §4.1. However, some complications probably occur.

4.5.6 According to Eq.(4.4.10), find $\tilde{q}$ and q. It is also allowed to extract them from $\tilde{h}$ and h mentioned in 4.5.5 above.

4.5.7 To enumerate certain type of cubic maps on the Klein bottle or even more general nonorientable surfaces, and further on the torus, or even more general orientable surfaces looks with penetration hopeful from what has been discussed in §4.4.

Chapter 5

Eulerian Maps

§5.1 Planar Eulerian maps

A map is called *Eulerian* if all the valencies of its vertices are even (or say, all vertices are *even*). Let $\mathcal{U}$ be the set of all the rooted planar Eulerian maps with the convention that the vertex map ϑ is in $\mathcal{U}$ for convenience.

Further, $\mathcal{U}$ is divided into 3 classes: $\mathcal{U}_0$, $\mathcal{U}_1$ and $\mathcal{U}_2$, *i.e.*,

$$\mathcal{U} = \mathcal{U}_0 + \mathcal{U}_1 + \mathcal{U}_2 \tag{5.1.1}$$

such that $\mathcal{U}_0 = \{\vartheta\}$, or simply write $\{\vartheta\} = \vartheta$, and

$$\mathcal{U}_1 = \{U | \forall U \in \mathcal{U} \text{ with } a = e_r(U) \text{ being a loop}\}.$$

Lemma 5.1.1 *Any Eulerian map (not necessarily planar) has no cut-edge.*

Proof By contradiction. Assume that a Eulerian map M has a cut-edge $e = (u, v)$ such that $M = M_1 \cup e \cup M_2$, $M_1 \cap M_2 = \emptyset$, where M_1 and M_2 are submaps of M with the property that M_1 is incident to u and M_2, to v. From the Eulerianity of M, u and v are the unique odd vertex in M_1 and M_2, respectively. This contradicts to that both M_1 and M_2 are a submap of M because the number of odd vertices in a map is always even. □

Lemma 5.1.2 *Let $\mathcal{U}_{\langle 1\rangle} = \{U - a | \forall U \in \mathcal{U}_1\}$ where $a = e_r(U)$ is the root-edge. Then, we have*

$$\mathcal{U}_{\langle 1\rangle} = \mathcal{U} \odot \mathcal{U} \tag{5.1.2}$$

where $\odot$ *is the* 1*v-production.*

Proof Because for $U \in \mathcal{U}_1$, the root-edge a is a loop, we see that $U - a = U_1 \dot{+} U_2$ where U_1 and U_2 are in the inner and outer domain of a respectively. Of course, it can be checked that both U_1 and U_2 are maps in $\mathcal{U}$. Thus, the set on the left hand side of (5.1.2) is a subset of that on the right.

On the other hand, for any $U = U_1 \dot{+} U_2$, $U_1, U_2 \in \mathcal{U}$, we may uniquely construct a map U' by adding a loop at the common vertex of U_1 and U_2. The root-edge of U' is chosen to be the loop such that U_1 and U_2 are respectively in its inner and outer domains. It is easily checked that U' is a Eulerian map and hence $U' \in \mathcal{U}_1$. However, $U = U' - a \in \mathcal{U}_{\langle 1 \rangle}$. That implies the set on the right hand side of (5.1.2) is a subset of that on the left as well. □

For any map $U \in \mathcal{U}_2$, we see that the root-edge a of U has to be a link. From Lemma 5.1.1, if $U \bullet a = U_1 \dot{+} U_2$ such that the root-vertex is the common vertex, then the valencies of the vertices in both U_1 and U_2 are even. Further for any $U \in \mathcal{U}$, if $U = U_1 \dot{+} U_2$, then the valencies of the common vertex between U_1 and U_2 are both even as well.

Lemma 5.1.3 *Let* $\mathcal{U}_{(2)} = \{U \bullet a | \forall U \in \mathcal{U}_2\}$ *where* a *is the root-edge of* U. *Then, we have*

$$\mathcal{U}_{(2)} = \mathcal{U} - \vartheta \tag{5.1.3}$$

where $\vartheta = \mathcal{U}_0$ *for simplicity.*

Proof From what has just been discussed, the set on the left hand side of (5.1.3) is a subset of that on the right.

Conversely for any $U \in \mathcal{U} - \vartheta$, we may always construct a map U' by splitting the root-vertex into v_1 and v_2 with the new edge $a' = (o_1, o_2)$ as the root-edge of U' such that the valencies of o_1 and o_2 in U' are both even. However, we see that $U = U' \bullet a' \in \mathcal{U}_{(2)}$. That implies the set on the right hand side of (5.1.3) is a subset of that on the left. □

For a map $U \in \mathcal{U} - \vartheta$, assume the valency of the root-vertex o is $2k$, $k \geq 1$,without loss of generality. The map U' obtained by splitting the root-vertex into o_1 and o_2 with the new edge $a' = (o_1, o_2)$ such that the valency $\rho(o_1;U') = 2i$ and hence $\rho(o_2;U') = 2k - 2i + 2$ is denoted by $U_{[2i]}, i = 1, 2, \cdots, k$.

From Lemma 5.1.1, we see that the procedure works and that all the resultant maps $U_{[2i]}$ are also Eulerian maps.

Lemma 5.1.4 *For U_2, we have*

$$|\mathcal{U}_2| = \sum_{U \in \mathcal{U} - \vartheta} \left|\left\{U_{[2i]} \middle| \ i = 1, 2, \cdots, m(U)\right\}\right| \tag{5.1.4}$$

where $2m(U)$ is the valency of the root-vertex of U.

Proof First, we show that for any $U \in \mathcal{U}_2$, it appears in the set on the right hand side of (5.1.4) only once. Assume that $a = (o, v)$ is the root-edge of U and that $\rho(o) = 2s$ and $\rho(v) = 2t$. Let U' be the map obtained by contracting the root-edge a, *i.e.*, $U' = U \bullet a$. Then, there is the only possibility that $U = U'_{[2s]}$ in the set on the right hand side of (5.1.4).

Then, we show that for any map U in the set on the right hand side of (5.1.4), it appears also only once in $\mathcal{U}_2$. This is obvious from Lemma 5.1.3 because all elements are distinguished and they are all maps in $\mathcal{U}_2$ by considering the Eulerianity with the root-edges being links. □

In what follows, we see what kind of equation should be satisfied by the enufunction u of rooted planar Eulerian maps with vertex partition. Write

$$u = \sum_{U \in \mathcal{U}} x^{2m(U)} \underline{y}^{\underline{n}(U)} \tag{5.1.5}$$

where $2m(U)$ is the valency of the root-vertex as mentioned above and $\underline{n}(U) = (n_2(U), \cdots, n_{2i}(U), \cdots), n_{2i}(U)$ is the number of nonrooted vertices of valency $2i$, $i \geq 1$.

Theorem 5.1.1 *The function u defined in* (5.1.5) *satisfies the following functional equation:*

$$u = 1 + x^2 u^2 + x^2 \int_y \left(y^2 \delta_{x^2, y^2}\left(u(\sqrt{z})\right)\right) \tag{5.1.6}$$

where $u(z) = u|_{x=z} = u(z, \underline{y})$.

Proof The contribution of $\mathcal{U}_0$ to u is

$$u_0 = 1 \tag{5.1.7}$$

since $2m(\vartheta) = 0$ and $\underline{n}(\vartheta) = \underline{0}$.

From Lemma 5.1.2, the contribution of $\mathcal{U}_1$ to u is

$$\begin{aligned} u_1 &= x^2 \sum_{U \in \mathcal{U}_{\langle 1 \rangle}} x^{2m(U)} \underline{y}^{\underline{n}(U)} \\ &= x^2 u^2. \end{aligned} \tag{5.1.8}$$

The contribution of $\mathcal{U}_2$ to u is denoted by u_2. Let

$$\tilde{u}(t) = \sum_{u \in \mathcal{U}_2} x^{2m(U)} z^{2j(U)} \underline{y}^{\underline{\tilde{n}}(U)}$$

where $2j(U)$ is the valency of the nonrooted end of the root-edge and $\underline{\tilde{n}}(U) = (\tilde{n}_2(U), \tilde{n}_4(U), \cdots, \tilde{n}_{2i}(U), \cdots)$, $\tilde{n}_{2i}(U)$ is the number of vertices of valency $2i$ except for the two ends of the root-edge. It is easily seen that

$$\underline{\tilde{n}}(U) = \underline{n}(U) - e_{2j(U)}$$

where $e_{2j(U)}$ is the vector with all the components 0 except only for the $j(U)$-th which is 1. In addition, it can be verified that

$$u_2 = \int_y \tilde{u}(y). \tag{5.1.9}$$

From Lemma 5.1.4, we have

$$\begin{aligned} \tilde{u}(z) &= \sum_{U \in \mathcal{U} - \vartheta} \left(\sum_{i=1}^{m(U)} x^{2i} z^{2m(U) - 2i + 2} \right) \underline{y}^{\underline{n}(U)} \\ &= \frac{x^2}{z^2} \sum_{U \in \mathcal{U} - \vartheta} z^{2m(U)+2} \frac{1 - \left(\frac{x}{z}\right)^{2m(U)}}{1 - \left(\frac{x}{z}\right)^2} \underline{y}^{\underline{n}(U)} \\ &= x^2 z^2 \sum_{U \in \mathcal{U}} \frac{z^{2m(U)} - x^{2m(U)}}{z^2 - x^2} \underline{y}^{\underline{n}(U)} \\ &= x^2 z^2 \delta_{z^2, x^2} \left(u(\sqrt{t}) \right). \end{aligned}$$

Then by (5.1.9), we find that

$$u_2 = x^2 \int_y y^2 \delta_{y^2,x^2}\Big(u(\sqrt{t})\Big). \tag{5.1.10}$$

From (5.1.7), (5.1.8) and (5.1.10), the theorem is obtained. □

If we write $u(x,y) = u(x,\hat{y})$ where $\hat{\underline{y}} = (y^2, y^4, \cdots, y^{2i}, \cdots)$, then $f_{\mathrm{El}} = u(x,\sqrt{y})$ is the enufunction of this type of maps with x and y corresponding to the two parameters: the valency of the root vertex and the size, respectively.

Lemma 5.1.5 *The function f_{El} satisfies the following quadratic equation:*

$$\begin{aligned} x^2y(1-x^2)f_{\mathrm{El}}^2 - (1-x^2+x^2y)f_{\mathrm{El}} \\ +(1-x^2)+x^2yf^* = 0 \end{aligned} \tag{5.1.11}$$

where $f^* = f_{\mathrm{El}}(1,y)$.

Proof First, the contributions of $\mathcal{U}_0$ and $\mathcal{U}_1$ to f_{El} can be easily found to be $f_{\mathcal{U}_0}(x,y) = 1$ and $f_{\mathcal{U}_1}(x,y) = x^2 f_{\mathrm{El}}^2$, respectively.

Then, from Lemma 5.1.4 the contribution of $\mathcal{U}_2$ to f_{El} is

$$\begin{aligned} f_{\mathcal{U}_2}(x,y) &= y\sum_{U\in\mathcal{U}}\Big(\sum_{i=1}^{m(U)} x^{2i}\Big)y^{n(U)} \\ &= x^2y\sum_{U\in\mathcal{U}}\frac{1-x^{2m(U)}}{1-x^2}y^{n(U)} \\ &= x^2y\frac{f^*-f_{\mathrm{El}}}{1-x^2}. \end{aligned}$$

As a consequence we have

$$\begin{aligned} f_{\mathrm{El}} &= f_{\mathcal{U}_0}(x,y) + f_{\mathcal{U}_1}(x,y) + f_{\mathcal{U}_2}(x,y) \\ &= 1 + x^2yf_{\mathrm{El}}^2 + x^2y\frac{f^*-f_{\mathrm{El}}}{1-x^2} \end{aligned}$$

which is equivalent to (5.1.11). □

The discriminant of Eq.(5.1.11) is

$$\delta(x,y) = (yX - X + 1)^2 - 4yX(X-1) \times(X - 1 - yxf^*) \qquad (5.1.12)$$

where $X = x^2$.

If X is considered as a power series ξ of y such that $X = \xi$ is a double root of $\delta(x,y)$, then we may find the following two equations:

$$\xi^2y^2 + 2\xi(\xi-1)y + (\xi-1)^2 -4\xi(\xi-1)^2y + 4y^2\xi(\xi-1)f^* = 0;$$

$$2\xi y^2 - 2(2\xi-1)y + 2(\xi-1) -4(\xi-1)(3\xi-1)y + 4y^2\xi(3\xi-2)f^* = 0$$

with ξ as a parameter. From them, we extract the parametric expressions:

$$y = \frac{-(\xi-1)(\xi-2)}{\xi^2};$$

$$f^* = \frac{1+\xi-\xi^2}{(\xi-2)^2}. \qquad (5.1.13)$$

Theorem 5.1.2 *The number of rooted planar Eulerian maps of size s is*

$$E_s = \begin{cases} 1, & s = 0; \\ \dfrac{3\cdot 2^{s-1}(2s)!}{s!(s+2)!}, & s \geq 1. \end{cases} \qquad (5.1.14)$$

Proof From the first relation of (5.1.13), we see that $\xi = 1 - y\frac{\xi^2}{\xi-2}$. Then, it is allowed to employ Corollary 1.5.2 for finding the coefficient of y^s in f^* as the power series of y as

$$\partial_\xi^s f^* = \frac{1}{s!}\frac{\mathrm{d}^{s-1}}{\mathrm{d}\xi^{s-1}}\left(\frac{\xi^{2s}}{(2-\xi)^s}\frac{\mathrm{d}f^*}{\mathrm{d}\xi}\right)\bigg|_{\xi=1}$$

$$= \frac{3\cdot 2^{s-1}(2s)!}{s!(s+2)!}$$

for $s \geq 1$. The case $s = 0$ is trivial. □

§5.2 Tutte formula

Let $U \in \mathcal{U}$ be a rooted Eulerian planar map. For a vertex $v = (x_1, x_2, \cdots, x_l)$ of U, we can construct another map denoted by $U\{v\}$ which is obtained by deleting v and adding l new edges Ky_i, $i = 1, 2, \cdots, l$, such that l new vertices are produced as

$$\begin{cases} v_1 = (x_1, y_1, \alpha\beta y_l); \\ v_i = (x_i, y_i, \alpha\beta y_{i-1}),\ 2 \leq i \leq l. \end{cases} \tag{5.2.1}$$

This operation on U is called *holing* at v. The map which is obtained by holing every vertex from U is called the *holed map* of U. Write $\mathring{U}$ as the holed map of U, *i.e.*, $\mathring{U} = U\{v|\forall v \in V(U)\}$. It is easily checked that any holed map is cubic and planar. Generally, for the root-vertex it is always imagined that the *boundary* (the circuit which consists of all the new edge added in the holing at the vertex) of the *hole* (the face bounded only by new edges) is the *outerface* of $\mathring{U}$, while others are *innerfaces*.

Because all the valencies of vertices in a Eulerian map are even, we may assume that the valency of root-vertex is $2m$ and that the number of nonrooted vertices with valency $2i$ is $n_i, i \geq 1$. From the basic knowledge of graphs, we see that

$$\begin{cases} \epsilon(U) = m + \sum_{i\geq 1} in_i; \\ \nu(U) = 1 + \sum_{i\geq 1} n_i \end{cases} \tag{5.2.2}$$

where $\epsilon(U)$ and $\nu(U)$ are the size and the order of U respectively.

Write $\mathcal{U}(m; \underline{n})$ as the set of all the Eulerian planar maps with valency of the root-vertex being $2m$ and the number of nonrooted vertices of valency $2i$ being $n_i, i \geq 1$. Here, $\underline{n} = (n_1, n_2, \cdots)$.

On the other hand, we may construct a kind of cubic planar map in the following procedure.

First, make n holes in the plane such that the boundaries of any two of them have no point in common and one of them includes the infinity. Then, label them by $1, 2, \cdots, n$ such that 1 is that with the infinity. The complement of the union of all the holes is called a *band.*

Second, mark $2l_i$ points as vertices on each hole $i, i = 1, 2, \cdots, n$. Let S be the set of holes labeled by $2, 3, \cdots, n$.

Third, choose a vertex v on the boundary of hole 1 and another vertex u. Draw a curve from v to u such that no point is in common with the boundary of any hole other than 1 and that u is on its boundary.

Fourth, cut along the curve and ignore the vertex v and the vertex u both.

C1 When u is also on the boundary v is on, then the plane is divided into two bands by the two closed curves each of which is formed by the curves from v to u and a part of the boundary of hole 1. In each of the two bands, the hole whose boundary is one of the two closed curves is labeled 1 as well.

The subsets of S which consist of all the holes in the inner domains of the two closed curves are denoted by P and $\bar{P}$. Of course, we have

$$S = P + \bar{P}. \tag{5.2.3}$$

C2 When u is on the boundary of a holes j other than that v is on, *i.e.*, $j \neq 1$, then the holes 1 and j are unified into 1 by cutting along a line connecting u and v without inner point in common with the boundaries of 1 and j, and then by ignoring u and v to obtain a band which has one hole less than those the original has.

Fifth, do the Third on each band until nothing can be done.

The procedure is called *slicing* in Tutte's terminology.

Theorem 5.2.1 *The slicing always terminates at the time when all the vertices are paired. The resultant which consists of all the* *boundaries of holes* $1, 2, \cdots, n$ *with all the curves drawn is a cubic planar map.*

Proof Prove the first statement by contradiction. Suppose the slicing terminates when vertex u and v have not paired yet. Then, u and v have to be on distinct bands. Because each band is only allowed to have even number of vertices that are not paired, on the band u is on has another vertex w which is not paired. However, on the band the slicing is allowed to go on by doing the Third with u and w chosen. This is a contradiction.

Prove the second statement by induction on the number of vertices. It is easy to check when the number is small. In general, from the Fourth on slicing, the following two cases should be considered.

Case 1 For C1, write the two cubic planar maps corresponding to those obtained by slicing on the respective two bands as M_1 and M_2. Then the whole map is formed from M_1 and M_2 by identifying an inner segment of one edge on the outerface boundary of M_1 and an inner segment of one edge on that of M_2. Because of the cubicness of M_1 and M_2, M is cubic as well. Further, it can be seen that M is planar iff so are M_1 and M_2. Then by the induction hypothesis M is planar.

Case 2 For C2, because the band obtained in the Fourth has two vertices less than that the original has, we see that the map M corresponding to the band is cubic and planar by hypothesis. However, the map corresponding to the original band is obtained by identifying two inner segments of different edges on the boundary of the outerface of M, from the cubicness and the planarity of M, the map is cubic and planar as well.

As a consequence, the theorem is obtained. □

A map obtained by slicing on a band of the plane is called a *slicification*. Because any slicification is a cubic planar map, we may always deal with it as a rooted map. Let $\mathcal{S}(m, \underline{n})$ be the set of all the slicifications with the *outer* hole (that is labeled by 1) of valency $2m$ and the number of holes, labeled by those other than 1, of vertices $2i$ being $n_i \geq 0, i \geq 1$, $\underline{n} = (n_1, n_2, \cdots)$.

Lemma 5.2.1 *For $m \geq 1$ and $n_i \geq 0,\ i \geq 1$, we have*

$$|\mathcal{U}(m;\underline{n})| = |\mathcal{S}(m;\underline{n})| \tag{5.2.4}$$

where $\underline{n} = (n_1, n_2, \cdots,)$.

Proof For $U \in \mathcal{U}(m;\underline{n})$, we have $\mathring{U}$ is the holed map of U. By following the slicing in pairing vertices as shown by the edges of U. From the planarity of U, the slicing terminates at the time when all edges have been cut. Then, we find that the slicification is isomorphic to $\mathring{U}$.

Conversely, for any $M \in \mathcal{S}(m;\underline{n})$ by contracting all the holes into points as vertices, we find a Eulerian planar map U. It can be checked that the holed map $\mathring{U}$ of U is also isomorphic to M. □

For convenience, let $M \in \mathcal{S}$, the set of all slicifications of the plane, be with k holes: $1, 2, \cdots, k$. On the boundary of i, let $2l_i$ be the number of vertices. Write

$$l = \sum_{i=1}^{k} l_i. \tag{5.2.5}$$

Then, M has $2l$ vertices all together. Let

$$\gamma(l_1, l_2, \cdots, l_k) \tag{5.2.6}$$

be the number of all the slicifications with k holes as described above. We may also write $\gamma(l_1; S)$ instead of (5.2.6) where $S = \{2, 3, \cdots, k\}$ is the set of all the holes except the root one 1.

Lemma 5.2.2 *The number $\gamma(l_1, \cdots, l_k)$ satisfies the following recursive relation:*

$$\begin{aligned}\gamma(l_1; S) &= \sum_{P \subseteq S} \sum_{j=0}^{l_1 - 1} \gamma(j; P)\gamma(l_1 - j - 1; \bar{P}) \\ &\quad + \sum_{r=2}^{k} 2l_r \gamma(l_1 + l_r - 1; S - \{r\}),\ k \geq 2 \\ \gamma(l_1; \emptyset) &= \frac{(2l_1)!}{l_1!(l-1)!},\ k = 1\end{aligned} \tag{5.2.7}$$

where $\gamma(0,P)$ is naturally defined to be 0.

Proof For the case $k=1$, from §4.3 it corresponds to the number of outerplanar cubic maps and hence it is known to be the Catalan number.

In general, by the procedure described above, we discuss in two cases.

Case 1 When both u and v are on the boundary of the hole 1. Let the $2l_1$ vertices be numbered by $v_1, v_2, \cdots, v_{2l_1}$. We may assume $v=v_1$ without loss of generality. In order to obtain the bands with all holes having an even number of vertices, u has to be v_{2j}, $j=0,1,\cdots,l_1-1$. Of course, $j=0$ contributes nothing to $\gamma(l_1,S)$. That contributes to $\gamma(l_1,S)$ in sum as

$$\sum_{P\subseteq S}\sum_{j=0}^{l_1-1}\gamma(j;P)\gamma(l_1-j-1;\bar{P})$$

which is the first term on the right hand side of (5.2.7).

Case 2 When u is on the boundary of hole $r\neq 1$, because there are $2l_r$ possibilities for the choice of u, the contribution of this case to $\gamma(l_1;S)$ is

$$\sum_{r=2}^{k}2l_r\gamma(l_1+l_r-1;S-\{r\}).$$

That is the last term on the right hand side of (5.2.7). Therefore, the lemma is found. □

According to the relation (5.2.7), we are allowed to determine

$$\gamma(l_1,\cdots,l_k)=\frac{(l-1)!}{(l-k+2)!}\prod_{i=1}^{k}\frac{(2l_i)!}{l_i!(l_i-1)!} \tag{5.2.8}$$

where l is given in (5.2.5).

Theorem 5.2.2 *The number of rooted Eulerian planar maps of order ν and size ϵ with the number of nonrooted vertices of valency $2i$ being q_i and the valency of the root-vertex being $2m$ is*

$$\frac{(\epsilon-1)!}{(\epsilon-\nu+2)!}\frac{(2m)!}{m!(m-1)!}\prod_{i\geq 1}\frac{1}{q_i!}\left(\frac{(2i-1)!}{i!(i-1)!}\right)^{q_i} \tag{5.2.9}$$

for ν, ϵ and $m \geq 1$.

Proof From Lemma 5.2.2, by considering that in C2 of the Fourth of slicing, the vertex has $2n_i, i \geq 2$, choices and that non-rooted vertices of the same valency are with the symmetry, (5.2.8) soon becomes (5.2.9). The lemma is obtained. □

§5.3 Eulerian planar triangulations

Let $\mathcal{U}_{\text{Tri}}$ be the set of all rooted Eulerian planar near triangulations with the outerfaces as the exception. Of course, the simplest one is the triangle. However, the vertex map ϑ is treated as a degenerate map for convenience.

Lemma 5.3.1 *For any $U \in \mathcal{U}_{\text{Tri}}$, no edge of U is a loop or an isthmus (cut-edge in other words).*

Proof By contradiction. Assume edge $e = (u, v)$ of U is an isthmus. Then, $U - e = U_1 + U_2$, where U_1 and U_2 are maps. However, by the Eulerianity of U, each of them has only one vertex, u or v, of odd valency. This is a contradiction.

If $e = (u, v), u = v$, is a loop of U, from the triangularity of U, the loop with its inner domain still forms a Eulerian planar near-triangulation $U_{e(\text{in})}$ of root-face valency 2. However, from the near-triangularity the valency of outerface should be divisible by 3. This is a contradiction. □

For $U \in \mathcal{U}_{\text{Tri}}$, let

$$\hat{U} = U - \{\mathcal{P}r, \mathcal{P}\alpha\beta(\mathcal{P}r), (\mathcal{P}\alpha\beta)^2(\mathcal{P}r)\}$$

be the map obtained by deleting the three edges incident with

$$\mathcal{P}r,\ (\mathcal{P}\alpha\beta)\mathcal{P}r,\ (\mathcal{P}\alpha\beta)^2\mathcal{P}r,$$

i.e., those on the boundary of the nonrooted face which is incident with the root-edge, from U. Then, $\mathcal{U}_{\text{Tri}}$ can be divided into four classes: $\mathcal{U}_{\text{tri}_i}, i = 0, 1, 2,$ and 3 such that $\mathcal{U}_{\text{Tri}_0} = \vartheta$, the set of only one map which is the vertex map, and

$$\mathcal{U}_{\mathrm{Tri}_i} = \{U|\forall U \in \mathcal{U}_{\mathrm{Tri}}, \hat{U} \text{ has } i \text{ connected components}\} \quad (5.3.1)$$

for $i = 1, 2,$ and 3. Thus,

$$\mathcal{U}_{\mathrm{Tri}} = \sum_{i=0}^{3} \mathcal{U}_{\mathrm{Tri}_i}. \quad (5.3.2)$$

Further, write

$$\hat{\mathcal{U}}_{\mathrm{Tri}_i} = \{\hat{U}|\forall U \in \mathcal{U}_{\mathrm{Tri}_i}\} \quad (5.3.3)$$

for $i = 1, 2,$ and 3.

Lemma 5.3.2 *For $\mathcal{U}_{\mathrm{Tri}_1}$, we have*

$$\hat{\mathcal{U}}_{\mathrm{Tri}_1} = \mathcal{U}_{\mathrm{Tri}} - \mathcal{U}_{\mathrm{Tri}}(3) - \vartheta \quad (5.3.4)$$

where $\mathcal{U}_{\mathrm{Tri}}(3) = \{U|\forall U \in \mathcal{U}_{\mathrm{Tri}},\ m(U) = 3\}$ in which $m(U)$ is the valency of the root-face in U.

Proof From what was mentioned in the proof of Lemma 5.3.1, we see that the set on the right hand side of (5.3.4) is the set of all rooted Eulerian planar near-triangulations with the outface valency not less than 6, and hence we write it as $\mathcal{U}_{\mathrm{Tri}}(\geq 6)$.

For any $U \in \hat{\mathcal{U}}_{\mathrm{Tri}_1}$, let $U = \hat{W}$,

$$W = (\mathcal{X}_{\alpha\beta}, \mathcal{P}) \in \mathcal{U}_{\mathrm{Tri}_1}$$

with its root r. From the Eulerianity and the triangularity of W, it is easily checked that $U \in \mathcal{U}_{\mathrm{Tri}_1}$. Since there has to be a segment

$$\langle(\mathcal{P}\alpha\beta)^{-1}r, \mathcal{P}^2 r, \mathcal{P}\alpha\beta(\mathcal{P}^2 r),\ (\mathcal{P}\alpha\beta)^2\mathcal{P}^2(\mathcal{P}^2 r), (\mathcal{P}\alpha\beta)^3\mathcal{P}^2(\mathcal{P}^2 r), \mathcal{P}\alpha\beta r\rangle$$

on the outerface boundary of U whose root is $\mathcal{P}\alpha\beta r$, we have $U \in \mathcal{U}_{\mathrm{Tri}}(\geq 6)$. That implies $\hat{\mathcal{U}}_{\mathrm{Tri}_1} \subseteq \mathcal{U}_{\mathrm{Tri}}(\geq 6)$.

Conversely, for any $U = (\mathcal{X}_{\alpha\beta}, \mathcal{P}) \in \mathcal{U}_{\mathrm{Tri}}(\geq 6)$ with its root r, we are allowed to construct a map $W = (\mathcal{X}'_{\alpha\beta}, \mathcal{Q})$ with its root x by adding three edges Kx, Ky and Kz on U such that

$$\mathcal{Q}t = \begin{cases} x, \ t = \alpha\beta(\mathcal{P}\alpha\beta)^{-5}r; \\ y, \ t = x; \\ (\mathcal{P}\alpha\beta)^{-4}r, \ t = y, \end{cases}$$

$$\mathcal{Q}t = \begin{cases} \alpha\beta y, \ t = \alpha\beta(\mathcal{P}\alpha\beta)^{-3}r; \\ z, \ t = \alpha\beta y; \\ (\mathcal{P}\alpha\beta)^{3}r, \ t = z, \end{cases}$$

$$\mathcal{Q}t = \begin{cases} \alpha\beta z, \ t = \alpha\beta(\mathcal{P}\alpha\beta)^{-1}r; \\ \alpha\beta x, \ t = \alpha\beta z; \\ r, \ t = \alpha\beta x \end{cases}$$

and for all other cases $\mathcal{Q}t = \mathcal{P}t$. It is also seen that $W \in \mathcal{U}_{\mathrm{Tri}}$ and $U = \hat{W}$. Further, since $\hat{W}$ is connected, we have $W \in \mathcal{U}_{\mathrm{Tri}_1}$ and $U = \hat{W} \in \hat{\mathcal{U}}_{\mathrm{Tri}_1}$. This implies $\mathcal{U}_{\mathrm{Tri}}(\geq 6) \subseteq \hat{\mathcal{U}}_{\mathrm{Tri}_1}$. □

Let $U \in \mathcal{U}_{\mathrm{Tri}_2}$ with root r. Write o, u, and v as the vertices incident with r, βr and $\beta\mathcal{P}r$ respectively. Of course, o is the root-vertex, or say the *first vertex*, u is the nonrooted end of the root-edge, or say the *second vertex*, and v is called the *third vertex* of U.

Lemma 5.3.3 *For $U \in \mathcal{U}_{\mathrm{Tri}_2}$, if the third vertex is connected to neither the first nor the second in $\hat{U}$, then $\hat{U}$ is a map itself.*

Proof Because the first vertex o and the second vertex u are in the same component of $\hat{U}$, the third vertex v has to be an inner vertex of U. Since v is connected to neither o nor u, it has to be a cut-vertex of U. From the triangularity, the outerface boundary of the component v is in has to be without an edge. That implies the valency of v is 2 in U. The lemma follows. □

From Lemma 5.3.3, $\mathcal{U}_{\mathrm{Tri}_2}$ can be divided into 3 classes named by $\mathcal{U}^{(1)}_{\mathrm{Tri}_2}, \mathcal{U}^{(2)}_{\mathrm{Tri}_2}$, and $\mathcal{U}^{(3)}_{\mathrm{Tri}_2}$ such that

$$\mathcal{U}_{\mathrm{Tri}_2} = \sum_{i=1}^{3} \mathcal{U}^{(i)}_{\mathrm{Tri}_2} \tag{5.3.5}$$

where

$$\mathcal{U}_{\mathrm{Tri}_2}^{(i)} = \{U|\forall U \in \mathcal{U}_{\mathrm{Tri}_2}\text{such that the } i\text{-th vertex is not connected to any of the other two in } \hat{U}\} \quad (5.3.6)$$

for $i = 1, 2, 3$.

Lemma 5.3.4 *For $\mathcal{U}_{\mathrm{Tri}_2}$, we have*

$$\hat{\mathcal{U}}_{\mathrm{Tri}_2}^{(1)} = \mathcal{U}_{\mathrm{Tri}} \times (\mathcal{U}_{\mathrm{Tri}} - \vartheta); \quad (5.3.7)$$

$$\hat{\mathcal{U}}_{\mathrm{Tri}_2}^{(2)} = \mathcal{U}_{\mathrm{Tri}} \times (\mathcal{U}_{\mathrm{Tri}} - \vartheta); \quad (5.3.8)$$

$$\hat{\mathcal{U}}_{\mathrm{Tri}_2}^{(3)} = (\mathcal{U}_{\mathrm{Tri}} - \vartheta) \times \vartheta \quad (5.3.9)$$

where ϑ is the vertex map.

Proof First, to prove (5.3.7) and (5.3.8). Because $\hat{U}$ has two components here such that one component of $\hat{U}$ is incident with only the first vertex, or the second vertex and the other is incident with the other two among the three vertices for $U \in \mathcal{U}_{\mathrm{Tri}}^{(2)}$. The former is allowed to be any map in $\mathcal{U}_{\mathrm{Tri}}$ and the latter is only not allowed to be the vertex map in $\mathcal{U}_{\mathrm{Tri}}$. Therefore, the two formulae are true.

Then, to prove (5.3.9). From Lemma 5.3.3, by the similar discussion as above for (5.3.7–8), (5.3.9) can also be found. □

Lemma 5.3.5 *For $\mathcal{U}_{\mathrm{Tri}_3}$, we have*

$$\hat{\mathcal{U}}_{\mathrm{Tri}_3} = \mathcal{U}_{\mathrm{Tri}} \times \mathcal{U}_{\mathrm{Tri}} \times \mathcal{U}_{\mathrm{Tri}} \quad (5.3.10)$$

where $\times$ is the Cartesian product.

Proof Because all the three of the first, the second and the third vertices are cut-vertices, each component of $\hat{U}, U \in \mathcal{U}_{\mathrm{Tri}_3}$, is allowed to be any map in $\mathcal{U}_{\mathrm{Tri}_3}$. The lemma follows. □

The enufunction of $\mathcal{U}_{\mathrm{Tri}}$ is written as

$$f = f_{\mathcal{U}_{\mathrm{Tri}}}(x, y) = \sum_{U \in \mathcal{U}_{\mathrm{Tri}}} x^{m(U)} y^{n(U)} \quad (5.3.11)$$

where $m(U)$ is the valency of the root-face, and $n(U)$ is the number of non-rooted faces.

In what follows, we evaluate the contributions $f_{\mathcal{U}_{\mathrm{Tri}_i}}$ of $\mathcal{U}_{\mathrm{Tri}_i}$ to f for $i = 0, 1, 2$ and 3.

Of course, $i = 0$ is the trivial case, and we have

$$f_{\mathcal{U}_{\mathrm{Tri}_0}} = 1. \tag{5.3.12}$$

For $\mathcal{U}_{\mathrm{Tri}_1}$, from Lemma 5.3.2 we find

$$\begin{aligned} f_{\mathcal{U}_{\mathrm{Tri}_1}} &= x^{-3}y^3 \sum_{u \in \hat{\mathcal{U}}_{\mathrm{Tri}_1}} x^{m(U)}y^{n(U)} \\ &= x^{-3}y^3(f - f^{\triangle} - 1) \end{aligned} \tag{5.3.13}$$

where $f^{\triangle} = f_{\mathcal{U}_{\mathrm{Tri}(3)}}$, the enufunction of rooted Eulerian planar triangulations with the number of nonrooted faces as the parameter.

According to Lemma 5.3.4, we see that

$$f_{\mathcal{U}_{\mathrm{Tri}_2}} = f_{\mathcal{U}^{(1)}_{\mathrm{Tri}_2}} + f_{\mathcal{U}^{(2)}_{\mathrm{Tri}_2}} + f_{\mathcal{U}^{(3)}_{\mathrm{Tri}_2}} \tag{5.3.14}$$

where

$$f_{\mathcal{U}^{(1)}_{\mathrm{Tri}_2}} = y^2 \sum_{U \in \hat{\mathcal{U}}^{(1)}_{\mathrm{Tri}_2}} x^{m(U)}y^{n(U)}$$

by (5.3.7),

$$= y^2 f(f-1),$$

$$f_{\mathcal{U}^{(2)}_{\mathrm{Tri}_2}} = y^2 f(f-1)$$

as well by (5.3.8), and

$$f_{\mathcal{U}^{(3)}_{\mathrm{Tri}_2}} = y^2 \sum_{U \in \mathcal{U}^{(3)}_{\mathrm{Tri}_2}} x^{m(v)}y^{n(v)}$$

by (5.3.9),

$$= y^2(f-1).$$

For $\mathcal{U}_{\mathrm{Tri}_3}$, from (5.3.10) in Lemma 5.3.5 we soon obtain that

$$f_{\mathcal{U}_{\mathrm{Tri}_3}} = x^3 y f^3.$$

Theorem 5.3.1 *The enufunction f of $\mathcal{U}_{\mathrm{Tri}}$ defined by (5.3.11) satisfies the following cubic equation:*

$$x^3yf^3 + 2y^2f^2 + \left(\left(\frac{y}{x}\right)^3 - y^2 - 1\right)f$$

$$+1 - y^2 - \left(\frac{y}{x}\right)^3\left(f^{\Delta} + 1\right) = 0 \qquad (5.3.15)$$

where f^{Δ} is the coefficient of x^3 in f.

Proof From (5.3.12–14), we soon obtain that

$$f = 1 + x^{-3}y^3(f - f^{\Delta} - 1)$$

$$+2y^2f(f-1) + x^3yf^3.$$

By grouping the terms according to the power of f, the equation (5.3.15) is derived. □

Although the cubic equation (5.3.15) looks not easy to solve, we can extract its solution in an indirect way.

Since for any Eulerian planar triangulation of order ν and size ϵ, we always have

$$\epsilon = 3\nu - 6$$

from the Euler formula for maps, according to (5.2.9) the number of rooted Eulerian planar triangulations of order ν with q_i nonrooted vertices of valency $2i, i \geq 1$, is

$$N_{\mathrm{Etr}}(\nu, m; \underline{q}) = \frac{(3\nu-7)!(2m)!}{(2\nu-4)!m!(m-1)!}$$

$$\times \prod_{i\geq 1} \frac{1}{q_i!}\left(\frac{(2i-1)!}{i!(i-1)!}\right)^{q_i} \qquad (5.3.16)$$

where $\underline{q} = (q_1, q_2, \cdots)$ and $2m$ is the valency of root-vertex.

Let $\mathcal{N}_p$ be the set of all partitions of integer $p > 0$, *i.e.*,

$$\mathcal{N}_p = \left\{\underline{q} \geq 0 \middle| \sum_{i\geq 1} q_i = p\right\}. \qquad (5.3.17)$$

Then the number of rooted Eulerian planar triangulations of order ν is

$$N_{\mathrm{Etr}}(\nu) = \sum_{m\geq 1}\sum_{\underline{q}\in\mathcal{N}_{\nu-1}} N_{\mathrm{Etr}}(\nu, m; \underline{q}). \tag{5.3.18}$$

Theorem 5.3.2 *The enufunction $f^{\triangle}$ of rooted Eulerian planar triangulations with the number n of nonrooted faces as the parameter is determined by*

$$\partial_y^n f^{\triangle} = \begin{cases} N_{\mathrm{Etr}}\left(\frac{n+5}{2}\right), & n = 1 (\mathrm{mod}\ 2); \\ 0, & n = 0 (\mathrm{mod}\ 2); \\ 1, & n = 0 \end{cases} \tag{5.3.19}$$

where ∂_y^n is defined as in (1.5.1).

Proof Since the coefficient $\partial_y^n f^{\triangle}$ of y^n in $f^{\triangle}$ is the number of rooted Eulerian planar triangulations of n inner faces, by considering that $n+1 = 2\nu - 4$ from Euler formula for planar maps we have $\nu = \frac{n+5}{2}$. From (5.3.18), the theorem is obtained. □

However, how to determine $f^{\triangle}$ by solving the equation (5.3.16) directly is still a problem.

§5.4 Regular Eulerian planar maps

A map is said to be *regular*, if all the valencies of vertices are a constant number. If the constant is l, it is also called an *l-regular* map. Among all regular maps, the 4-regular planar maps are particularly important at least in rectilinear embeddings of graphs which is motivated by VLSI circuit design, the Gauss crossing problem in graph theory, and the knot problem in topology. Of course, all 4-regular maps are Eulerian.

A *nearly 4-regular* map is a rooted map such that all vertices probably except the root-vertex are of valency 4. Because the number of *odd vertices* (of valency to be an odd number) is always even in a map, the valency of root-vertex in any nearly 4-regular map has to be even. Thus, any nearly 4-regular maps are Eulerian as well.

Let $\mathcal{M}_{\rm nr}$ be the set of all nearly 4-regular planar maps. Denote by $2m(M)$ and $n(M)$ the valency of the root-vertex and the number of inner faces of M, $M \in \mathcal{M}_{\rm nr}$, respectively.

Lemma 5.4.1 *For any $M \in \mathcal{M}_{\rm nr}$, if $M = M_1 \dot{+} M_2$ such that both M_1 and M_2 are submaps of M, then the valencies of the common vertex (the rooted one in all cases here) in M_1 and M_2 are always even.*

Proof By contradiction. Assume o is the common vertex of M_1 and M_2 and the valency of o in M_1 is odd. From the 4-regularity, all other vertices in M_1 have to be even. Then, M_1 has only one vertex o which is odd. This is a contradiction to M_1 being a map. □

The simplest map in $\mathcal{M}_{\rm nr}$ is the loop map $(K_x, (x, \alpha\beta x))$. However, the vertex map ϑ is defined to be in $\mathcal{M}_{\rm nr}$ for convenience. Let

$$\mathcal{M}_{{\rm nr}_1} = \{\vartheta\};$$

$$\mathcal{M}_{{\rm nr}_2} = \{M | \forall M \in \mathcal{M}_{\rm nr}, e_r(M) \text{ is a link}\};$$
$$\mathcal{M}_{{\rm nr}_3} = \{M | \forall M \in \mathcal{M}_{\rm nr}, e_r(M) \text{ is a loop}\}. \tag{5.4.1}$$

Then, $\mathcal{M}_{\rm nr}$ are classified into the following form:

$$\mathcal{M}_{\rm nr} = \mathcal{M}_{{\rm nr}_1} + \mathcal{M}_{{\rm nr}_2} + \mathcal{M}_{{\rm nr}_3}. \tag{5.4.2}$$

Lemma 5.4.2 *Let $\mathcal{M}_{({\rm nr})_2} = \{M \bullet a | \forall M \in \mathcal{M}_{{\rm nr}_2}\}$ where $a = e_r$ is the root-edge of M. Then, we have*

$$\mathcal{M}_{({\rm nr})_2} = \mathcal{M}_{\rm nr} - \mathcal{M}_{\rm nr}(2) - \vartheta \tag{5.4.3}$$

where $\mathcal{M}_{\rm nr}(2) = \{M | \forall M \in \mathcal{M}_{\rm nr}, 2m(M) = 2\}$.

Proof For any $M \in \mathcal{M}_{({\rm nr})_2}$, let $M = M' \bullet a'$ where $a' = e_r(M')$, the root-edge of M', $M' \in \mathcal{M}_{{\rm nr}_2}$. Because a' is a link and M' is not allowed to be the link map, the valency of the root-vertex is not less than 4 and hence

$$M = M' \bullet a' \in \mathcal{M}_{\rm nr} - \mathcal{M}_{\rm nr}(2) - \vartheta.$$

Thus,

$$\mathcal{M}_{({\rm nr})_2} \subseteq \mathcal{M}_{\rm nr} - \mathcal{M}_{\rm nr}(2) - \vartheta.$$

Conversely, for any $M \in \mathcal{M}_{\rm nr} - \mathcal{M}_{\rm nr}(2) - \vartheta$, because the valency of the root-vertex is even not less than 4, we may denote the root-vertex by

$$(r, \mathcal{P}r, ..., \mathcal{P}^{2m(M)-1}r)$$

with $2m(M) \geq 4$ where $r = r(M)$ is the root of M. Thus, from Lemma 5.4.1 the only map $M' \in \mathcal{M}_{\rm nr_2}$ can be constructed by splitting the root-vertex $o = v_r$ into o_1 and o_2 with an edge $e_{r'} = Kr' = (o_2, o_1)$ such that

$$o_1 = (r, \mathcal{P}r, \mathcal{P}^2 r, \alpha\beta r'),$$

and hence

$$o_2 = (r', \mathcal{P}^3 r, ..., \mathcal{P}^{2m(M)-1}r),$$

and that $M = M' \bullet a'$ where $a' = e_{r'}$ is the root-edge of M' which is a link. Thus, $M \in \mathcal{M}_{(\rm nr)_2}$. This implies

$$\mathcal{M}_{\rm nr} - \mathcal{M}_{\rm nr}(2) - \vartheta \subseteq \mathcal{M}_{(\rm nr)_2}.$$

□

From the procedure of the proof of Lemma 5.4.2, we see that there are 1-to-1 correspondences between the sets on the two sides of (5.4.3) and between $\mathcal{M}_{\rm nr_2}$ and $\mathcal{M}_{(\rm nr)_2}$. Therefore, we also have

$$\left|\mathcal{M}_{\rm nr_2}\right| = \left|\mathcal{M}_{\rm nr} - \mathcal{M}_{\rm nr}(2) - \vartheta\right|. \tag{5.4.4}$$

For $M \in \mathcal{M}_{\rm nr_3}$, because the root-edge a of M is a loop, we have

$$M = M_1 \dot{+} M_2 \tag{5.4.5}$$

such that M_1 and M_2 are both submaps of M.

Lemma 5.4.3 *Let $\mathcal{M}_{\langle nr \rangle_3} = \{M - a | \forall M \in \mathcal{M}_{\rm nr_3}\}$ with a being the root-edge of M. Then, we have*

$$\mathcal{M}_{\langle \rm nr \rangle_3} = \mathcal{M}_{\rm nr} \odot \mathcal{M}_{\rm nr} \tag{5.4.6}$$

where $\odot$ is the $1v$-production.

Proof For any $M \in \mathcal{M}_{\langle \rm nr \rangle_3}$, since M has the form (5.4.5), from the near-regularity both M_1 and M_2 are members of $\mathcal{M}_{\rm nr}$. That implies $\mathcal{M}_{\langle \rm nr \rangle_3}$ is a subset of the set on the right hand side of (5.4.6).

Conversely, for any $M \in \mathcal{M}_{\mathrm{nr}} \odot \mathcal{M}_{\mathrm{nr}}$, since there exist M_1 and $M_2 \in \mathcal{M}_{\mathrm{nr}}$ such that $M = M_1 \dot{+} M_2$, we may construct a map M' by add a loop a' at the common vertex o (the root-vertex of M_1 and M_2) such that one of M_1 and M_2 is in the inner domain and the other in the outer domain of the loop a' which is the root-edge of M'. Because $M' \in \mathcal{M}_{\mathrm{nr}_2}$ and $M = M' - a'$, we see $M \in \mathcal{M}_{\langle \mathrm{nr} \rangle_3}$. That implies $\mathcal{M}_{\mathrm{nr}} \odot \mathcal{M}_{\mathrm{nr}} \subseteq \mathcal{M}_{\langle \mathrm{nr} \rangle_3}$. □

We are now going to evaluate the contributions of $\mathcal{M}_{\mathrm{nr}_i}, i = 1, 2, 3$, to the enufunction

$$f = f_{\mathrm{nr}} = \sum_{M \in \mathcal{M}_{\mathrm{nr}}} x^{2m(M)} y^{n(M)} \tag{5.4.7}$$

where $2m(M)$ and $n(M)$ are the root-vertex valency and the inner face number respectively.

For $\mathcal{M}_{\mathrm{nr}_1}$, it is trivial that

$$f_{\mathcal{M}_{\mathrm{nr}_1}} = 1 \tag{5.4.8}$$

because there is only one map, the loop map in $\mathcal{M}_{\mathrm{nr}_1}$. By (5.4.4), we have

$$\begin{aligned} f_{\mathcal{M}_{\mathrm{nr}_2}} &= x^{-2} \sum_{M \in \mathcal{M}_{(\mathrm{nr})_2}} x^{2m(M)} y^{n(M)} \\ &= x^{-2}(f - x^2 f^* - 1) \end{aligned} \tag{5.4.9}$$

where f^* is the coefficient of x^2 in f, or in other words, the enufunction of rooted nearly 4-regular maps of root-vertex valency 2 with the number of inner faces as the parameter.

Then by Lemma 5.4.3, we obtain

$$\begin{aligned} f_{\mathcal{M}_{\mathrm{nr}_3}} &= x^2 y \sum_{M \in \mathcal{M}_{(\mathrm{nr})_3}} x^{2m(M)} y^{n(M)} \\ &= x^2 y f^2. \end{aligned} \tag{5.4.10}$$

Theorem 5.4.1 *The enufunction f defined by* (5.4.7) *satisfies the quadratic equation as*

$$x^4 y f^2 + (1 - x^2) f + x^2 - x^2 f^* - 1 = 0 \tag{5.4.11}$$

where f^* is the coefficient of x^2 in f.

Proof According to (5.4.2), we have

$$f = f_{\mathcal{M}_{\mathrm{nr}_1}} + f_{\mathcal{M}_{\mathrm{nr}_2}} + f_{\mathcal{M}_{\mathrm{nr}_3}}.$$

Then from (5.4.8–10), we derive

$$f = 1 + x^{-2}(f - x^2 f^* - 1) + x^2 y f^2.$$

Multiplying by x^2 the two sides and by rearranging the terms, we may soon find (5.4.11). □

The discriminant of (5.4.11) is

$$D = (1 - x^2)^2 - 4x^4 y(x^2 - x^2 f^* - 1) \tag{5.4.12}$$

from which we build up two parametric relations:

$$\begin{aligned} D_{x=\xi} &= (1-\xi^2)^2 - 4\xi^4 y(\xi^2 - \xi^2 f^* - 1) = 0; \\ \left.\frac{\partial D}{\partial x}\right|_{x=\xi} &= -4(1-\xi^2)\xi - 16\xi^3 y(\xi^2 - \xi^2 f^* - 1) \\ &\quad -8\xi^4(\xi - \xi f^*) = 0. \end{aligned} \tag{5.4.13}$$

Further, by dividing the two sides of the second in (5.4.13) by -4ξ, after rearrangement, (5.4.13) becomes

$$\begin{aligned} &(1-\xi^2)^2 + 4\xi^4(1-\xi^2)y + 4\xi^6 y f^* = 0; \\ &(1-\xi^2) + 2\xi^2(3\xi^2 - 2)y - 6\xi^4 y f^* = 0. \end{aligned} \tag{5.4.14}$$

From doing that 3 times the first plus $2\xi^2$ times the second, we have

$$(1-\xi^2)(2\xi^2 + 3 - 3\xi^2) + 4\xi^4 y = 0,$$

and hence

$$y = \frac{(\xi^2 - 1)(3 - \xi^2)}{4\xi^4}. \tag{5.4.15}$$

From (5.4.15) and one in (5.4.14), we find

$$f^* = \frac{(1-\xi^2)(\xi^2 - 2)}{\xi^2(3 - \xi^2)}. \tag{5.4.16}$$

By introducing the substitution

$$\eta = 1 - \xi^2, \tag{5.4.17}$$

from (5.4.15–16) we have

$$y = -\frac{\eta(2+\eta)}{4(1-\eta)^2};$$

$$f^* = -\frac{\eta(1+\eta)}{(1-\eta)(2+\eta)}. \tag{5.4.18}$$

Because $\eta = -\frac{4(1-\eta)^2}{2+\eta} y$, and hence

$$\phi(\eta) = \frac{-4(1-\eta)^2}{2+\eta}, \ \phi(0) = -2 \neq 0,$$

it is allowed to employ Corollary 1.5.1 for finding

$$\begin{aligned}
\partial_y^n f^* &= \frac{1}{n!}\frac{\mathrm{d}^{n-1}}{\mathrm{d}\eta^{n-1}}\left(\phi^n \frac{\mathrm{d}f^*}{\mathrm{d}\eta}\right)\Big|_{\eta=0} \\
&= \frac{(-1)^{n+1}}{n!}\frac{\mathrm{d}^{n-1}}{\mathrm{d}\eta^{n-1}}\left(\frac{4^n(1-\eta)^{2n}2(1+2\eta)}{(2+\eta)^n(1-\eta)^2(2+\eta)^2}\right)\Big|_{\eta=0} \\
&= \frac{(-1)^{n+1}2^{2n+1}}{n!}\frac{\mathrm{d}^{n-1}}{\mathrm{d}\eta^{n-1}}\left(\frac{(1-\eta)^{2n-2}(1+2\eta)}{(2+\eta)^{n+2}}\right)\Big|_{\eta=0} \\
&= \frac{(-1)^{n+1}2^{2n+1}}{n!}\left((1+2\eta)\frac{\mathrm{d}^{n-1}}{\mathrm{d}\eta^{n-1}}\frac{(1-\eta)^{2n-2}}{(2+\eta)^{n+2}}\right)\Big|_{\eta=0} \\
&\quad +2(n-1)\frac{\mathrm{d}^{n-2}}{\mathrm{d}\eta^{n-2}}\left(\frac{(1-\eta)^{2n-2}}{(2+\eta)^{n+2}}\right)\Big|_{\eta=0} \\
&= \frac{(-1)^{n+1}2^{2n+1}}{n!}\left(\sum_{i=0}^{n-1}\binom{2n-2}{n-i-1}\binom{n+i+1}{i}\right. \\
&\quad \times\frac{(-1)^{n-1}(n-1)!}{2^{n+i+2}} \\
&\quad +2(n-1)\sum_{i=0}^{n-2}\binom{2n-2}{n-i-2}\binom{n+i+1}{i}
\end{aligned}$$

$$\times \frac{(-1)^{n-2}(n-2)!}{2^{n+i+2}}\Bigg)$$

$$= \frac{2^{n-1}}{n}\left(\frac{(2n-2)!}{(n+1)!(n-1)!}\left(\frac{3}{2}\right)^{n-3}\frac{8n^2-3n-2}{2}\right.$$

$$\left.-\frac{2(2n-2)!}{(n+1)!(n-1)!}\left(\frac{3}{2}\right)^{n-2}\frac{4n+1}{2}\right)$$

$$= \frac{2^{n-1}}{n}\left(\frac{3}{2}\right)^{n-3}\frac{(2n-2)!}{(n+1)!(n-2)!}\frac{9n}{2(n-1)}$$

$$= 3^{n-1}\frac{2(2n-2)!}{(n+1)!(n-1)!}. \tag{5.4.19}$$

Lemma 5.4.4 *For $M \in \mathcal{M}_{\rm nr}(2)$, let $\tilde{M}$ be the map obtained by deleting the root-vertex and adding an edge $(v_{(\mathcal{P}\alpha\beta)^{-1}r}, v_{\beta r})$ as the root-edge of $\tilde{M}$. Write*

$$\tilde{\mathcal{M}}_{\rm nr}(2) = \{\tilde{M} | \forall M \in \mathcal{M}_{\rm nr}(2) - L\} \tag{5.4.20}$$

where L is the loop map. Then we have

$$\tilde{\mathcal{M}}_{\rm nr}(2) = \mathcal{M}_{\rm nr}(4) \tag{5.4.21}$$

where $\mathcal{M}_{\rm nr}(4)$ is the set of all rooted 4-regular planar maps.

Proof For any $M \in \mathcal{M}_{\rm nr}(4)$, we may uniquely construct a map $M' \in \mathcal{M}_{\rm nr}(2)$ by deleting the root-edge $e_r = Kr$ and then adding a new vertex o with two edges Kr' and Ks such that

$$o = (r', \alpha\beta s),\ \mathcal{P}\alpha\beta r' = \mathcal{P}\alpha\beta r,\ \mathcal{P}s = \mathcal{P}r$$

where r' is chosen to be the root of M'. It is easily checked that

$$\tilde{M}' = M,\ M' \neq L$$

and hence $M \in \tilde{\mathcal{M}}_{\rm nr}(2)$. Thus, $\mathcal{M}_{\rm nr}(4) \subseteq \tilde{\mathcal{M}}_{\rm nr}(2)$.

Conversely, for any $\tilde{M} \in \tilde{\mathcal{M}}_{\rm nr}(2)$, by the procedure of constructing $\tilde{M}$ from

$$M \in \mathcal{M}_{\rm nr}(2) - L,$$

it is seen that $\tilde{M} \in \mathcal{M}_{\rm nr}(4)$. Thus, $\tilde{\mathcal{M}}_{\rm nr}(2) \subseteq \mathcal{M}_{\rm nr}(4)$. □

Theorem 5.4.2 *The number of rooted 4-regular planar maps with n inner faces is*

$$H_{\mathrm{nr}}(n) = 3^{n-1}\frac{2(2n-2)!}{(n+1)!(n-1)!} \tag{5.4.22}$$

for $n \geq 2$.

Proof From Lemma 5.4.4, by considering that $\tilde{M}$ has the same number of inner faces as M does, we see

$$H_{\mathrm{nr}}(n) = \partial_y^n f^*.$$

By (5.4.19) the theorem follows. □

§5.5 Notes

5.5.1 Although f_{El} in (5.1.11) is determined by (5.1.14) quite simply, it looks more complicated for finding f itself in (5.1.11). Based on what is obtained here, one is also allowed to discuss the enumerations of other kinds of Eulerian maps, rooted or nonrooted, planar or nonplanar, and so on. The form (5.1.14) can be seen in [Liu38] and then independently in [BeC2]. The former also covers an explicit formula with two parameters.

5.5.2 In view of the planar duality of Eulerian maps, the number of rooted bipartite planar maps of a given size can be found in [Liu41], where the enufunction with two parameters was not provided. Two years later it appeared in [Liu61]. Although the formula is still not simple enough, it is expressed as a summation with all terms being positive.

5.5.3 The formula (5.2.8) was first discovered by Tutte in [Tut5]. From theoretical point of view, (5.2.9) should lead to a solution of the functional equation (5.1.10). However, no one has derived it from solving the equation directly.

5.5.4 Several ways can be used to find a solution of the equation (5.3.15). One is considered to observe the planar dual case shown in §4.2. Others are suggested for investigating some relationships with what have been or can probably be found, such as the nonseparable case in both bipartite and Eulerian kinds.

5.5.5 Although the 4-regular planar maps are much related to quadrangulations on the disc, which was firstly discussed in [Bro1], the equation (5.4.11) is still new. The method looks much simpler. Based on this, we are also allowed to discuss non-separable, even 2- or 3- connected, cases. It does look simple enough for determining explicit formulae.

5.5.6 On the usages of 4-regular planar maps can be seen in [Liu60] for rectilinear embedding in VLSI, for the knot problem in topology [Liu59], for the Gaussian crossing problem in graph theory [Liu57], and for enumerations of some other kinds of maps [MuSc1].

5.5.7 Although the corresponding enumeration problems for varieties of Eulerian maps on surfaces can be naturally proposed, there do not appear to be many results in the literature up to now.

5.5.8 It would also have the algebraic way to enumerate general Eulerian maps according to [Jac3–4] and [JaV3].

Chapter 6

Nonseparable Maps

§6.1 Outerplanar nonseparable maps

An outerplanar map is said to be *nonseparable* if it has no cut-vertex. Let $\mathcal{O}_{\rm ns}$ be the set of all rooted outerplanar nonseparable maps with the convention of excluding the vertex map ϑ and the link map $L = (Kx, (x)(\alpha\beta x))$. We divide $\mathcal{O}_{\rm ns}$ into two classes: $\mathcal{O}_{\rm ns_0}$ and $\mathcal{O}_{\rm ns_1}$, *i.e.*,

$$\mathcal{O}_{\rm ns} = \mathcal{O}_{\rm ns_0} + \mathcal{O}_{\rm ns_1} \tag{6.1.1}$$

where $\mathcal{O}_{\rm ns_0}$ consists of only one map, *i.e.*, the loop map $L_1 = (Kr, (r, \alpha\beta r))$ and $\mathcal{O}_{\rm ns_1}$, of all others.

Let O_1 and O_2 be two maps in $\mathcal{O}_{\rm ns}$. Assume o_1 and o_2 are the root-vertices of O_1 and O_2 respectively. Write

$$o_2 = (r_2, \mathcal{P}r_2, \cdots, \mathcal{P}^{m(Q_2)-1}r_2) = (S_1, S_2) \tag{6.1.2}$$

and

$$o_1 = (r_1, \mathcal{P}r_1, \cdots, \mathcal{P}^{m(O_1)-1}r_1) = (S) \tag{6.1.3}$$

where S_1, S_2 and S are in linear order. Then, the map $O_1 \dot{+} O_2$ defined to be $O_1 \cup O_2$ provided that $O_1 \cap O_2 = \{o\}$, $o = o_1 = o_2$, and that the root-vertex o of $O_1 + O_2$ as

$$o = (S_1, S, S_2) \tag{6.1.4}$$

is called the *inner* 1v-*addition* of O_1 and O_2. Define

$$O_1 \dot{+} O_2 \dot{+} O_3 \dot{+} \cdots \dot{+} O_k = k \dot{\times} O \tag{6.1.5}$$

when $O_1 = O_2 = \cdots = O_k = O$. If $k = 0$, then $k \times\!\cdot\, O$ is defined to be nothing concerned.

Further, for two sets of maps, $\mathcal{Q}_1$ and $\mathcal{Q}_2$, let

$$\mathcal{Q}_1 \times\!\cdot\, \mathcal{Q}_2 = \{O_1 +\!\cdot\, O_2 | \forall O_1 \in \mathcal{Q}_1, \forall O_2 \in \mathcal{Q}_2\} \qquad (6.1.6)$$

be called the *inner* 1*v-product* of $\mathcal{O}_1$ and $\mathcal{O}_2$. Define

$$\mathcal{Q}_1 \times\!\cdot\, \mathcal{Q}_2 \times\!\cdot\, \cdots \times\!\cdot\, \mathcal{Q}_n = \mathcal{Q}^{\times n} \qquad (6.1.7)$$

when $\mathcal{Q}_1 = \mathcal{Q}_2 = \cdots = \mathcal{Q}_n$.

Lemma 6.1.1 *For any $O \in \mathcal{O}_{\mathrm{ns}_1}$, there exist an integer $k \geq 0$ and a map $O' \in \mathcal{O}_{\mathrm{ns}}$ such that*

$$O \bullet a = (k \times\!\cdot\, L_1) +\!\cdot\, O' \qquad (6.1.8)$$

where $a = e_r(O)$ is the root-edge of O.

Proof Because $O \in \mathcal{O}_{\mathrm{ns}_1}$ and $O \neq L_1$ from the convention, $O \bullet a$ is still an outerplanar map but not allowed to be the vertex map ϑ. From the outerplanarity of O, all the edges with the same two ends as those of the root-edge form a multiedge, or more precisely, a segment in the rotation at each of the two ends. The number of edges excluding the nonrooted edge on the root-face boundary, if any, in the multi-edge is $k+1, k \geq 0$. The remaining map of deleting the k loops in $O \bullet a$ is O', for which it is easily checked that $O' \in \mathcal{O}_{\mathrm{ns}}$ by considering that $O' \neq L$ from the nonseparability. Since the k loops in an inner face of O' form the map $k \times\!\cdot\, L$, we soon find (6.1.8). This is the lemma □

For $O' \in \mathcal{O}_{\mathrm{ns}}$, let

$$\mathcal{O}_{O'} = \{O | \forall O \in \mathcal{O}_{\mathrm{ns}_1},\ O \bullet a = (k \times\!\cdot\, L_1) +\!\cdot\, O',\ k \geq 0\}. \qquad (6.1.9)$$

Then the following lemma can be found.

Lemma 6.1.2 *For $\mathcal{O}_{\mathrm{ns}_1}$, we have*

$$\mathcal{O}_{\mathrm{ns}_1} = \sum_{O' \in \mathcal{O}_{\mathrm{ns}}} \mathcal{O}_{O'}. \qquad (6.1.10)$$

Proof First, prove that for any $O'_1, O'_2 \in \mathcal{O}_{\rm ns}$,

$$\mathcal{O}_{O'_1} \bigcap \mathcal{O}_{O'_2} = \emptyset$$

iff $O'_1 \neq O'_2$ in the rooted case. The necessity is obvious since otherwise $\mathcal{O}_{O'_1} = \mathcal{O}_{O'_2}$, a contradiction to the condition appears. Conversely, from (6.1.8) and considering the uniqueness of this expression, the sufficiency is soon found.

Then, prove

$$\mathcal{O}_{\rm ns_1} = \bigcup_{O' \in \mathcal{O}_{\rm ns}} \mathcal{O}_{O'}.$$

For any $O \in \mathcal{O}_{\rm ns_1}$, from Lemma 6.1.1 there is a map $O' \in \mathcal{O}_{\rm ns}$ such that $O \in \mathcal{O}_{O'}$. Conversely, for any $O \in \mathcal{O}_{O'}$, there is a map $O' \in \mathcal{O}_{\rm ns}$ such that $O \bullet a = (k \times\cdot L_1) +\cdot O'$. Because $O' \neq \vartheta$, $O \neq L_1$, from the uniqueness of O for k and O' given, it is seen that O is outerplanar and nonseparable as well. Thus, $O \in \mathcal{O}_{\rm ns_1}$. □

For $O' \in \mathcal{O}_{\rm ns}$, the root vertex of O' is written as

$$o' = (r', \mathcal{P}r', \cdots, \mathcal{P}^{m(O')-1}r')$$

in which the segment $\langle \mathcal{P}^{i-1}r', \mathcal{P}^i r' \rangle$ is called the *i-th angle* of O'. Denote by

$$O +\cdot_i O' = O +\cdot O' \tag{6.1.11}$$

the inner 1v-addition of O and O' provided O is incident with the i-th angle of O'. Here, the map O is always in the form $k \times\cdot L_1$. The operation appearing in (6.1.11) is also called *the i-th* of the inner 1v-addition.

Accordingly, write

$$\mathcal{O}_{O'}(i) = \{O | \forall O \in \mathcal{O}_1, O \bullet a = (k \times\cdot L) +\cdot_i O', \\ O' \in \mathcal{O}, k \geq 0\} \tag{6.1.12}$$

for $1 \leq i \leq m(O') - 1$.

Lemma 6.1.3 *For any $O' \in \mathcal{O}_{\rm ns}$, we have*

$$\mathcal{O}_{O'} = \sum_{i=1}^{m(O')-1} \mathcal{O}_{O'}(i) \tag{6.1.13}$$

where $\mathcal{O}_{O'}(i)$ *is defined in* (6.1.12).

Proof First, prove that for any $i, j, i \neq j$,

$$\mathcal{O}_{O'}(i) \bigcap \mathcal{O}_{O'}(j) = \emptyset.$$

Because $i \neq j$ and all the maps in $\mathcal{O}_{O'}(i)$ and $\mathcal{O}_{O'}(j)$ have the same root, from Theorem 1.1.5 the relation holds.

Then, prove

$$\mathcal{O}_{O'} = \bigcup_{i=1}^{m(O')-1} \mathcal{O}_{O'}(i).$$

From the definition of the i-th inner $1v$-addition, we soon see that the set on the right hand side is a subset of the set on the left. Conversely, from the definition of $O_{O'}$ in (6.1.9), any member of $\mathcal{O}_{O'}$ is also a member of the set on the other side. That implies the set on the left is a subset of that on the right. □

In what follows, we investigate the enufunction of $\mathcal{O}_{\mathrm{ns}}$ with vertex partition

$$f = f_{\mathcal{O}_{\mathrm{ns}}} = \sum_{O \in \mathcal{O}_{\mathrm{ns}}} x^{m(O)} \underline{y}^{\underline{n}(O)} \tag{6.1.14}$$

where $m(O)$ is the valency of root-vertex and $\underline{n} = (n_2(O), n_3(O), \cdots)$, $n_i(O)$ is the number of nonrooted vertices of valency i, $i \geq 2$, by evaluating the contributions $f_{\mathcal{O}_{\mathrm{ns}_0}}$ and $f_{\mathcal{O}_{\mathrm{ns}_1}}$ of $\mathcal{O}_{\mathrm{sep}_0}$ and $\mathcal{O}_{\mathrm{ns}_1}$ to f respectively.

Since the loop map has only one vertex and only one edge which contributes 2 to the power of x but nothing to the power of $\underline{y}$, we have

$$f_{\mathcal{O}_{\mathrm{ns}_0}} = x^2. \tag{6.1.15}$$

For $\mathcal{O}_{\mathrm{ns}_1}$, we have to evaluate the function

$$\tilde{f}(z) = \sum_{O \in \mathcal{O}_{\mathrm{ns}_1}} x^{m(O)} z^{s(O)} \underline{y}^{\underline{n}(O)}$$

where $s(O)$ is the valency of the nonrooted vertex incident to the root-edge. From Lemmas 6.1.1–3, we have

$$\tilde{f}(z) = \sum_{O' \in \mathcal{O}_{\mathrm{ns}}} \sum_{k \geq 0} \sum_{i=1}^{m(O')-1} x^{k+i+1} z^{m(O')+k-i+1} \underline{y}^{\underline{n}(O')}$$

$$
\begin{aligned}
&= \sum_{O' \in \mathcal{O}_{\mathrm{ns}}} \frac{xz}{1-xz} \left(\sum_{i=1}^{m(O')-1} x^i z^{m(O')-i} \right) \underline{y}^{\underline{n}(O')} \\
&= \frac{xz}{1-xz} \sum_{O' \in \mathcal{O}_{\mathrm{ns}}} \left(\frac{xz^{m(O')} - zx^{m(O')}}{z-x} \right) \underline{y}^{\underline{n}(O')} \\
&= \frac{xz}{1-xz} \partial_{x,z}\big(f(u)\big).
\end{aligned} \tag{6.1.16}
$$

Theorem 6.1.1 *The enufunction f defined by (6.1.14) satisfies the following functional equation:*

$$
f = x^2 + x \int_y \frac{y}{1-xy} \partial_{x,y}\big(f(u)\big) \tag{6.1.17}
$$

where $\partial_{x,y}$ is defined as in (1.6.8).

Proof From (6.1.1), we have

$$
f = f_{\mathcal{O}_{\mathrm{ns}_0}} + f_{\mathcal{O}_{\mathrm{ns}_1}}. \tag{6.1.18}
$$

By (6.1.16), we find

$$
\begin{aligned}
f_{\mathcal{O}_{\mathrm{ns}_1}} &= \int_y \tilde{f}(y) \\
&= \int_y \frac{xy}{1-xy} \partial_{x,y}(f(u)).
\end{aligned} \tag{6.1.19}
$$

Then, by substituting (6.1.15) and (6.1.19) into (6.1.18), the theorem is obtained. □

It looks not easy to solve the equation (6.1.17) directly for determining the enufunction. However, we can do it in the other way for face partition instead of vertex partition. In order to do this, a new kind of decomposition of $\mathcal{O}_{\mathrm{ns}}$ has to be investigated.

Let $\mathcal{O}_{\mathrm{sf}}$ be the set obtained from $\mathcal{O}_{\mathrm{ns}}$ with the replacement of the loop map L_1 by the link map L. Then, $\mathcal{O}_{\mathrm{sf}}$ is classified into $\mathcal{O}_{\mathrm{sf}_0}$ and $\mathcal{O}_{\mathrm{sf}_1}$, *i.e.*,

$$
\mathcal{O}_{\mathrm{sf}} = \mathcal{O}_{\mathrm{sf}_0} + \mathcal{O}_{\mathrm{sf}_1} \tag{6.1.20}
$$

where $\mathcal{O}_{\mathrm{sf}_0}$ consists of only the link map L and $\mathcal{O}_{\mathrm{sf}_1}$, of all the others in $\mathcal{O}_{\mathrm{sf}}$.

The enufunction $h = h_{\mathcal{O}_{\rm sf}}$ considered here is

$$h_{\mathcal{O}_{\rm sf}}(x,y) = \sum_{O\in\mathcal{O}_{\rm sf}} x^{m(O)}\underline{y}^{\underline{n}(O)} \tag{6.1.21}$$

where $m(O)$ is the valency of root-face and $\underline{n}(O) = (n_1(O), n_2(O), \cdots), n_i(O)$ is the number of inner faces of valency $i, i\geq 1$.

Let $\mathcal{O}_{\langle{\rm sf}\rangle_1} = \{O - a|\forall O \in \mathcal{O}_{{\rm sf}_1}\}$ where $a = e_r$. Then, for any $O' \in \mathcal{O}_{\langle{\rm sf}\rangle_1}$, assume $O' = O - a$ and the inner face incident with the root-edge a of O is of valency $k, k \geq 1$. From the outerplanarity of O, we see

$$O' = O_1\hat{+}O_2\hat{+}\cdots\hat{+}O_k \tag{6.1.22}$$

where $O_j \cap O_{j+1} = v_{(\mathcal{P}\alpha\beta)^j\mathcal{P}\beta r}, j = 1, 2, \cdots, k-1$, and $O_j \in \mathcal{O}_{\rm sf}$ with

$$r(O_1) = \mathcal{P}\beta r,\ r(O_j) = (\mathcal{P}\alpha\beta)^{j-1}\mathcal{P}\beta r$$

for $j = 2, 3, \cdots, k$. A map O which has the form (6.1.22) is called the *chain* $1v$*-addition* of $O_1, O_2, \cdots$, and O_k.

Lemma 6.1.4 *Write* $\hat{\mathcal{O}}_k = \{O_1\hat{+}\cdots\hat{+}O_k|\forall O_j \in \mathcal{O}_{\rm sf}, i = 1, 2, \cdots, k\}$. *Then, we have*

$$\mathcal{O}_{\langle{\rm sf}\rangle_1} = \sum_{k\geq 1}\hat{\mathcal{O}}_k \tag{6.1.23}$$

where $\hat{\mathcal{O}}_k = \mathcal{O}_1\hat{\times}\cdots\hat{\times}\mathcal{O}_k = \mathcal{O}^{\hat{\times}k}$ *for* $\mathcal{O} = \mathcal{O}_i$, $i = 1, 2, \cdots, k$, *is also called the chain* $1v$*-production.*

Proof For any $O' \in \mathcal{O}_{\langle{\rm sf}\rangle_1}$, from what has just been mentioned above, there exists an integer $k \geq 1$ such that $O' \in \hat{\mathcal{O}}_k$ and hence $\mathcal{O}_{\langle{\rm sf}\rangle_1}$ is a subset of the set on the right hand side of (6.1.23).

Conversely, for any O', a member of the set on the right of (6.1.23), write $O' \in \hat{\mathcal{O}}_k$ for a $k, k \geq 1$. Since O' has the form (6.1.22), we may construct a map O by adding an edge Kr being from $v_{\beta r(O_k)}$ to $v_{r(O_1)}$ with $v_{\beta r(Q_k)} = v_r$ and $\mathcal{P}r = \beta r(O_k)$. Of course, $O' = O - a,\ a = Kr$. It is easily checked that $O \in \mathcal{O}_{{\rm sf}_1}$. This implies $O' \in \mathcal{O}_{\langle sf\rangle_1}$. Hence, the set on the left of (6.1.23) is a subset of $\mathcal{O}_{\langle{\rm sf}\rangle_1}$. □

According to Lemma 6.1.4, we may find the contribution of $\mathcal{O}_{\mathrm{sf}_1}$ to h is

$$\begin{aligned} h_{\mathcal{O}_{\mathrm{sf}_1}} &= x \sum_{O' \in \mathcal{O}_{\langle sf \rangle_1}} y_{k(O')+1} x^{m(O')-k(O')} \underline{y}^{\underline{n}(O')} \\ &= \sum_{k \geq 1} x y_{k+1} x^{-k} \left(\sum_{O \in \mathcal{O}_{\mathrm{sf}}} x^{m(O)} \underline{y}^{\underline{n}(O)} \right)^k \\ &= x \sum_{k \geq 1} y_{k+1} x^{-k} h^k. \end{aligned} \tag{6.1.24}$$

Theorem 6.1.2 *The enufunction h defined by* (6.1.21) *satisfies the functional equation as*

$$h = x^2 + x \int_y \frac{y^2 h}{x - yh} \tag{6.1.25}$$

in any way.

Proof Since $\mathcal{O}_{\mathrm{sf}_0}$ has a single member which is the link map, the contribution of $\mathcal{O}_{\mathrm{sf}_0}$ to h is

$$h_{\mathcal{O}_{\mathrm{sf}_0}} = x^2. \tag{6.1.26}$$

For $\mathcal{O}_{\mathrm{sf}_1}$, by (6.1.24) we have

$$\begin{aligned} \int_y^{-1} h_{\mathcal{O}_{\mathrm{sf}_1}} &= x \sum_{k \geq 1} \left(\frac{y}{x} h \right)^k \\ &= \frac{xyh}{x - yh}. \end{aligned} \tag{6.1.27}$$

Then by (6.1.26–27), we obtain that

$$\begin{aligned} h &= h_{\mathcal{O}_{\mathrm{sf}_0}} + h_{\mathcal{O}_{\mathrm{sf}_1}} \\ &= x^2 + x \int \frac{yh}{x - yh}. \end{aligned}$$

This is just (6.1.25). The theorem follows. □

From Theorem 6.1.2 we find

$$\frac{h}{x} = x + \sum_{k\geq 1} y_{k+1}\left(\frac{h}{x}\right)^k \tag{6.1.28}$$

which is allowed to employ Corollary 1.5.2. Then we have

$$\begin{aligned}\frac{h}{x} &= x + \sum_{m\geq 1} \frac{1}{m!}\frac{\mathrm{d}^{m-1}}{\mathrm{d}\xi^{m-1}}\left(\sum_{k\geq 1} y_{k+1}\xi^k\right)^m \Bigg|_{\xi=x} \\ &= x + \sum_{m\geq 1} \frac{1}{m!} \sum_{\underline{n}\in\mathcal{R}_{s-m+1}} \frac{m!s!}{\underline{n}!(s-m+1)!} \\ &\quad \times x^{s-m+1} \underline{y}^{\underline{n}}\end{aligned} \tag{6.1.29}$$

where

$$m = \sum_{i\geq 2} n_i,\ s = \sum_{k\geq 1} kn_{k+1}$$

and

$$\mathcal{R}_{s-m+1} = \{\underline{n}|0 \leq n_i \leq m\}.$$

Theorem 6.1.3 *The number of distinct rooted nonseparable outerplanar maps with q edges, the valency m of root-face, and n_i non rooted faces of valency $j, j \geq 2$, is*

$$N_{\mathcal{O}_{\mathrm{sf}}}(q, \underline{n}) = \frac{1}{q}\left(\frac{q!}{(m-1)!\underline{n}!}\right) \tag{6.1.30}$$

where $\underline{n} = (n_2, n_3, \cdots)$.

Proof From (6.1.29), we have

$$h = x^2 + \sum_{m\geq 1} \frac{1}{m!} \sum_{\underline{n}\in\mathcal{R}_{s-m+1}} \frac{m!s!}{n!(s-m+1)!} x^{s-m+2} \underline{y}^{\underline{n}}.$$

That implies the number of such maps with m nonrooted faces, the valency of root-face being $s-m+2$ and n_i nonrooted faces of valency i is

$$\frac{s!}{\underline{n}!(s-m+1)!}. \tag{6.1.31}$$

Since

$$2q = (s - m + 2) + \sum_{k \geq 2} kn_k$$
$$= (s - m + 2) + s + m$$
$$= 2s + 2,$$

we have $q = s + 1$. By considering that $s - m + 2$ is the valency of the root-face, from (6.1.31) the theorem is obtained. □

§6.2 Eulerian nonseparable maps

Although general Eulerian planar maps have been investigated in Chapter 5, nonseparable Eulerian maps are still necessary to be considered for finding a functional equation satisfied by the enufunction or the relationships between general and nonseparable cases in order to enumerate the corresponding maps.

Let $\mathcal{M}_{\mathrm{Ens}}$ be the set of all rooted nonseparable Eulerian planar maps. For any $M \in \mathcal{M}_{\mathrm{Ens}}$, let $2m(M)$ be the valency of root-vertex and $n_{2i}(M)$ be the number of nonrooted vertices of valency $2i$ for $i \geq 1$.

The smallest map in $\mathcal{M}_{\mathrm{Ens}}$ is the loop map $L_1 = (Kr, (r, \alpha\beta r))$. Then, $\mathcal{M}_{\mathrm{Ens}}$ is divided into two classes: $\mathcal{M}_{\mathrm{Ens}_0}$ and $\mathcal{M}_{\mathrm{Ens}_1}$ where $\mathcal{M}_{\mathrm{Ens}_0}$ is the set of only one map which is the loop map L_1 and $\mathcal{M}_{\mathrm{Ens}_1}$, of all others in $\mathcal{M}_{\mathrm{Ens}}$, *i.e.*,

$$\mathcal{M}_{\mathrm{Ens}} = \mathcal{M}_{\mathrm{Ens}_0} + \mathcal{M}_{\mathrm{Ens}_1}. \tag{6.2.1}$$

For two maps $M_1 = (\mathcal{X}_1, \mathcal{P}_1)$ and $M_2 = (\mathcal{X}_2, \mathcal{P}_2)$ with the roots r_1 and r_2 respectively, the composition of M_1 and M_2 into $M = (\mathcal{X}, \mathcal{P})$ is called the *root addition* of M_1 and M_2 and written by

$$M = M_1 +\!|_r M_2. \tag{6.2.2}$$

if $\mathcal{X} = \mathcal{X}_1 + (\mathcal{X}_2 - Kr_2)$ and $\mathcal{P}$ is induced from $\mathcal{P}_1$ and $\mathcal{P}_2$ with

$$(r_1, \mathcal{P}r_2, \mathcal{P}_2r_2, \cdots, \mathcal{P}_2^{\rho(v_2)-1}r_2, \mathcal{P}_1r_1, \cdots, \mathcal{P}_1^{\rho(v_{r_1})-1}r_1)$$

and

$$(\beta r_1, \mathcal{P}_1\beta r_1, \cdots, \mathcal{P}_1^{\rho(v_{\beta r_1})-1}\beta r_1, \mathcal{P}_2\beta r_2, \cdots, \mathcal{P}_2^{\rho(v_{\beta r_2})-1}\beta r_2)$$

as the root-vertex and the nonrooted end of the root-edge of M while $r = r_1$ is the root of M. It is easily checked that the operation of root addition does not satisfy the commutative law. Moreover, it can be shown that the operation satisfies the associative law. This allows us to write

$$\sum_{i=1}^{k}\!|_r M_i = M_1 +\!|_r \sum_{i=2}^{k}\!|_r M_i$$

$$= \sum_{i=1}^{k-1}\!|_r M_i +\!|_r M_k. \tag{6.2.3}$$

For $M \in \mathcal{M}_{\mathrm{Ens}_1}$, let $r = r(M)$ be the root with the two ends of the root-edge being

$$v_r = (r, T), \tag{6.2.4}$$

where

$$T = \langle \mathcal{P}r, \mathcal{P}^2 r, \cdots, \mathcal{P}^{\rho(v_r)-1} r\rangle$$

and

$$v_{\beta r} = (\alpha\beta r, S), \tag{6.2.5}$$

where

$$S = \langle \mathcal{P}\alpha\beta r, \mathcal{P}^2\alpha\beta r, \cdots, \mathcal{P}^{\rho(v_{\beta r})-1}\alpha\beta r\rangle.$$

If for an integer $j \geq 1$, there exist two integers $m \geq 1$ and $n \geq 1$ such that

$$(\mathcal{P}\alpha\beta)^m \mathcal{P}^j r = \mathcal{P}^n \alpha\beta r, \tag{6.2.6}$$

then T and S are said to be *f-splitable* because

$$T = \langle T_1, T_2\rangle,\ S = \langle S_2, S_1\rangle$$

where

$$T_1 = \langle \mathcal{P}r, \cdots, \mathcal{P}^{j-1} r\rangle,\ T_2 = \langle \mathcal{P}^j r, \cdots, \mathcal{P}^{\rho(v_r)-1} r\rangle,$$

$$S_2 = \langle \mathcal{P}\alpha\beta r, \cdots, \mathcal{P}^{n-1}\alpha\beta r\rangle,\ S_1 = \langle \mathcal{P}^n\alpha\beta r, \cdots, \mathcal{P}^{\rho(v_\beta)-1}\alpha\beta r\rangle.$$

Here, $\{T_1, S_1\}$ and $\{T_2, S_2\}$ are called *f-split pairs*. Of course, $j \neq 1$ because there always exists an integer l such that $(\mathcal{P}\alpha\beta)^l \mathcal{P}r = \alpha\beta r$ and $(\mathcal{P}\alpha\beta)^{l+1}\mathcal{P}r = \mathcal{P}r$. However, $j = \rho(v_1) - 1$ is allowed possibly to occur.

If for every $j > 1$, there do not exist $m \geq 1$ and $n \geq 1$ such that (6.2.6) holds, then T and S are called *f-unsplitable*.

If T and S are f-splitted into

$$T = \langle T_1, T_2, \cdots, T_k \rangle, \ S = \langle S_k, \cdots, S_2, S_1 \rangle$$

such that every f-split pair $\{T_i, S_i\}$, $i = 1, \cdots, k$, is f-unsplitable, then the number k is called the *root-index* of M. Let M_i be the maximal submap of

$$M - \bigcup_{j=1, j\neq i}^{k} \{T_j, S_j\}$$

involved with $\{T_i, S_i\}$ for $i = 1, 2, \cdots, k$. Since $M \in \mathcal{M}_{\text{Ens}_1}$, from the Eulerianity for any i the lengths of T_i and S_i have the same parity and the number of T_i with odd length is odd among all $T_i, i = 1, 2, \cdots, k$.

Lemma 6.2.1 *For $M \in \mathcal{M}_{\text{Ens}_1}$, we have*

$$M = \sum_{i=1}^{k}{}_r M_i \tag{6.2.7}$$

such that k is the root-index of M, there is an odd number of $M_i \in \mathcal{M}_{\text{Ens}_1}$ and all others, not Eulerian, among all $M_i, 1 \leq i \leq k$.

Proof From what are just discussed, we see that M has the form (6.2.7). By considering the Eulerianity of M, S_i and T_i of M_i have the same parity and the number of T_i with odd lengths is odd. Because the root-edge of M is not a loop, so is that of M_i and hence those with odd length of T_i is a member of $\mathcal{M}_{\text{Ens}_1}$. The lemma follows. □

According to (6.2.7), we see that

$$M \bullet a = \sum_{i=1}^{k}{}^{\cdot} M_i \bullet a_i \tag{6.2.8}$$

where a, a_i are the root-edges of $M, M_i, i = 1, 2, \cdots, k$, respectively, and the summation with a dot is the inner $1v$-addition defined as in §6.1. Meanwhile, we have to pay attention to that the root-vertices of $M \bullet a$, $M_i \bullet a_i$ are with the forms

$$(S, T), \ (S_i, T_i) \tag{6.2.9}$$

for $i = 1, 2, \cdots, k$, respectively and that the following relations

$$s = \sum_{i=1}^{k} s_i = 1 \,(\mathrm{mod}\ 2);$$

$$t = \sum_{i=1}^{k} t_i = 1 \,(\mathrm{mod}\ 2);$$

$$s_i = t_i \,(\mathrm{mod}\ 2) \tag{6.2.10}$$

are satisfied where s, t, s_i, t_i are the lengths of S, T, S_i, T_i, $i = 1, 2, \cdots, k$, respectively.

Lemma 6.2.2 *Let* $\mathcal{M}_{(\mathrm{Ens})_1}(s,t) = \{M \bullet a | \forall M \in \mathcal{M}_{\mathrm{Ens}_1}$ with the root-vertex $(S,T)\}$. *Then we have*

$$\mathcal{M}_{(\mathrm{Ens})_1}(s,t) = \sum_{k\geq 1} \sum_{(\underline{s},\underline{t})\in\Omega} \prod_{i=1}^{k}\cdot \mathcal{M}_{\mathrm{Ens}}(s_i, t_i) \tag{6.2.11}$$

where $s, t, (\underline{s}, \underline{t}) = (s_1, \cdots, s_k, t_1, \cdots, t_k)$ *are as described in* (6.2.10),

$$\Omega = \{(\underline{s},\underline{t}) | \forall (\underline{s},\underline{t}) \text{ satisfied by } (6.2.10)\};$$

$$\mathcal{M}_{\mathrm{Ens}}(s_i, t_i) = \{M | \forall M \in \mathcal{M}_{\mathrm{Ens}} \text{ with the root-vertex } (S_i, T_i)\}$$

and the operator Π *with a dot represents the inner* 1v*-product as defined in* §6.1.

Proof Let the set on the right hand side of (6.2.11) be denoted by $\tilde{\mathcal{M}}$ for convenience.

For any $M \in \mathcal{M}_{(\mathrm{Ens})_1}(s,t)$, we can construct a map M' obtained by splitting the root vertex into o_1 and o_2 with an edge Kr' where

$$o_1 = (r', T) \text{ and } o_2 = (\alpha\beta r', S)$$

from M such that $M = M' \bullet a', M' \in \mathcal{M}_{\mathrm{Ens}_1}$. From Lemma 6.2.1, M' has the form shown in (6.2.7) and hence M is with the form (6.2.8) satisfying (6.2.9) and (6.2.10). By the Eulerianity and the nonseparability of M', any map in the terms on the right hand side of (6.2.8) is allowed to be in $\mathcal{M}_{\text{Ens}}$ which includes the loop map L_1. From the relations (6.2.9–10), $M \in \tilde{\mathcal{M}}$. Thus, $\mathcal{M}_{(\text{Ens})_1}(s,t) \subseteq \tilde{\mathcal{M}}$.

Conversely for any $M \in \tilde{\mathcal{M}}$, since M has the form as shown in (6.2.8) in the replacement of $M \bullet a, M_i \bullet a_i$ by M, M_i respectively with (6.2.9–10), we may construct maps M', M_i' from M, M_i by splitting the root-vertex $o' = (S,T)$ into $o_1' = (r_1', T)$ and $o_2' = (\alpha\beta r', S)$ with an edge Kr', and the root-vertex $o_i = (S_i, T_i)$ into $o_i' = (r_i', T_i)$ and $o_{i2}' = (\alpha\beta r_i', S_i)$ with an edge Kr_i' for $i = 1, 2, \cdots, k$, respectively. From the Eulerianity and nonseparability of M_i with the relations (6.2.9–10), $M \in \mathcal{M}_{(\text{Ens})_1}$. Because $M = M' \bullet a'$, we have $M \in \mathcal{M}_{(\text{Ens})_1}(s,t)$. Thus, $\tilde{\mathcal{M}} \subseteq \mathcal{M}_{(\text{Ens})_1}(s,t)$. □

In what follows we evaluate the contributions of $\mathcal{M}_{\text{Ens}_0}$ and $\mathcal{M}_{\text{Ens}_1}$ to the enufunction

$$f = f_{\mathcal{M}_{\text{Ens}}}(x, \underline{y}) = \sum_{m \in \mathcal{M}_{\text{Ens}}} x^{2m(M)} \underline{y}^{\underline{n}(M)} \tag{6.2.12}$$

where $2m(M)$ is the valency of the root-vertex and $\underline{n}(M) = (n_2(M), n_4(M), \cdots), n_{2i}(M)$ is the number of nonrooted vertices of valency $2i, i \geq 1$.

Because $\mathcal{M}_{\text{Ens}_0}$ consists of only one map, the loop map $L_1 = (Kr, (r, \alpha\beta r))$, the contribution of $\mathcal{M}_{\text{Ens}_0}$ to f is

$$f_{\mathcal{M}_{\text{Ens}_0}} = x^2 \tag{6.2.13}$$

by considering that L_1 has the valency of the root-vertex being 2 without nonrooted vertex.

For $\mathcal{M}_{\text{Ens}_1}$, we should evaluate the function

$$\tilde{f}(x,z) = \sum_{M \in \mathcal{M}_{\text{Ens}_1}} x^{2m(M)} z^{2l(M)} \underline{y}^{\underline{n}(M)}. \tag{6.2.14}$$

From Lemma 6.2.2, we have

$$\begin{aligned}\tilde{f}(x,z) &= xz \sum_{M \in \mathcal{M}_{(\text{Ens})_1}} x^{2m(M)-s(M)} z^{s(M)} \underline{y}^{\underline{n}(M)} \\ &= xz \sum_{k \geq 1} \sum_{0 \leq l \leq \lfloor k/2 \rfloor} \binom{k}{2l+1}\end{aligned}$$

$$\times\left(\sum_{M\in\mathcal{M}_{\mathrm{Ens}}} x^{2m(M)}\sum_{i=1}^{m(M)}\left(\frac{z}{x}\right)^{2i-1}\underline{y}^{\underline{n}(M)}\right)^{2l+1}$$

$$\times\left(\sum_{M\in\mathcal{M}_{\mathrm{Ens}}} x^{2m(M)}\sum_{i=1}^{m(M)-1}\left(\frac{z}{x}\right)^{2i}\underline{y}^{\underline{n}(M)}\right)^{k-2l-1}$$

$$= xz\sum_{k\geq 1}\sum_{0\leq l\leq \lfloor k/2\rfloor}\binom{k}{2l+1}$$

$$\times\left(\sum_{M\in\mathcal{M}_{\mathrm{Ens}}}\frac{xz(x^{2m(M)}-z^{2m(M)})}{x^2-z^2}\underline{y}^{\underline{n}(M)}\right)^{2l+1}$$

$$\times\left(\sum_{M\in\mathcal{M}_{\mathrm{Ens}}}\frac{z^2x^{2m(M)}-x^2z^{2m(M)}}{x^2-z^2}\underline{y}^{\underline{n}(M)}\right)^{k-2l-1}$$

$$= xz\sum_{k\geq 1}\sum_{0\leq l\leq \lfloor k/2\rfloor}\binom{k}{2l+1}$$

$$\times(xz\delta_{x^2,z^2}f)^{2l+1}(\partial_{x^2,z^2}f)^{k-2l-1}$$

$$= \frac{xy}{2}\sum_{k\geq 1}\Big((\partial_{x^2,y^2}f + xz\delta_{x^2,z^2}f)^k$$

$$-(\partial_{x^2,z^2}f - xz\delta_{x^2,z^2}f)^k\Big)$$

$$= \frac{x^2z^2\delta_{x^2,z^2}f}{(1-\partial_{x^2,z^2}f)^2-(xz\delta_{x^2,z^2}f)^2} \tag{6.2.15}$$

where $f = f(\sqrt{u}) = f_{\mathcal{M}_{\mathrm{Ens}}}(\sqrt{u},\underline{y})$.

Theorem 6.2.1 *The enufunction f defined by* (6.2.12) *satisfies the following functional equation:*

$$f = x^2 + x^2\int_y \frac{y^2\delta_{x^2,y^2}f}{y(1-\partial_{x^2,y^2}f)^2-(xy\delta_{x^2,y^2}f)^2}. \tag{6.2.16}$$

Proof From (6.2.1) we see that

$$f = f_{\mathcal{M}_{\mathrm{Ens}_0}} + f_{\mathcal{M}_{\mathrm{Ens}_1}}.$$

By (6.2.14–15), noticing that

$$f_{\mathcal{M}_{\mathrm{Ens}_1}} = \int_y \tilde{f}(x, y),$$

we soon find that

$$f = x^2 + \int_z \frac{x^2 z^2 \delta_{x^2,z^2} f}{(1 - \partial_{x^2,z^2} f)^2 - (xz\delta_{x^2,z^2} f)^2}$$

which is equivalent to (6.2.16). The theorem is obtained. □

Because it looks much difficult for solving equation (6.2.16) directly to determine the enufunction f with vertex partition, we first investigate the enufunction with few parameters for rooted Eulerian nonseparable maps as

$$f = f_{\mathcal{M}_{\mathrm{Ens}}} = \sum_{M \in \mathcal{M}_{\mathrm{Ens}}} x^{2m(M)} y^{n(M)} z^{q(M)} \tag{6.2.17}$$

where $2m(M), n(M)$ and $q(M)$ are the valency of the root-vertex, the number of nonrooted vertices and the number of nonrooted faces of M.

Theorem 6.2.2 *The enufunction f defined by* (6.2.17) *satisfies the following equation:*

$$f = x^2 z + x^2 y \frac{f - f^*}{x^2(1 + f^*)^2 - (f + 1)^2} \tag{6.2.18}$$

where $f^* = f(1, y, z)$.

Proof First, the contribution $f_{\mathcal{M}_{\mathrm{Ens}_0}}$ of $\mathcal{M}_{\mathrm{Ens}_0}$ to f is $x^2 z$ by considering that the loop map has only one nonrooted face.

Then, the contribution of $\mathcal{M}_{\mathrm{Ens}_1}$ is

$$\begin{aligned} f_{\mathcal{M}_{\mathrm{Ens}_1}} &= \sum_{M \in \mathcal{M}_{\mathrm{Ens}_1}} x^{2m(M)} y^{n(M)} z^{q(M)} \\ &= xy \sum_{M \in \mathcal{M}_{(\mathrm{Ens})_1}} x^{2m(M) - s(M)} y^{n(M)} z^{q(M)} \end{aligned}$$

where $s(M)$ is one less than the valency of the nonrooted end of the root-edge. By Lemma 6.2.2 we have

$$\begin{aligned}
f_{\mathcal{M}_{(\mathrm{Ens})_1}} &= xy\sum_{k\geq 1}\sum_{0\leq l\leq \lfloor k/2\rfloor}\binom{k}{2l+1} \\
&\quad\times\left(\sum_{M\in\mathcal{M}_{\mathrm{Ens}}} x^{2m(M)}\sum_{i=1}^{m(M)}\left(\frac{1}{x}\right)^{2i-1} y^{n(M)}z^{q(M)}\right)^{2l+1} \\
&\quad\times\left(\sum_{M\in\mathcal{M}_{\mathrm{Ens}}} x^{2m(M)}\sum_{i=1}^{m(M)-1} x^{-2i}y^{n(M)}z^{q(M)}\right)^{k-2l-1} \\
&= xy\sum_{k\geq 1}\sum_{0\leq l\leq \lfloor k/2\rfloor}\binom{k}{2l+1} \\
&\quad\times\left(\sum_{M\in\mathcal{M}_{\mathrm{Ens}}} \frac{x(1-x^{2m(M)})}{1-x^2} y^{n(M)}z^{q(M)}\right)^{2l+1} \\
&\quad\times\left(\sum_{M\in\mathcal{M}_{\mathrm{Ens}}} \frac{x^2-x^{2m(M)}}{1-x^2} y^{n(M)}z^{q(M)}\right)^{k-2l-1} \\
&= xy\sum_{k\geq 1}\frac{1}{2}\left((x\Delta+\Delta_1)^k-(x\Delta-\Delta_1)^k\right) \\
&= \frac{xy}{2}\left(\frac{\Delta_1+\Delta}{1-\Delta_1-x\Delta}-\frac{\Delta_1-x\Delta}{1-\Delta_1+x\Delta}\right) \\
&= \frac{x^2y\Delta}{(1-\Delta_1)^2-x^2\Delta^2}
\end{aligned}$$

where

$$\Delta = \frac{f^*-f}{1-x^2};\ \Delta_1 = \frac{x^2f^*-f}{1-x^2}.$$

Since

$$(1-\Delta_1)^2-x^2\Delta^2 = \frac{x^2(1+f^*)-(1+f)^2}{x^2-1},$$

after simplification we find

$$f_{\mathcal{M}_{\mathrm{Ens}_1}} = x^2y\frac{f-f^*}{x^2(1+f^*)-(1+f)^2}.$$

In consequence $f = f_{\mathcal{M}_{\mathrm{Ens}_0}} + f_{\mathcal{M}_{\mathrm{Ens}_1}}$ has the form (6.2.18). The theorem is obtained. □

If (6.2.18) has $y = z$, then the following cubic equation is found:

$$f = x^2 y \left(1 + \frac{f - f^*}{x^2(1+f^*)^2 - (1+f)^2}\right). \tag{6.2.19}$$

However, no simpler formula of f^* has been derived from (6.2.19) yet up to now.

§6.3 Planar nonseparable maps

Let $\mathcal{M}_{\mathrm{ns}}$ be the set of all rooted nonseparable planar maps with the convention that the loop map $L_1 = (Kr, (r, \alpha\beta r))$ is included but the link map $L = (Kr, (r)(\alpha\beta r))$ is not for convenience.

Then, $\mathcal{M}_{\mathrm{ns}}$ is divided into two parts $\mathcal{M}_{\mathrm{ns}_0}$ and $\mathcal{M}_{\mathrm{ns}_1}$, *i.e.*,

$$\mathcal{M}_{\mathrm{ns}} = \mathcal{M}_{\mathrm{ns}_0} + \mathcal{M}_{\mathrm{ns}_1} \tag{6.3.1}$$

such that $\mathcal{M}_{\mathrm{ns}_0}$ consists of only the loop map L_1.

Lemma 6.3.0 *A map $M \in \mathcal{M}_{\mathrm{ns}}$, $M \neq L_1$ iff its dual $M^* \in \mathcal{M}_{\mathrm{ns}}$.*

Proof By contradiction. Assume $M = (\mathcal{X}_{\alpha,\beta}, \mathcal{P}) \in \mathcal{M}_{\mathrm{ns}}$, $M \neq L_1$ and its dual $M^* = (\mathcal{X}_{\beta,\alpha}, \mathcal{P}\alpha\beta) \notin \mathcal{M}_{\mathrm{ns}}$. Let

$$v^* = (x, \mathcal{P}\alpha\beta x, \cdots, (\mathcal{P}\alpha\beta)^m x)$$

be a cut-vertex of M^*. Then we have a face $f^* = (x, \mathcal{P}x, \cdots, \mathcal{P}^n x)$ on M^* such that there exists an integer $j, 1 \leq j \leq n$, on f^* satisfying $\mathcal{P}^j x = (\mathcal{P}\alpha\beta)^i x$ for some $i, 1 \leq i \leq m$, *i.e.*, $v_x^* = v_{\mathcal{P}^j x}^* = v^*$. However, f^* is a vertex of M which has the face v^* having the symmetry and hence f^* is a cut-vertex of M. A contradiction to the assumption appears. The necessity is true.

Conversely, from the duality the sufficiency is true as well. □

For any $M \in \mathcal{M}_{\mathrm{ns}}$, let $m(M)$ be the valency of the root-vertex and $\underline{n}(M) = (n_1(M), n_2(M), \cdots), n_i(M)$ be the number of nonrooted vertices of valency $i, i \geq 1$.

From the nonseparability, the root-edge $a = (v_1, v_{\beta r})$ of any map M in $\mathcal{M}_{\mathrm{ns}_1}$ is always a link. The map $M \bullet a$ obtained by contracting the root-edge a in M has the same number of faces as M does.

Lemma 6.3.1 *For any $M \in \mathcal{M}_{\mathrm{ns}_1}$ there is an integer $k \geq 1$ with*

$$M \bullet a = \sum_{i=1}^{k} \cdot M_i \tag{6.3.2}$$

such that all M_i are allowed to be any map in $\mathcal{M}_{\mathrm{ns}}$ and that $M_i, i = 1, 2, \cdots, k$, does not have the form (6.3.2) *for $k > 1$.*

Proof In fact, from what were mentioned in §6.2, we see that k is the root-index of M and that all $M_i, 1 \leq i \leq k$, do not have the form (6.3.2) for $k > 1$. From the nonseparability of M, by considering that all vertices of M_i except for the root-vertex are the same as those of M for $i = 1, 2, \cdots, k$, since M_i does not have the form (6.3.2) for $k > 1$, the root-vertex is not a cut-vertex for $i = 1, 2, \cdots, k$. That implies all $M_i, 1 \leq i \leq k$, are allowed to be any map in $\mathcal{M}_{\mathrm{ns}}$ including the loop map. The lemma follows. □

Now let us write

$$\mathcal{M}_k = \Big\{ \sum_{i=1}^{k} \cdot M_i \Big| \ \forall M_i \in \mathcal{M}_{\mathrm{ns}}, 1 \leq i \leq k \Big\}, \tag{6.3.3}$$

and

$$\mathcal{M}_{(\mathrm{ns})_1} = \{ M \bullet a | \forall M \in \mathcal{M}_{\mathrm{ns}_1} \}, \tag{6.3.4}$$

where $a = e_r$, the root-edge of M.

Lemma 6.3.2 *For $\mathcal{M}_{\mathrm{ns}_1}$, we have*

$$\mathcal{M}_{(\mathrm{ns})_1} = \sum_{k \geq 1} \mathcal{M}_k;$$

$$\mathcal{M}_k = \mathcal{M}_{\mathrm{ns}}^{\times k} \tag{6.3.5}$$

where $\times$ is the inner $1v$-production.

Proof By the definition of inner $1v$-product in §6.2, the last form of (6.3.5) is easily seen.

From Lemma 6.3.1 we can find that

$$\mathcal{M}_{(\mathrm{ns})_1} = \bigcup_{k \geq 1} \mathcal{M}_k.$$

Moreover, for any $i, j, i \neq j$, we always have

$$\mathcal{M}_i \bigcap \mathcal{M}_j = \emptyset.$$

Therefore, the first form of (6.3.5) is true. □

Based on the two lemmas above, we are allowed to evaluate the contributions of $\mathcal{M}_{\mathrm{ns}_0}$ and $\mathcal{M}_{\mathrm{ns}_1}$ to the enufunction $f_{\mathcal{M}_{\mathrm{ns}}}$ of $\mathcal{M}_{\mathrm{ns}}$ with vertex partition, *i.e.*,

$$f = f_{\mathcal{M}_{\mathrm{ns}}} = \sum_{M \in \mathcal{M}_{\mathrm{ns}}} x^{m(M)} \underline{y}^{\underline{n}(M)}, \tag{6.3.6}$$

where $m(M)$ is the valency of root-vertex and

$$\underline{n}(M) = (n_1(M), n_2(M), \cdots)$$

with $n_i(M)$ being the number of nonroot-vertices of valency $i, i \geq 1$.

Since $\mathcal{M}_{\mathrm{ns}_0}$ consists of only the loop map, which has the root-vertex of valency 2 without nonrooted vertex, we have

$$f_{\mathcal{M}_{\mathrm{ns}_0}} = x^2. \tag{6.3.7}$$

For $\mathcal{M}_{\mathrm{ns}_1}$, we have to evaluate the function

$$\tilde{f}(x, z) = \sum_{M \in \mathcal{M}_{\mathrm{ns}_1}} x^{m(M)} z^{s(M)} \underline{y}^{\underline{n}(M)} \tag{6.3.8}$$

where $s(M)$ is the valency of the nonrooted vertex $v_{\beta r}$ incident with the root-edge e_r of M.

By considering that for $M \in \mathcal{M}_{\mathrm{ns}_1}$, $m(M \bullet a) = (m(M) - 1) + (s(M) - 1)$, we soon find

$$\tilde{f}(x, z) = xz \sum_{\tilde{M} \in \mathcal{M}_{\mathrm{ns}_1}} x^{m(\tilde{M}) - s(\tilde{M})} z^{s(\tilde{M})} \underline{y}^{\underline{n}(\tilde{M})}$$

where $s(\tilde{M})$ is the contribution of the valency of the nonrooted end of the root-edge of M to the valency of the root-vertex of $\tilde{M} = M \bullet a$, $M \in$

$\mathcal{M}_{\mathrm{ns}_1}$. Because $s(\tilde{M})$ is allowed to be any number between 1 and $m(\tilde{M})-1$, from Lemma 6.3.2 we have

$$\begin{aligned}\tilde{f}(x,z) &= xz\sum_{k\geq 1}\left(\sum_{M\in\mathcal{M}_{\mathrm{ns}}} x^{m(M)}\sum_{i=1}^{m(M)-1}\left(\frac{z}{x}\right)^i \underline{y}^{n(M)}\right)^k \\ &= xz\sum_{k\geq 1}\left(\sum_{M\in\mathcal{M}_{\mathrm{ns}}} \frac{zx^{m(M)}-xz^{m(M)}}{x-z}\underline{y}^{n(M)}\right)^k \\ &= xz\sum_{k\geq 1}(\partial_{x,y}f)^k \\ &= \frac{xz\partial_{x,z}f}{1-\partial_{x,z}f}\end{aligned}$$

where $f=f(u)=f_{\mathcal{M}_{\mathrm{ns}}}(u,\underline{y})$, and hence

$$\begin{aligned}f_{\mathcal{M}_{\mathrm{ns}_1}} &= \int_y \tilde{f}(x,y) \\ &= x\int_y \frac{y\partial_{x,y}f}{1-\partial_{x,y}f}.\end{aligned} \tag{6.3.9}$$

Theorem 6.3.1 *The enufunction of $\mathcal{M}_{\mathrm{ns}}$ defined by (6.3.6) satisfies the following functional equation:*

$$f = x^2 + x\int_y \frac{y\partial_{x,y}f}{1-\partial_{x,y}f}. \tag{6.3.10}$$

Proof Since $f_{\mathcal{M}_{\mathrm{ns}}} = f_{\mathcal{M}_{\mathrm{ns}_0}} + f_{\mathcal{M}_{\mathrm{ns}_1}}$, from (6.3.7) and (6.3.9) the theorem is obtained. □

Although the equation (6.3.10) looks simpler, no solution has been found yet. However, a much simpler formula has been extracted for the enufunction of $\mathcal{M}_{\mathrm{ns}}$ with two parameters: the order and the *dual order* (*i.e.*, the number of faces).

In order to do that we have to investigate the enufunction of $\mathcal{M}_{\mathrm{ns}}$ with three parameters

$$h = h_{\mathcal{M}_{\mathrm{ns}}} = \sum_{M\in\mathcal{M}_{\mathrm{ns}}} x^{m(M)}y^{n(M)}z^{q(M)} \tag{6.3.11}$$

where $m(M), n(M)$ and $q(M)$ are the valency of root-vertex, the number of nonrooted vertices and the number of nonrooted faces respectively.

Following the same decomposition of $\mathcal{M}_{\rm ns}$ as mentioned above, we may evaluate the corresponding contributions of $\mathcal{M}_{\rm ns_0}$ and $\mathcal{M}_{\rm ns_1}$ to h which is defined in (6.3.11).

For $\mathcal{M}_{\rm ns_0}$ it is trivial to have

$$h_{\mathcal{M}_{\rm ns_0}} = x^2 z \tag{6.3.12}$$

since the loop map has the root-vertex of valency 2 with only one nonrooted face and without nonrooted vertex.

For $\mathcal{M}_{\rm ns_1}$ we have

$$\begin{aligned} h_{\mathcal{M}_{\rm ns_1}} &= xy \sum_{k\geq 1} \left(\sum_{M\in\mathcal{M}_{\rm ns}} \sum_{i=1}^{m(M)-1} x^i y^{n(M)} z^{q(M)} \right)^k \\ &= xy \sum_{k\geq 1} \left(\sum_{M\in\mathcal{M}_{\rm ns}} \frac{x - x^{m(M)}}{1-x} y^{n(M)} z^{q(M)} \right)^k \\ &= xy \sum_{k\geq 1} \left(\frac{xh^* - h}{1-x} \right)^k \\ &= \frac{xy(xh^* - h)}{1 - x - xh^* + h} \end{aligned} \tag{6.3.13}$$

where $h^* = h_{\mathcal{M}_{\rm ns}}(1, y, z)$, the enufunction of $\mathcal{M}_{\rm ns}$ with the number of nonrooted vertices and the number of nonrooted faces as the two parameters.

Theorem 6.3.2 *The enufunction h defined by (6.3.11) satisfies the following quadratic equation:*

$$\begin{aligned} h^2 + (1 - x + xy - x^2 z - xh^*)h& \\ -x^2((2-x)^2 z + (y - zx)h^*) &= 0 \end{aligned} \tag{6.3.14}$$

where h^* is the same as defined in (6.3.13).

Proof According to (6.3.12–13), we find

$$h = x^2 z + \frac{xy(xh^* - h}{1 - x - xh^* + h}.$$

By the rearrangement of coefficients in powers of h, Eq. (6.3.14) is soon obtained. □

Because the discriminant of the equation (6.3.14) is

$$D(x) = (1 - x + xy - x^2 z - xh^*)^2 + 4x^2((1-x)z + (y - zx)h^*),$$

we may construct the following two equations with $x = \xi(y, z)$ as a parameter:

$$D(\xi) = (1 - \xi + \xi y - \xi^2 z - \xi h^*)^2 + 4\xi^2\Big((1-\xi)z + (y - z\xi)h^*\Big) = 0;$$

$$\frac{\mathrm{d}D}{\mathrm{d}\xi} = 2(1 - \xi + \xi y + \xi^2 z - \xi h^*) \times (y - 1 + 2\xi z - h^*) + 4\xi\Big(\xi(2 - 3\xi)z + (2y - 3z\xi)h^*\Big) = 0. \tag{6.3.15}$$

Since both of the equations in (6.3.15) are quadratic in h^*, by one of them we may find h^* as a function of y and z with the parameter ξ. Then, from the other in (6.3.15), y can be expressed as a function of z with ξ. By introducing a new parameter η to express y or z, through a suitable substitution of considering the duality between y and z, we may have

$$y = \xi(1-\eta)^2;$$
$$z = \eta(1-\xi)^2;$$
$$h^* - z = \xi\eta(1 - \xi - \eta). \tag{6.3.16}$$

By considering that $(x_1, x_2) = (\xi, \eta)$, $(t_1, t_2) = (y, z)$, $\phi_1 = \frac{1}{(1-\xi)^2}$, $\phi_2 = \frac{1}{(1-\eta)^2}$ in (1.5.20–21), we have

$$\Delta(\xi, \eta) = \det \begin{pmatrix} 1 - \dfrac{\xi}{\phi_1}\dfrac{\partial \phi_1}{\partial \xi} & -\dfrac{\eta}{\phi_1}\dfrac{\partial \phi_1}{\partial \eta} \\ -\dfrac{\xi}{\phi_2}\dfrac{\partial \phi_2}{\partial \xi} & 1 - \dfrac{\eta}{\phi_2}\dfrac{\partial \phi_2}{\partial \eta} \end{pmatrix}$$

$$= \det \begin{pmatrix} \dfrac{1-3\xi}{1-\xi} & 0 \\ 0 & \dfrac{1-3\eta}{1-\eta} \end{pmatrix}$$

$$= \frac{(1-3\xi)(1-3\eta)}{(1-\eta)(1-\xi)}.$$

By employing Theorem 1.5.2, we obtain

$$\partial_{(y,z)}^{(n,q)} h^* = \partial_{(\xi,\eta)}^{(n,q)} \left(\frac{\xi\eta(1-\xi-\eta)(1-3\xi)(1-3\eta)}{(1-\xi)^{n+1}(1-\eta)^{q+1}} \right)$$

$$= \left(\frac{\mathrm{d}^n}{\mathrm{d}\xi^n} \frac{\xi(1-3\xi)}{(1-\xi)^{n+1}} \bigg|_{\xi=0} \right) \left(\frac{\mathrm{d}^q}{\mathrm{d}\eta^q} \frac{\eta(1-3\eta)}{(1-\eta)^{q+1}} \bigg|_{\eta=0} \right)$$

$$- \left(\frac{\mathrm{d}^n}{\mathrm{d}\xi^n} \frac{\xi^2(1-3\xi)}{(1-\xi)^{n+1}} \bigg|_{\xi=0} \right) \left(\frac{\mathrm{d}^q}{\mathrm{d}\eta^q} \frac{\eta(1-3\eta)}{(1-\eta)^{q+1}} \bigg|_{\eta=0} \right)$$

$$- \left(\frac{\mathrm{d}^n}{\mathrm{d}\xi^n} \frac{\xi(1-3\xi)}{(1-\xi)^{n+1}} \bigg|_{\xi=0} \right) \left(\frac{\mathrm{d}^q}{\mathrm{d}\eta^q} \frac{\eta^2(1-3\eta)}{(1-\eta)^{q+1}} \bigg|_{\eta=0} \right).$$

Moreover, since

$$\frac{\mathrm{d}^l}{\mathrm{d}\theta^l} \frac{\theta(1-3\theta)}{(1-\theta)^{l+1}} \bigg|_{\theta=0} = -\frac{(l-2)(2l-2)!}{(l-1)!};$$

$$\frac{\mathrm{d}^l}{\mathrm{d}\theta^l} \frac{\theta^2(1-3\theta)}{(1-\theta)^{l+1}} \bigg|_{\theta=0} = -\frac{(l-4)(2l-3)!}{(l-2)!},$$

we soon find

$$\partial_{(y,z)}^{(n,q)} h^* = \frac{(2n+q-2)!(2q+n-2)!}{n!(2n-1)!(2q-1)!} \tag{6.3.17}$$

for $n \geq 0, q \geq 1$ except for $n = 0, q = 1$.

Theorem 6.3.3 *The number of rooted distinct nonseparable planar maps with order m and dual order p (i.e., the number of faces) is*

$$N_{\mathcal{M}_{\mathrm{ns}}}(m,p) = \frac{(2m+p-5)!(2p+m-5)!}{(m-1)!(p-1)!(2m-3)!(2p-3)!} \tag{6.3.18}$$

for $m \geq 1, p \geq 2$.

Proof Because the coefficient of $y^n z^q$ in h^* is the number of rooted nonseparable planar maps with $n \geq 0$ nonrooted vertices and $q \geq 1$ nonrooted faces except for $n = 0$ and $q = 1$, by considering that $m = n + 1$ and $p = q + 1$, from (6.3.17) the theorem is obtained for $m \geq 1$ and $p \geq 2$ but $m = 1$ and $p = 2$.

However, it is checked that for $m = 1$ and $p = 2$, the value of (6.3.18) is 1, which is just corresponding to the only one map, the loop map in $\mathcal{M}_{\text{ns}}$, the theorem is then proved. □

§6.4 Nonseparable maps on surfaces

A nonseparable map which has all its vertices, on the root-face boundary, of valency 3 is said to be the *cubic boundary*.

For $M = (\mathcal{X}, \mathcal{P}) \in \mathcal{M}_{\text{ns}}$, the set of all rooted nonseparable planar maps, let $m(M)$ be the valency of root-vertex of M. The root-vertex is denoted by

$$v_r = (r, \mathcal{P}r, \cdots, \mathcal{P}^{m(M)-1}r). \tag{6.4.1}$$

We may extend M into a cubic boundary map X by splitting the vertex v_r into $m(M)$ ends:

$$v_i = v_{\mathcal{P}^{i-1}r},\ i = 1, 2, \cdots, m(M) \tag{6.4.2}$$

and adding $m(M)$ edges:

$$e_i = Kx_{i-1},\ i = 1, 2, \cdots, m(M) \tag{6.4.3}$$

such that

$$v_i = (x_i, \mathcal{P}^i r, \alpha\beta x_{i+1}) \tag{6.4.4}$$

for $i = 0, 1, \cdots, m(M) - 1$, $x_{m(M)} = x_0 = x$ being the root of X. It can be seen that the root-face of X is

$$f_0 = (x, x_{m(M)-1}, x_{m(M)-2}, \cdots, x_1) \tag{6.4.5}$$

because

$$\mathcal{P}\alpha\beta x = x_{m(M)-1}, (\mathcal{P}\alpha\beta)^2 x = x_{m(M)-2}, \cdots,$$
$$(\mathcal{P}\alpha\beta)^{m(M)-1} x = x_1$$

from (6.4.4). That implies all the $m(M)$ vertices defined by (6.4.4) form the root-face boundary of X. Of course, they are all of valency 3. Thus X is a cubic boundary map and is called a *cubic boundary extension* of M.

Lemma 6.4.1 *Let $\mathcal{M}_{\mathrm{ns}}(m,n)$ and $\mathcal{B}(m,n)$ be the sets of all the rooted nonseparable planar maps of size n with the root-vertex of valency m and all the cubic boundary maps of size n with the root-face valency m for $m,n \geq 1$ respectively. Then we have*

$$\left|\mathcal{M}_{\mathrm{ns}}(m,n)\right| = \left|\mathcal{B}(m,m+n)\right| \tag{6.4.6}$$

in any way.

Proof For any $M \in \mathcal{M}_{\mathrm{ns}}(m,n)$, by the procedure described above there is a unique cubic boundary map which is the cubic boundary extension X of M. It is easy to check that $X \in \mathcal{B}(m,m+n)$.

Conversely, for any $X \in \mathcal{B}(m,m+n)$ we may assume that the vertices on the boundary of the root-face are given by (6.4.2) and (6.4.4). Then, by deleting the edges as described in (6.4.3) and identifying all the ends of edges as shown in (6.4.3) into a vertex denoted by v_r as in (6.4.1) as well, we can uniquely obtain a map M which is in $\mathcal{M}_{\mathrm{ns}}(m,n)$ from the nonseparability of X.

In consequence, there is a 1-to-1 correspondence between $\mathcal{M}_{\mathrm{ns}}(m, n)$ and $\mathcal{B}(m,m+n)$ and hence the lemma is true. □

According to Lemma 6.4.1, in order to determine the cardinality of $\mathcal{B}(m,m+n)$ it is only necessary to determine that of $\mathcal{M}_{\mathrm{ns}}(m,n)$. This task can just be done by following what was discussed in §6.3.

Now let us write

$$f = f_{\mathcal{M}_{\mathrm{ns}}}(x,y) = \sum_{M\in\mathcal{M}_{\mathrm{ns}}} x^{m(M)}y^{n(M)} \tag{6.4.7}$$

where $m(M)$ is the valency of the root-vertex and $n(M)$, the size of M.

By the similar procedure, based on Lemma 6.3.2 we can find the enufunction defined by (6.4.7) satisfies the following equation as

$$f^2 + \Big((1-x)(1+xy) - xf^*\Big)f$$

$$-x^2(1-x)y(1+f^*)=0 \tag{6.4.8}$$

where $f^* = f(1,y)$ is the corresponding enufunction with the size as the parameter.

In fact $f = h(x,y,y)$, which was defined by (6.3.11) from the Euler formula for planar maps, and hence the equation (6.4.8) is the result of the equation (6.3.14) in the case $y = z$.

From this point of view, for Eq.(6.4.8) we have the following parametric expressions:

$$\begin{aligned} y &= \xi(1-\xi)^2; \\ f^* - y &= \xi^2(1-2\xi). \end{aligned} \tag{6.4.9}$$

Thus, we are now allowed to employ Corollary 1.5.1 for finding

$$\begin{aligned} f^* - y &= \sum_{n\geq 1} \frac{y^n}{n!}\frac{\mathrm{d}^{n-1}}{\mathrm{d}\xi^{n-1}}\left(\frac{2\xi(1-3\xi)}{(1-\xi)^{2n}}\right)\Bigg|_{\xi=0} \\ &= \sum_{n\geq 1}\frac{y^n}{n!}\left(\frac{2(n-1)(3n-3)!}{(2n-1)!}\right. \\ &\qquad \left. - \frac{3(n-1)(n-2)(3n-4)!}{(2n-1)!}\right) \\ &= \sum_{n\geq 1}\frac{2(3n-3)!}{n!(2n-1)!}y^n. \end{aligned} \tag{6.4.10}$$

Although by substituting (6.4.9) into (6.4.8) for eliminating y and f^*, f can be solved to be a function of x and ξ and hence by the first of (6.4.9) f as a function of x and y is soon found, we are not allowed to write in detail here from the limitation of spaces and assume that

$$N_{\mathrm{ns}}(m,n) = \partial_{(x,y)}^{(m,n)} f \tag{6.4.11}$$

which is given beforehand as the number of rooted nonseparable planar map with size n and the root-vertex of valency m for $m, n \geq 1$.

Let $N_{\mathrm{cb}}(m,n)$ be the number of cubic boundary maps with size n and the root-face boundary of valency m. Then from Lemma 6.4.1, we have

$$N_{\mathrm{cb}}(m,n) = N_{\mathrm{ns}}(m,n-m) \tag{6.4.12}$$

for $n \geq 2m$ which is naturally satisfied.

For any map $M = (\mathcal{M}_{\alpha,\beta}, \mathcal{P})$ on a nonorientable surface of genus $\tilde{g} \geq 1$, according to the theory of polyhedra described in [Liu58], there is a minimal submap R which has only one face on the surface. Such a submap R of M on the surface of genus g (or nonorientable genus $\tilde{g}$) is said to be g(or $\tilde{g}$)-*essential*. Here, we only discuss the nonorientable case to save space.

Lemma 6.4.2 *A submap R of a map on the surface of genus $\tilde{g}$ is $\tilde{g}$-essential iff R has no vertex of valency 1 with the dimension $\tilde{g}$ of the cycle space such that no cycle is a face boundary except only probably for R itself.*

Proof Because R is $\tilde{g}$-essential, we see that R has exactly one face on a nonorientable surface of genus $\tilde{g}$. For R of genus $\tilde{g}$, R has the dimension of the cycle space at least $\tilde{g}$. By the minimality and the theory appearing in [Liu58], the dimension has to be exactly $\tilde{g}$. Since R has exactly one face, any cycle except only for R itself is not allowed to be a face boundary. Of course, from the minimality, R does not have a vertex of valency 1. This is the necessity.

For the sufficiency, we have to investigate the relationship between the order and the size of R. For evaluating the difference between the order and the size of R, we are allowed to assume no vertex have valency 2. Let the order and the size of R be $\nu(R)$ and $\epsilon(R)$ respectively. Because R has the dimension $\tilde{g}$ of its cycle space, we see that for any spanning tree of R, R has exactly $\tilde{q}$ cotree edges. That implies

$$\epsilon(R) = \nu(R) - 1 + \tilde{g}.$$

Thus, for the difference $\Delta(R) = \nu(R) - \epsilon(R)$, we have

$$\Delta(R) = 1 - \tilde{g}. \tag{6.4.13}$$

Since R has only one face, from (6.4.13) by Euler formula R is a map on the nonorientable surface of genus $\tilde{g}$. Then by considering that any edge on a cycle is not allowed to delete for maintaining as a map on the surface of genus $\tilde{g}$ and that any edge which is not on a cycle (and here a cut-edge) is not allowed to deleted for having only one face except only for an *articulate edge* (an edge with an end of valency 1), we see that R has to be $\tilde{g}$-essential. □

If a map on the nonorientable surface of genus $\tilde{g}$ has a $\tilde{g}$-essential submap marked, then it is called *$\tilde{g}$-oriented.* Two $\tilde{g}$-oriented maps are said to be the *same* if they are isomorphic in correspondence with the marked $\tilde{g}$-essential submaps;*distinct*, otherwise here.

For a map M on the surface of genus $\tilde{g}$ let R be the marked $\tilde{g}$-essential submap with the only face

$$(x_1, x_2, \cdots, x_{2\epsilon(R)}). \tag{6.4.14}$$

By cutting along R on M, we may obtain another map M' such that all the edges on R are replaced by

$$K\tilde{x}_1, K\tilde{x}_2, \cdots, K\tilde{x}_{2\epsilon(R)} \tag{6.4.15}$$

where

$$K\tilde{x}_i = \{x_i, \tilde{x}_i, \beta x_i, \beta\tilde{x}_i\},\ i = 1, 2, \cdots, 2\epsilon(R),$$

and that a new face is added as

$$(\tilde{x}_1, \tilde{x}_2, \cdots, \tilde{x}_{2\epsilon(R)}) \tag{6.4.16}$$

with all faces on M remaining unchanged. Such a map M' is called the *$\tilde{g}B$-map* of M for R which is the $\tilde{g}$-essential submap. It is easily verified that the $\tilde{1}$B-map introduced in §3.4 is just the special case of $g = 1$ for a $\tilde{g}$B-map.

Lemma 6.4.3 *All $\tilde{g}B$-maps of a map M on the surface of non-orientable genus $\tilde{g}$ are planar maps.*

Proof Because the Euler characteristic of M is the same as that of $\tilde{M}$, which is obtained by contracting all the edges on a spanning tree of R, therefore $\tilde{g}$B-maps of M have the same Euler characteristic as those of $\tilde{M}$ do. However, it is easily checked that for any $\tilde{g}$B-map M' of $\tilde{M}$,

$$\nu(M') = 2\tilde{g} - 1 + \nu(\tilde{M}),$$

$$\epsilon(M') = \epsilon(\tilde{M}) + \tilde{g},\ \phi(M') = \phi(\tilde{M}) + 1.$$

That implies the Euler characteristic

$$E(M') = E(\tilde{M}) + 2\tilde{g} - 1\tilde{g} + 1$$

$$= 1 - \tilde{g} + \tilde{g} + 1$$
$$= 2.$$

Therefore M' is a planar map. The lemma is true. □

Let $\mathcal{M}_{\text{nst}}(m;\tilde{g})$ be the set of all $\tilde{g}$-oriented nonsplitable maps with the marked submap of size $m(\geq \tilde{g}$, of course). Then

$$\mathcal{M}_{\text{nst}}(m;\tilde{g}) = \sum_{n\geq m} \mathcal{M}_{\text{nst}}(m,n;\tilde{g}) \tag{6.4.17}$$

where $\mathcal{M}_{\text{nst}}(m,n;\tilde{g})$ is such a subset of $\mathcal{M}_{\text{nst}}(m;\tilde{g})$ that all its maps are of size n. Of course, only $n \geq m$ does make sense because maps in $\mathcal{M}_{\text{nst}}(m;\tilde{g})$ have at least m edges.

Lemma 6.4.4 *Let $\mathcal{M}_{\text{nsp}}(m,n)$ be the set of all rooted nonseparable planar map of size n with the root-face of valency m for $n \geq m \geq 1$. Then, we have*

$$\left|\mathcal{M}_{\text{nst}}(m,n;\tilde{g})\right| = N_{1\text{f}}(m;\tilde{g})\left|\mathcal{M}_{\text{nsp}}(2m,n+m)\right| \tag{6.4.18}$$

for $m \geq \tilde{g} \geq 1$ where $N_{1\text{f}}(m;\tilde{g})$ is the number of rooted maps without articulate vertex with m edges, only one face and the cycle space of dimension $\tilde{g}$ on the surface of nonorientable genus $\tilde{g}$.

Proof For any $M \in \mathcal{M}_{\text{nst}}(m;\tilde{g})$, we may uniquely find its $\tilde{g}$B-map M'. From Lemma 6.4.3, M' is planar. Because M is nonsplitable, M' has to be nonseparable for the same reason shown in §3.4–5. By counting the numbers of edges in M' and on the outerface boundary of M', it is soon seen that $M' \in \mathcal{M}_{\text{nsp}}(2m, m+n)$. That implies the number on the left hand side of (6.4.18) is not less than that on the right.

Conversely, for any map $M \in \mathcal{M}_{\text{nsp}}(2m, m+n)$, there are just $N_{1\text{f}}(m;\tilde{g})$ possibilities for forming a map M' of size n with the $\tilde{g}$-essential submap of size m on the surface of nonorientable genus $\tilde{g}$ because any map without articulate vertex, with m edge, only one face and the cycle space of dimension $\tilde{g}$ is a $\tilde{g}$-essential submap of a map on the surface of nonorientable genus $\tilde{g}$. By the reason described in §3.4–5 again, M' is nonsplitable. From simply counting, we see that

$M' \in \mathcal{M}_{\rm nst}(m;\tilde{g})$. Then, the number on the right of (6.4.8) is not less than that on the left. □

This lemma suggests us to enumerate the only one face maps without articulate vertex with the cycle space of dimension $\tilde{g}$ on the surface of genus $\tilde{g}$. This task will be completed in §8.5.

If $\tilde{g} = 1$, then Lemma 6.4.1 can also be employed to enumerate a kind of oriented nonsplitable map on the projective plane by following similarly what was discussed in §4.4.

§6.5 Notes

6.5.1 The equation (6.1.17) was firstly described in [Liu30], and (6.1.25–30), in [Liu31] where some other kinds of outerplanar maps and the relationships between outerplanar maps and plane trees were also provided. A number of simpler formulae can be seen in [DoY1] as well. One might like to enumerate Hamiltonian planar maps through outerplanar maps by suitable identifying the boundaries of two such maps.

6.5.2 A direct method has not yet be found for solving the functional equation (6.1.17) to obtain an explicit form of the enufunction of outerplanar maps with vertex partition.

6.5.3 One might find the functional equation (6.2.16) in [Liu23, Liu53] where no detailed proof was presented. However, a number of applications can be found. How to solve (6.2.16) directly is still an open problem.

6.5.4 In fact, the cubic equation (6.2.19) is equivalent to what was provided in [Liu41] from the planar duality. By extracting the relation between general and nonseparable cases, based on (5.2.13) the following parametric expressions have been found:

$$y = \frac{(\xi - 1)(1 + \xi + \xi^2)^2}{\xi^2(2 - \xi)^2};$$

$$1 + f^* = \frac{1 + \xi - \xi^2}{(2 - \xi)^2}, \tag{6.5.1}$$

from which f^* can be determined by employing the Lagrangian inversion. However, no simpler formula has been obtained up to now.

6.5.5 The formula (6.3.18) is owed to Brown and Tutte [BrT1]. However, the decomposition used in §6.3 is different from that of Brown and Tutte. In fact, it can be seen as the dual form by considering that a planar map is nonseparable iff so is its planar dual.

6.5.6 The functional equation (6.3.10) has not been solved yet even not complicated in form among those we have known by now.

6.5.7 The formula (6.4.10) for enumerating rooted nonseparable planar maps have been found in several ways, for instance, those appearing in [Liu17, WaL3 *etc.*]. However, it is here shown as a direct result of Brown and Tutte [BrT1].

6.5.8 It looks still possible to employ a kind of decomposition of maps on surfaces for finding a functional equation satisfied by certain enufunctions. However, there seems to be nobody yet following this idea for enumerating nonseparable maps (rooted first) on general surfaces orientable, or non-orientable.

Chapter 7

Simple Maps

§7.1 Loopless maps

A map is said to be *loopless* if it has no edge which is a loop. Let $\mathcal{M}_{\rm nl}$ be the set of all rooted loopless planar maps. By a *nearly loopless* map is meant a planar map in which only the rooted edge is a loop. The set of all nearly loopless maps is denoted by $\mathcal{M}_{\rm nnl}$.

A nearly loopless map with its root-face (or in other words, outer-face) boundary consisting of only one edge (which is the loop) is called an *inner map*. And, $\mathcal{M}_{\rm in}$ stands for the set of all inner maps.

Lemma 7.1.1 *Let $\mathcal{M}_{\langle\rm in\rangle} - \{M - a | \forall M \in \mathcal{M}_{\rm in}\}$ where $a = e_r$ is the root-edge of M. Then we have*

$$\mathcal{M}_{\langle\rm in\rangle} = \mathcal{M}_{\rm nl} \tag{7.1.1}$$

where $\mathcal{M}_{\rm nl}$ is the set of all rooted loopless planar maps.

Proof For any $M \in \mathcal{M}_{\langle\rm in\rangle}$, because there is an $M' \in \mathcal{M}_{\rm in}$ such that $M = M' - a'$, a' is the root-edge of M' and M' has only one loop which is a', we see that $M \in \mathcal{M}_{\rm nl}$. Thus $\mathcal{M}_{\langle\rm in\rangle} \subseteq \mathcal{M}_{\rm nl}$.

Conversely, for any $M \in \mathcal{M}_{\rm nl}$, we may always construct a map M' by adding a loop a' as the outer-face boundary of M' from M and hence $M' \in \mathcal{M}_{\rm in}$. However, it is easily checked that $M = M' - a'$. This implies $M \in \mathcal{M}_{\langle\rm in\rangle}$ and hence $\mathcal{M}_{\rm nl} \subseteq \mathcal{M}_{\langle\rm in\rangle}$. □

For two maps M_1 and M_2, the 1-addition of M_1 and M_2 is denoted by $M_1 \dot{+} M_2$ as defined in (2.1.3).

However, the root of M is chosen to be the same as that of M_2 without loss of generality. For two sets $\mathcal{M}_1$ and $\mathcal{M}_2$, the 1-product of them is denoted by $\mathcal{M}_1 \odot \mathcal{M}_2$ as defined in (2.1.4).

Lemma 7.1.2 *For* $\mathcal{M}_{\text{nnl}}$ *we have*

$$\mathcal{M}_{\text{nnl}} = \mathcal{M}_{\text{nl}} \bigodot \mathcal{M}_{\text{in}} \tag{7.1.2}$$

where $\odot$ *is the* 1- *production.*

Proof For $M \in \mathcal{M}_{\text{nnl}}$, because the root-edge of M is a loop the inner part of M with the loop itself is denoted by M_2 and the outer part of M is M_1. It is easily checked that $M_2 \in \mathcal{M}_{\text{in}}$ and $M_1 \in \mathcal{M}_{\text{nnl}}$. Since $M = M_1 \dot{+} M_2$, we have

$$M \in \mathcal{M}_{\text{nnl}} \bigodot \mathcal{M}_{\text{in}}.$$

Thus $\mathcal{M}_{\text{nnl}} \subseteq \mathcal{M}_{\text{nl}} \odot \mathcal{M}_{\text{in}}$.

Conversely, for $M \in \mathcal{M}_{\text{nl}} \odot \mathcal{M}_{\text{in}}$, because $M = M_1 \dot{+} M_2$ has only one loop, the root-edge which is the same as that of M_2, by considering $M_1 \in \mathcal{M}_{\text{nl}}$ and $M_2 \in \mathcal{M}_{\text{in}}$ we see that $M \in \mathcal{M}_{\text{nnl}}$. Thus $\mathcal{M}_{\text{nl}} \odot \mathcal{M}_{\text{in}} \subseteq \mathcal{M}_{\text{nnl}}$. □

Now $\mathcal{M}_{\text{nl}}$ is divided into three classes $\mathcal{M}_{\text{nl}}^{(\text{I})}, \mathcal{M}_{\text{nl}}^{(\text{II})}$, and $\mathcal{M}_{\text{nl}}^{(\text{III})}$, *i.e.*,

$$\mathcal{M}_{\text{nl}} = \mathcal{M}_{\text{nl}}^{(\text{I})} + \mathcal{M}_{\text{nl}}^{(\text{II})} + \mathcal{M}_{\text{nl}}^{(\text{III})}, \tag{7.1.3}$$

where $\mathcal{M}_{\text{nl}}^{(\text{I})}$ consists of a single map which is the vertex map ϑ and

$$\mathcal{M}_{\text{nl}}^{(\text{II})} = \{M | \forall M \in \mathcal{M}_{\text{nl}} - \vartheta \text{ with the root- edge not in a multi-edge}\}.$$

Of course, $\mathcal{M}_{\text{nl}}^{(\text{III})}$ consists of all the maps whose root-edges are all in multi-edges.

Lemma 7.1.3 *Let* $\mathcal{M}_{(\text{nl})}^{(\text{II})} = \{M \bullet a | \forall M \in \mathcal{M}_{\text{nl}}^{(\text{II})}\}$ *where* a *is the root-edge of* M. *Then*

$$\mathcal{M}_{(\text{nl})}^{(\text{II})} = \mathcal{M}_{\text{nl}}, \tag{7.1.4}$$

the set of all rooted loopless planar maps.

Proof For any $M \in \mathcal{M}^{(\mathrm{II})}_{(\mathrm{nl})}$ there is a map $M' \in \mathcal{M}^{(\mathrm{II})}_{(\mathrm{nl})}$ such that $M = M' \bullet a'$ where a' is the multi-edge of M', M is loopless as well and hence $M \in \mathcal{M}_{\mathrm{nl}}$.

Conversely, for any $M \in \mathcal{M}_{\mathrm{nl}}$ we may obtain another map M' by splitting the root-vertex into two vertices with the root-edge a' joining the two vertices.

Of course, a' has to be a link which is not in a multi-edge. Since $M = M' \cdot a$, $M' \in \mathcal{M}^{(\mathrm{II})}_{\mathrm{nl}}$, we see that $M \in \mathcal{M}^{(\mathrm{II})}_{(\mathrm{nl})}$. □

For a map $M \in \mathcal{M}_{\mathrm{nl}}$, we may find another map $\nabla_i M$, which is called the *i-th decrement,* by splitting the root-vertex o of M into v_{1i} and v_{2i} with the new edge $(o_{1i}, o_{2i}) = Kr_i$ such that

$$o_{2i} = (\alpha\beta r_i, r, \mathcal{P}r, \cdots, \mathcal{P}^{i-1}r)$$

and

$$o_{1i} = (r_i, \mathcal{P}^i r, \mathcal{P}^{i+1}r, \cdots, \mathcal{P}^{m(M)-1}r)$$

where r_i is chosen to be the root of $\nabla_i M$ for $i = 0, 1, \cdots, m(M)$ and $m(M)$ is the valency of the root-vertex o of M.

Lemma 7.1.4 *For* $\mathcal{M}^{(\mathrm{II})}_{\mathrm{nl}}$ *we have*

$$\mathcal{M}^{(\mathrm{II})}_{\mathrm{nl}} = \sum_{M \in \mathcal{M}_{\mathrm{nl}}} \left\{\nabla_i M \mid 0 \le i \le m(M)\right\}. \tag{7.1.5}$$

Proof Let the set on the right hand side of (7.1.5) be denoted by $\mathcal{M}'$ for convenience.

For $M \in \mathcal{M}^{(\mathrm{II})}_{\mathrm{nl}}$, let $\tilde{M} = M \bullet a$. From lemma 7.1.3, $\tilde{M} \in \mathcal{M}_{\mathrm{nl}}$. Then, there exists an i such that $M = \nabla_i \tilde{M}$. In fact, $i = t - 1$ where t is the valency of the nonrooted end of the root-edge of M. Hence, $M \in \mathcal{M}'$.

Conversely, for any $M \in \mathcal{M}'$, there exists a map $\tilde{M} \in \mathcal{M}_{\mathrm{nl}}$ such that $M = \nabla_i M$ for some $i, 0 \le i \le m(\tilde{M})$. Since $\tilde{M}$ has no loop, the root-edge of M is not in a multi-edge. That implies $M \in \mathcal{M}^{(\mathrm{II})}_{\mathrm{nl}}$. □

For $M \in \mathcal{M}_{\rm nl}^{\rm (III)}$, because the root-edge a of M is in a multi-edge, assume $l \geq 2$ is the number of edges in the multi-edge. Then, there are $l-1$ loops in $\tilde{M} = M \bullet a$. And further, $\tilde{M}$ has the form

$$\tilde{M} = M_1 \dot{+} M_2 \dot{+} \cdots \dot{+} M_{l-1} \dot{+} M_0, \tag{7.1.6}$$

where $\dot{+}$ is the *inner* 1*v-addition* defined in §6.1.

Lemma 7.1.5 *Let* $\mathcal{M}_{\rm (nl)}^{\rm (III)} = \{M \bullet a | \forall M \in \mathcal{M}_{\rm nl}^{\rm (III)}\}$. *Then we have*

$$\mathcal{M}_{\rm (nl)}^{\rm (III)} = \sum_{k \geq 1} \mathcal{M}_{\rm in}^{\dot{\times} k} \dot{\times} \mathcal{M}_{\rm nl}, \tag{7.1.7}$$

where $\dot{\times}$ *is the inner* 1*v-production defined in* §6.1.

Proof For any $M \in \mathcal{M}_{\rm (nl)}^{\rm (III)}$, because M has the form shown in (7.1.6) for $k = l-1 \geq 1$ and $M_i, i = 1, 2, \cdots, k$, are all inner maps with M_0 which is allowed to be any map in $\mathcal{M}_{\rm nl}$. Therefore, M is a member of the set on the right hand side of (7.1.7).

For any M in the set on the right of (7.1.7), since M has the form as shown in (7.1.6) such that all $M_i \in \mathcal{M}_{\rm in}, i = 1, 2, \cdots, k = l-1$, and $M_0 \in \mathcal{M}_{\rm nl}$, we may assume that the root-vertex is as

$$o_r = (S_0, S_{l-1}, \cdots, S_2, S_1, T_1, T_2, \cdots, T_{l-1}, T_0)$$

where $(S_i, T_i), S_i \neq \emptyset, T_i \neq \emptyset, i = 0, 1, 2, \cdots, l-1$, are the root-vertices of M_i respectively. Let $\tilde{M}$ be the map obtained by splitting the root-vertex o_r into two vertices o_1 and o_2 with the edge $(o_1, o_2) = K\tilde{r}$ such that

$$o_1 = (\tilde{r}, T_1, T_2, \cdots, T_{l-1}, T_0)$$

and

$$o_2 = (\alpha\beta\tilde{r}, S_o, S_{l-1}, S_2, S_1).$$

It is easily checked that the root-edge $\tilde{a} = (o_1, o_2)$ is in a multi-edge from $l \geq 2$ and hence $\tilde{M} \in \mathcal{M}_{\rm nl}^{\rm (III)}$. Because $M = \tilde{M} \bullet \tilde{a}$, we see that

$$M \in \mathcal{M}_{\rm (nl)}^{\rm (III)}.$$

In consequence, the lemma is true. □

For two maps M_1 and M_2, let $a_1 = (p_1, q_1)$ and $a_2 = (p_2, q_2)$ be the root-edges of M_1 and M_2 respectively. Write

$$p_i = (r_i, S_i) \text{ and } q_i = (\alpha\beta r_i, T_i)$$

for $i = 1, 2$. Then, the map M obtained by identifying the root-edges a_1 and a_2 into $a = (p, q)$ such that

$$p = (r, S_1, S_2) \text{ and } q = (\alpha\beta r, T_2\ T_1)$$

is said to be the *inner* 1*e-sum* from M_1 into M_2. Or denote $M = M_1 +\!\rangle M_2$. In correspondence, for two sets $\mathcal{M}_1$ and $\mathcal{M}_2$, the set

$$\mathcal{M}_1 \times\!\rangle \mathcal{M}_2 = \{M_1 +\!\rangle M_2 | \forall M_1 \in \mathcal{M}_1, \forall M_2 \in \mathcal{M}_2\} \tag{7.1.8}$$

is called the *inner* 1*e-product* from $\mathcal{M}_1$ into $\mathcal{M}_2$. Because both inner 1*e*-sum and inner 1*e*-product are associate, we are allowed to write

$$\sum_{i=1}^{k}\!\rangle M_i = \sum_{i=1}^{k-1}\!\rangle M_i +\!\rangle M_k = M_1 +\!\rangle \sum_{i=2}^{k}\!\rangle M_i \tag{7.1.9}$$

or $k \times\!\rangle M$ when $M_1 = M_2 = \cdots M_k = M$, and

$$\prod_{i=1}^{k}\!\rangle \mathcal{M}_1 = \mathcal{M}_1 \times\!\rangle \prod_{i=2}^{k}\!\rangle \mathcal{M}_i \tag{7.1.10}$$

or $\mathcal{M}^{\times\rangle k}$ when $\mathcal{M}_1 = \mathcal{M}_2 = \cdots = \mathcal{M}_k = \mathcal{M}$.

Lemma 7.1.6 *For $\mathcal{M}_{\rm nl}^{\rm (III)}$, we have*

$$\mathcal{M}_{\rm nl}^{\rm (III)} = \mathcal{M}_{\rm nl}^{\rm (II)}\ \langle\times \sum_{k\geq 1}$$

$$\times \left(\sum_{M \in \mathcal{M}_{\rm in}} \left\{ \nabla_i M |\ 1 \leq i \leq m(M) \quad 1 \right\} \right)^{\times\rangle k} \tag{7.1.11}$$

where $m(M)$ is the valency of the root-vertex of M.

Proof For $M \in \mathcal{M}_{\rm nl}^{\rm (III)}$ assume that the multi-edge which contains the root-edge a of M have k nonrooted edges. Because $M \bullet a$ is in the form as (7.1.6) with $k = l - 1$, from Lemma 7.1.5 it can be seen that M is a member of the set on the right hand side of (7.1.11).

Conversely, for any map M in the set on the right of (7.1.11), because there exists an integer $k \geq 1$ such that

$$\begin{aligned} M &= M_0 \langle + \Big(M_1 + \rangle \cdots + \rangle M_k \Big) \\ &= \left(\sum_{i=1}^{k} M_i \right) + \rangle M_0 \end{aligned}$$

where $M_0 \in \mathcal{M}_{\mathrm{nl}}^{(\mathrm{II})}$, all $M_i, 1 \leq i \leq k$, are loopless maps with the multiedges the root-edges is in and containing exactly one nonrooted edge which is on the root-face boundary each, we see that M is a map in $\mathcal{M}_{\mathrm{nl}}^{(\mathrm{III})}$ by considering the inner $1e$-sum. □

According to the lemmas above, we are allowed to evaluate the contributions of $\mathcal{M}_{\mathrm{nl}}^{(\mathrm{I})}$, $\mathcal{M}_{\mathrm{nl}}^{(\mathrm{II})}$ and $\mathcal{M}_{\mathrm{nl}}^{(\mathrm{III})}$ to the enufunction

$$g_{\mathcal{M}_{\mathrm{nl}}}(x, \underline{y}) = \sum_{M \in \mathcal{M}_{\mathrm{nl}}} x^{m(M)} \underline{y}^{\underline{n}(M)} \tag{7.1.12}$$

of $\mathcal{M}_{\mathrm{nl}}$ with vertex partition where $m(M)$ is the valency of the root-vertex of M and $\underline{n}(M) = (n_1(M), n_2(M), \cdots)$, $\underline{y} = (y_1, y_2, \cdots)$, $n_i(M)$ is the number of nonrooted vertices of valency $i, i \geq 1$, in M.

First, it is trivial that

$$g_{\mathcal{M}_{\mathrm{nl}}^{(\mathrm{I})}} = 1 \tag{7.1.13}$$

since $\mathcal{M}_{\mathrm{nl}}^{(\mathrm{I})}$ consists of only the vertex map ϑ which has $m(\vartheta) = 0$ and $\underline{n}(\vartheta) = \underline{0}$.

Then based on Lemma 7.1.4, we have

$$\begin{aligned} g_{\mathcal{M}_{\mathrm{nl}}^{(\mathrm{II})}} &= \sum_{M \in \mathcal{M}_{\mathrm{nl}}} \left(\sum_{i=0}^{m(M)} x^{m(M)-i+1} y_{i+1} \right) \underline{y}^{\underline{n}(M)} \\ &= \int_y \sum_{M \in \mathcal{M}_{\mathrm{nl}}} \left(\sum_{i=0}^{m(M)} x^{m(M)-i+1} y^{i+1} \right) \underline{y}^{\underline{n}(M)} \\ &= \int_y xy \sum_{M \in \mathcal{M}_{\mathrm{nl}}} \frac{x^{m(M)+1} - y^{m(M)+1}}{x - y} \underline{y}^{\underline{n}(M)} \end{aligned}$$

$$= x \int_y y\delta_{x,y}(zg) \tag{7.1.14}$$

where $g = g_{\mathcal{M}_{\rm nl}}(z, \underline{y})$.

From Lemma 7.1.6, in order to determine $g_{\mathcal{M}_{\rm nl}^{(\rm III)}}$, we have to know $g_{\mathcal{M}_{\rm in}}$. From Lemma 7.1.1, it is easily seen that

$$g_{\mathcal{M}_{\rm in}} = x^2 g_{\mathcal{M}_{\rm nl}}(x, \underline{y}) \tag{7.1.15}$$

since the loop contributes x^2 to the enufunction of $\mathcal{M}_{\rm in}$.

Further, from Lemma 7.1.6 and (7.1.4), we have

$$g_{\mathcal{M}_{\rm nl}^{(\rm III)}} = x \int_y \left(y\delta_{x,y}(zg) \sum_{k\geq 1} \left(\Delta_{\mathcal{M}_{\rm in}}\right)^k \right)$$

$$= x \int_y \left(y\delta_{x,y}(zg) \frac{\Delta_{\mathcal{M}_{\rm in}}}{1 - \Delta_{\mathcal{M}_{\rm in}}} \right) \tag{7.1.16}$$

where by paying attention to that the root-edge of M has no contribution to the valency of its ends,

$$\Delta_{\mathcal{M}_{\rm in}} = \frac{1}{xy} \sum_{M\in\mathcal{M}_{\rm in}} \left(\sum_{i=1}^{m(M)-1} x^{m(M)-i+1} y^{i+1} \right) \underline{y}^{\underline{n}(M)}$$

$$= \frac{1}{xy} \sum_{M\in\mathcal{M}_{\rm in}} \frac{y^2 x^{m(M)+1} - x^2 y^{m(M)+1}}{x - y} \underline{y}^{\underline{n}(M)}$$

$$= \partial_{x,y} g_{\mathcal{M}_{\rm in}}(z)$$

by (7.1.15)

$$= \partial_{x,y}(z^2 g) \tag{7.1.17}$$

in which $g = g_{\mathcal{M}_{\rm nl}}(z, \underline{y})$ defined in (7.1.12). Substituting (7.1.17) into (7.1.16), by (1.6.9) we soon find

$$g_{\mathcal{M}_{\rm nl}^{(\rm III)}} = \int_y \frac{\partial^2_{x,y}(z^2 g)}{1 - \partial_{x,y}(z^2 g)}. \tag{7.1.18}$$

Theorem 7.1.1 *The enufunction g of $\mathcal{M}_{\rm nl}$ with vertex partition given by (7.1.12) satisfies the following equation:*

$$g = 1 + \int_y \frac{\partial_{x,y}(z^2 g)}{1 - \partial_{x,y}(z^2 g)} \tag{7.1.19}$$

where $g = g_{\mathcal{M}_{\rm nl}}(z, \underline{y})$ *in the parentheses as in* (7.1.14) *and* (7.1.17).

Proof Since $g = g^{(\rm I)}_{\mathcal{M}_{\rm nl}} + g^{(\rm II)}_{\mathcal{M}_{\rm nl}} + g^{(\rm III)}_{\mathcal{M}_{\rm nl}}$, from (7.1.13), (7.1.14) and (7.1.18) we have

$$g = 1 + x\int_y y\delta_{x,y}(zg) + \int_y \frac{\partial^2_{x,y}(z^2 g)}{1 - \partial_{x,y}(z^2 g)}$$

by (1.6.9),

$$= 1 + \int_y \left(\partial_{x,y}(z^2 g) + \frac{\partial^2_{x,y}(z^2 g)}{1 - \partial_{x,y}(z^2 g)}\right)$$

$$= 1 + \int_y \frac{\partial_{x,y}(z^2 g)}{1 - \partial_{x,y}(z^2 g)}.$$

This is the theorem. □

Although the equation (7.1.19) has no explicit solution to be found, we are allowed to determine the enufunction of $\mathcal{M}_{\rm nl}$ with two parameters: the size and the valency of root-vertex. Let

$$h_{\mathcal{M}_{\rm nl}} = \sum_{M \in \mathcal{M}_{\rm nl}} x^{m(M)} y^{n(M)} \tag{7.1.20}$$

when $m(M)$ and $n(M)$ are respectively the valency of root-vertex and the size of M.

Lemma 7.1.7 *The enufunction* h *defined by* (7.1.20) *satisfies the equation as*

$$x^2 y h^2 + (1 - x - xyh^*)h - (1 - x) = 0 \tag{7.1.21}$$

where $h^* = h(1, y)$, *i.e., the enufunction of* $\mathcal{M}_{\rm nl}$ *with the size as the parameter.*

Proof By considering that $h_{\mathcal{M}^{(\rm I)}_{\rm nl}} = 1$;

$$h_{\mathcal{M}^{(\rm II)}_{\rm nl}} = \sum_{M \in \mathcal{M}_{\rm nl}} \left(y \sum_{i=0}^{m(M)} x^{m(M)-i+1} \right) y^{n(M)}$$

$$= \sum_{M \in \mathcal{M}_{\rm nl}} xy \left(\frac{1 - x^{m(M)+1}}{1-x} \right) y^{n(M)}$$

$$= \frac{xy(h^* - xh)}{1-x} \tag{7.1.22}$$

from Lemma 7.1.4;

$$h_{\mathcal{M}_{\rm in}} = x^2 yh \tag{7.1.23}$$

from Lemma 7.1.1; and

$$h_{\mathcal{M}_{\rm nl}^{(\rm III)}} = \frac{xy}{1-x}(h^* - xh) \sum_{k \geq 1} \times \left(\sum_{M \in \mathcal{M}_{\rm in}} \frac{x - x^{m(M)}}{1-x} y^{n(M)} \right)^k$$

$$= \frac{xy}{1-x}(h^* - xh) \frac{xh^*_{\rm in} - h_{\rm in}}{1 - x - (xh^*_{\rm in} - h_{\rm in})}$$

where $h_{\rm in} = h_{\mathcal{M}_{\rm in}}$, by (7.1.23),

$$= \frac{xy}{1-x} \frac{(h^* - xh)(xyh^* - x^2yh)}{1 - x - (xyh^* - x^2yh)}, \tag{7.1.24}$$

from (7.1.13), (7.1.22) and (7.1.24) we find

$$h = 1 + \frac{xy}{1-x} \left(\frac{(1-x)(h^* - xh)}{1 - x - xy(h^* - xh)} \right).$$

By simplification, (7.1.21) is obtained. □

According to the similar procedure used in §5.4 for solving the equation (5.4.11), we may also extract the following parametric expressions as follows:

$$y = \frac{\xi - 1}{\xi^4};$$

$$h^* = \xi^2(2 - \xi). \tag{7.1.25}$$

Theorem 7.1.2 *The number of distinct rooted loopless maps with size n is*

$$H_{\rm nl}(n) = \frac{6(4n+1)!}{(3n+3)!n!} \tag{7.1.26}$$

for $n \geq 0$.

Proof By employing Corollary 1.5.2, we see that

$$\partial_y^n h^* = \frac{1}{n!}\frac{\mathrm{d}^{n-1}}{\mathrm{d}\xi^{n-1}}\left(\xi^{4n+1}(4-3\xi)\right)\Big|_{\xi=1}$$
$$= \frac{1}{n!}\left(\frac{4(4n+1)!}{(3n+2)!} - \frac{3(4n+2)!}{(3n+3)!}\right)$$

from which $H_{\mathrm{nl}}(n) = \partial_y^n h^*$ is obtained just as in (7.1.26). □

Of course, from (7.1.26) the corresponding numbers of nearly loopless, inner planar maps, *etc.* can easily be found [Liu15, Liu17].

§7.2 Loopless Eulerian maps

Let $\mathcal{M}_{\mathrm{nlE}}$ be the set of rooted loopless Eulerian planar maps. For any $M \in \mathcal{M}_{\mathrm{nlE}}$, let $2m(M)$ be the valency of the root-vertex and $\underline{n}(M) = (n_2(M), n_4(M), \cdots, n_{2j}(M), \cdots)$ where $n_{2j}(M)$ is the number of nonrooted vertices of valency $2j, j \geq 1$, in M. First, we aim at finding an equation satisfied by the enufunction:

$$f_{\mathcal{M}_{\mathrm{nlE}}}(x, \underline{y}) = \sum_{M \in \mathcal{M}_{\mathrm{nlE}}} x^{2m(M)} \underline{y}^{\underline{n}(M)} \tag{7.2.1}$$

where $\underline{y} = (y_2, y_4, \cdots, y_{2j}, \cdots)$.

As mentioned in §7.1, an *inner Eulerian map* is such a rooted Eulerian planar map in which there is exactly one loop, of course the root-edge, that forms the outerface boundary.

Lemma 7.2.1 *Let* $\mathcal{M}_{\mathrm{in}}^{(\mathrm{E})}$ *be the set of all inner Eulerian maps. Then we have*

$$\mathcal{M}_{\langle\mathrm{in}\rangle}^{(\mathrm{E})} = \mathcal{M}_{\mathrm{nlE}}, \tag{7.2.2}$$

where

$$\mathcal{M}_{\langle\mathrm{in}\rangle}^{(\mathrm{E})} = \left\{M - a \mid \forall M \in \mathcal{M}_{\mathrm{in}}^{(\mathrm{E})}\right\}$$

with $a = Kr$, *the root-edge of* M.

Proof Similar to the proof of Lemma 7.1.1 but always with the Eulerianity of maps in mind here. □

A map which is rooted Eulerian and planar with exactly one loop, of course the root-edge, is called an *almost loopless Eulerian map*, or in other words, a *nearly loopless Eulerian map*, also.

Lemma 7.2.2 *Let $\mathcal{M}_{\mathrm{alE}}$ be the set of all almost loopless Eulerian maps. Then we have*

$$\mathcal{M}_{\mathrm{alE}} = \mathcal{M}_{\mathrm{nlE}} \odot \mathcal{M}_{\mathrm{in}}^{(\mathrm{E})} \tag{7.2.3}$$

where $\odot$ is the 1v-production, or in other words, the 1v-multiplication as shown in §2.1.

Proof Similar to the proof of Lemma 7.1.2 but with the attention on the Eulerianity as well. □

The set $\mathcal{M}_{\mathrm{nlE}}$ is divided into three parts: $\mathcal{M}_{\mathrm{nlE}}^{(\mathrm{I})} = \vartheta$, the vertex map,

$$\mathcal{M}_{\mathrm{nlE}}^{(\mathrm{II})} = \{M \mid \forall M \in \mathcal{M}_{\mathrm{nlE}} - \vartheta \text{ with the root-edge of } M \text{ not in a multi-edge}\},$$

and of course, $\mathcal{M}_{\mathrm{nlE}}^{(\mathrm{III})}$ consists of those with all the root-edges in multi-edges. That is

$$\mathcal{M}_{\mathrm{nlE}} = \mathcal{M}_{\mathrm{nlE}}^{(\mathrm{I})} + \mathcal{M}_{\mathrm{nlE}}^{(\mathrm{II})} + \mathcal{M}_{\mathrm{nlE}}^{(\mathrm{III})}. \tag{7.2.4}$$

Lemma 7.2.3 *Let $\mathcal{M}_{(\mathrm{nlE})}^{(\mathrm{II})} = \{M \bullet a \mid \forall M \in \mathcal{M}_{\mathrm{nlE}}^{(\mathrm{II})}\}$ with a as the root-edge as usual. Then we have*

$$\mathcal{M}_{(\mathrm{nlE})}^{(\mathrm{II})} = \mathcal{M}_{\mathrm{nlE}} - \vartheta. \tag{7.2.5}$$

Proof Pay attention to that the link map

$$L = \Big(Kr, (r)(\alpha\beta r)\Big)$$

is not Eulerian and that only one loopless map M such that $M \bullet a = \vartheta$. That is $M = L$. The procedure of proving Lemma 7.1.3 can also be applied here whenever the Eulerianity is always in mind. □

By employing the operation ∇_i introduced in §7.1 on a map, we may express $\mathcal{M}^{(\mathrm{II})}_{\mathrm{nlE}}$ in terms of maps from $\mathcal{M}_{\mathrm{nlE}} - \vartheta$.

Lemma 7.2.4 *For* $\mathcal{M}^{(\mathrm{II})}_{\mathrm{nlE}}$ *we have*

$$\mathcal{M}^{(\mathrm{II})}_{\mathrm{nlE}} = \sum_{M \in \mathcal{M}_{\mathrm{nlE}} - \vartheta} \left\{ \nabla_{2i+1} M \mid 0 \le i \le m(M) - 1 \right\} \tag{7.2.6}$$

where $2m(M)$ *is the valency of the root-vertex of* M.

Proof For any $M \in \mathcal{M}^{(\mathrm{II})}_{\mathrm{nlE}}$, let $\tilde{M} = M \bullet a$, where $a = Kr$. From Lemma 7.2.3, $\tilde{M} \in \mathcal{M}_{\mathrm{nlE}} - \vartheta$. Assume $a = (p, q)$ where $p = (r, S)$ and $q = (\alpha\beta r, T)$. Then $r(\tilde{M}) = \tilde{r} = \mathcal{P}\alpha\beta r$ and the root-vertex of $\tilde{M}$ is $\tilde{o} = (TS)$. Because M is Eulerian, each of S and T involves an odd number of edges. If T is incident with $2i + 1$ edges, then S has

$$2m(\tilde{M}) + 2 - (2i + 1) = 2m(\tilde{M}) - 2i + 1$$

edges. That implies $M = \nabla_{2i+1} \tilde{M}$ and hence M is a member of the set on the right hand side of (7.2.6).

Conversely, for M in the set on the right of (7.2.6), let $M = \nabla_{2i+1} \tilde{M}$ for some $\tilde{M} \in \mathcal{M}_{\mathrm{nlE}} - \vartheta$ and $i, 0 \le i \le m(\tilde{M}) - 1$. Because $2i + 1 + 1 = 2i + 2$ is the valency of the vertex which is the nonrooted end of the root-edge of M, the valency of the root-vertex of M has to be

$$2m(\tilde{M}) - 2 - 2i + 2 = 2m(\tilde{M}) - 2i.$$

Since both of them are even, from the Eulerianity of $\tilde{M}$, we see that M is Eulerian. Further, from the looplessness of $\tilde{M}$ and hence the root-edge of M not in any multi-edge, we have $M \in \mathcal{M}^{(\mathrm{II})}_{\mathrm{nlE}}$. □

By employing the inner $1v$-addition and the inner $1v$-production explained in §7.1, we are allowed to express

$$\mathcal{M}^{(\mathrm{III})}_{(\mathrm{nlE})} = \{ M \bullet a \mid \forall M \in \mathcal{M}^{(\mathrm{III})}_{\mathrm{nlE}} \}$$

for a being the root-edge of M in terms of those from $\mathcal{M}^{(\mathrm{E})}_{\mathrm{in}}$ and $\mathcal{M}_{\mathrm{nlE}}$.

Lemma 7.2.5 *For $\mathcal{M}^{(\mathrm{III})}_{(\mathrm{nlE})}$ we have*

$$\mathcal{M}^{(\mathrm{III})}_{(\mathrm{nlE})} = \mathcal{M}_{\mathrm{nlE}} \times\!\cdot \sum_{k\geq 1} \mathcal{M}^{(\mathrm{E})^{\times k}}_{\mathrm{in}} \tag{7.2.7}$$

where $\times\cdot$, $\cdot\times$ as well, is defined as in (7.1.7).

Proof For $M \in \mathcal{M}^{(\mathrm{III})}_{\mathrm{nlE}}$, write the root-vertex

$$o = (r, T_0, \alpha\beta l_k, T_k, \cdots, \alpha\beta l_1, T_1,$$

$$S_1, l_1, \cdots, S_k, l_k S_0) \tag{7.2.8}$$

of M such that $|T_1| = |S_1| = 0 (\mathrm{mod}\ 2)$. Let $M_0, M_1, \cdots, M_k$ be the maps extended from the root-vertices:

$$o_0 = (r, T_0, S_0), o_1 = (\alpha\beta l_1, T_1, S_1, l_1),$$

$$\cdots, o_k = (\alpha\beta l_k, T_k, S_k, l_k), \tag{7.2.9}$$

respectively. From the Eulerianity of M, we see that all the root-vertices $o_0, o_1, \cdots, o_k$ are of even valency in $M_0, M_1, \cdots, M_k$ and hence all the maps are Eulerian. Because $Kl_i, i = 1, 2, \cdots, k$, are all loops, all $M_i \in \mathcal{M}^{(\mathrm{E})}_{\mathrm{in}}$, $i = 1, 2, \cdots, k$. Since r is not a loop in M, M_0 is allowed to be any map in $\mathcal{M}_{\mathrm{nlE}}$. However,

$$M = M_1 \dot{+} M_2 \dot{+} \cdots \dot{+} M_k \dot{+} M_0$$

$$= M_0 \dot{+} M_1 \dot{+} M_2 \dot{+} \cdots \dot{+} M_k.$$

That implies M is a member of the set on the right hand side of (7.2.7).

Conversely, for M in the set on the right of (7.2.7), let

$$M = M_0 \dot{+} M_i \dot{+} M_2 \dot{+} \cdots \dot{+} M_k$$

for $M_i \in \mathcal{M}^{(\mathrm{E})}_{\mathrm{in}}, i = 1, 2, \cdots, k$, and $M_0 \in \mathcal{M}_{\mathrm{nlE}}$. Then, we may construct a map M' by splitting the root-vertex o of M as shown in (7.2.8) into p and q with the edge $(p, q) = Kr$ such that

$$p = (r', S_1, l_1, \cdots, S_k, l_k, S_0)$$

and

$$q = (\alpha\beta r', r, T_0, \alpha\beta l_k, T_k, \cdots, \alpha\beta l_1 T_1).$$

From the Eulerianity of $M_i, 0 \le i \le k$, it is allowed to choose S_1, T_1, if necessary, such that p and q are with even valencies. Because (p, q) is in a multi-edge, we see that $M' \in \mathcal{M}_{\text{nlE}}^{(\text{III})}$. However, it is easily checked

$$M = M' \bullet a',\ a' = (p, q).$$

Thus $M \in \mathcal{M}_{(\text{nlE})}^{(\text{III})}$, the set on the left of (7.2.7). □

The set $\mathcal{M}_{\text{nlE}}^{(\text{III})}$ has to be partitioned into two parts: $\mathcal{M}_A$ and $\mathcal{M}_B$ such that

$$\begin{aligned}\mathcal{M}_B = \{M | \ \forall M \in \mathcal{M}_{\text{nlE}}^{(\text{III})},&\\ Kr = (p, q), p = (r, S, l, J)&\\ \text{and } q = (\alpha\beta r, T, \alpha\beta l, R)\}.&\end{aligned}$$

Or in other words, all the maps in $\mathcal{M}_{\text{nlE}}^{(\text{III})}$ are with an edge $Kl, l \neq r$, in the multi-edge, which contains Kr, on the outerface boundary. Since

$$\mathcal{M}_{\text{nlE}}^{(\text{III})} = \mathcal{M}_A + \mathcal{M}_B, \tag{7.2.10}$$

the meaning of $\mathcal{M}_A$ is naturally determined from $\mathcal{M}_B$.

In order to decompose $\mathcal{M}_B$, from Lemma 5.1.1 we have to introduce a kind of non-Eulerian maps, called *bridge maps*, in each of which the root-edge $Kr = (p, q)$ is a cut-edge such that the extended submaps by $(\mathcal{P}\alpha\beta r, T)$ and $(\mathcal{P}r, S)$ as the root-vertices are loopless Eulerian maps, where

$$q = (\alpha\beta r, \mathcal{P}\alpha\beta r, T) \text{ and } p = (r, \mathcal{P}r, S).$$

Easily verified that if $\hat{\mathcal{M}}$ is the set of all bridge maps , then

$$\hat{\mathcal{M}} = \mathcal{M}_{\text{nlE}} \odot \{L\} \odot \mathcal{M}_{\text{nlE}} \tag{7.2.11}$$

where $L = (Kr, (r)(\alpha\beta r))$ is the link map.

Let $K = \{1, 2, \cdots, k\}$, the integer set distinguished from the Klein group which was employed in §1.1. For any $R \subseteq K$ and maps M_1, M_2, $\cdots$, M_k, write

$$M_R(M_j) = \begin{cases} \left\{\nabla_{2i+1} M_j | 0 \le i \le m(M_j) - 1\right\}, \ j \in R; \\ \\ \left\{\nabla_{2i} M_j | 1 \le i \le m(M_j) - 1\right\}, \ \text{otherwise}. \end{cases} \tag{7.2.12}$$

Lemma 7.2.6 *For $\mathcal{M}_B$ we have*

$$\mathcal{M}_B = \sum_{k\geq 1} \sum_{\substack{R\subseteq K \\ |R|=2l+1 \\ 0\leq l\leq \lfloor \frac{k-1}{2} \rfloor}} \left(\hat{\mathcal{M}} \langle \times \prod_{j=1}^{k} \rangle \right.$$

$$\left. \times \left(\sum_{M_j \in \mathcal{M}_{\text{in}}^{(\mathrm{E})}} M_R(M_j) \right) \right). \tag{7.2.13}$$

Proof The procedure of the proof is similar to that of Lemma 7.2.5, except that we have to pay attention to $\hat{\mathcal{M}}$ being a set of non-Eulerian maps and that the parity has to be considered as shown in (7.2.12). □

In order to decompose $\mathcal{M}_A$ we have to introduce another kind of non-Eulerian maps in each of which all vertices are of even valencies except only for the two ends of the rooted edge which is not a cut-edge. Such a non-Eulerian map is called a *via map*. If $\tilde{\mathcal{M}}$ denotes the set of all via maps, then it is easily seen that

$$\tilde{\mathcal{M}} = \sum_{M\in\mathcal{M}_{\text{nlE}}} \left\{ \nabla_{2i} M \mid 0 \leq i \leq m(M) \right\} - \hat{\mathcal{M}} \tag{7.2.14}$$

from reasoning similar to that in the proof of Lemma 7.2.4.

Lemma 7.2.7 *For $\mathcal{M}_A$, we have*

$$\mathcal{M}_A = \mathcal{M}_{\text{ev}} + \mathcal{M}_{\text{od}} \tag{7.2.15}$$

where

$$\mathcal{M}_{\text{ev}} = \sum_{k\geq 1} \sum_{\substack{R\subseteq K \\ |R|=2l \\ 0\leq l\leq \lfloor \frac{k}{2} \rfloor}} \left(\mathcal{M}_{\text{nlE}}^{(\text{II})} \langle \times \prod_{j=1}^{k} \rangle \right.$$

$$\left. \times \left(\sum_{M_j \in \mathcal{M}_{\text{in}}^{(\mathrm{E})}} M_R(M_j) \right) \right);$$

and

$$\mathcal{M}_{\mathrm{od}} = \sum_{k\geq 1} \sum_{\substack{R\subseteq K \\ |R|=2l+1 \\ 0\leq l\leq \lfloor \frac{k-1}{2} \rfloor}} \left(\tilde{\mathcal{M}} \langle \times \prod_{j=1}^{k} \rangle \right.$$

$$\left. \times \left(\sum_{M_j \in \mathcal{M}_{\mathrm{in}}^{(\mathrm{E})}} M_R(M_j) \right) \right). \tag{7.2.16}$$

Proof Similarly to the proof of Lemma 7.2.6. □

Now we are allowed to evaluate the contributions of $\mathcal{M}_{\mathrm{nlE}}^{(\mathrm{I})}$, $\mathcal{M}_{\mathrm{nlE}}^{(\mathrm{II})}$ and $\mathcal{M}_{\mathrm{nlE}}^{(\mathrm{III})}$ to the enufunction defined by (7.2.1).

Of course, since $\mathcal{M}_{\mathrm{nlE}}^{(\mathrm{I})}$ is in fact the vertex map, the contribution of $\mathcal{M}_{\mathrm{nlE}}^{(\mathrm{I})}$ is

$$f_{\mathcal{M}_{\mathrm{nlE}}^{(\mathrm{I})}} = 1. \tag{7.2.17}$$

Let us write

$$f = f(\sqrt{z}) = f_{\mathcal{M}_{\mathrm{nlE}}}(\sqrt{z}, \underline{y}) \tag{7.2.18}$$

for $z = x^2$. According to Lemma 7.2.4,

$$\begin{aligned} f_{\mathcal{M}_{\mathrm{nlE}}^{(\mathrm{II})}} &= \int_y \left(\sum_{M\in \mathcal{M}_{\mathrm{nlE}-\vartheta}} \underline{y}^{\underline{n}(M)} \sum_{j=0}^{m(M)-1} x^{2j+2} y^{2m(M)-2j} \right) \\ &= \int_y \left(x^2 y^2 \sum_{M\in \mathcal{M}_{\mathrm{nlE}-\vartheta}} \frac{x^{2m(M)} - y^{2m(M)}}{x^2 - y^2} \underline{y}^{\underline{n}(M)} \right) \\ &= x^2 \int_y \left(y^2 \delta_{x^2,y^2}(f-1) \right) \\ &= x^2 \int_y \left(y^2 \delta_{x^2,y^2} f \right) \end{aligned}$$

by (1.6.9),

$$= \int_y \partial_{x^2,y^2}(zf). \tag{7.2.19}$$

In order to determine $f_{\mathcal{M}_{\mathrm{nlE}}^{(\mathrm{III})}}$, we have to evaluate $f_{\mathcal{M}_A}$ and $\mathcal{M}_B$ based on Lemmas 7.2.6–7.

First, by (7.2.14) we see that

$$f_{\tilde{\mathcal{M}}+\hat{\mathcal{M}}} = \int_y \left(xy \sum_{M \in \mathcal{M}_{\mathrm{nlE}}} y^{\underline{n}(M)} \sum_{i=0}^{m(M)} x^{2i} y^{2m(M)-2i} \right)$$

$$= \int_y \left(\sum_{M \in \mathcal{M}_{\mathrm{nlE}}} \frac{x^{2m(M)+2} - xy^{2m(M)+2}}{x^2 - y^2} y^{\underline{n}(M)} \right)$$

$$= \int_y \left(xy \delta_{x^2,y^2}(zf) \right). \tag{7.2.20}$$

From the second of (7.2.16), (7.2.18) and (7.2.20),

$$f_{\mathcal{M}_{\mathrm{od}}+\mathcal{M}_B} = \int_y \left(xy\delta_{x^2y^2}(zf) \sum_{k\geq 1} \sum_{l=0}^{\lfloor \frac{k-1}{2} \rfloor} \binom{k}{2l+1} \right.$$

$$\times \left(\sum_{M \in \mathcal{M}_{\mathrm{in}}^{(\mathrm{E})}} y^{\underline{n}(M)} \sum_{i=0}^{m(M)-1} x^{2i+1} y^{2m(M)-2i-1} \right)^{2l+1}$$

$$\left. \times \left(\sum_{M \in \mathcal{M}_{\mathrm{in}}^{(\mathrm{E})}} y^{\underline{n}(M)} \sum_{i=1}^{m(M)-1} x^{2i} y^{2m(M)-2i} \right)^{k-2l-1} \right)$$

$$= \int_y \left(xy\delta_{x^2,y^2}(zf) \sum_{k\geq 1} \sum_{l=0}^{\lfloor \frac{k-1}{2} \rfloor} \binom{k}{2l+1} \right.$$

$$\left. \times (xy\delta_{x^2,y^2} f_{\mathrm{in}})^{2l+1} (\partial_{x^2,y^2} f_{\mathrm{in}})^{k-2l-1} \right)$$

$$= \int_y \left(\frac{xy\delta_{x^2,y^2}(zf)}{2} \left(\sum_{k\geq 1} \left(\partial_{x^2,y^2} f_{\mathrm{in}} + xy\delta_{x^2,y^2} f_{\mathrm{in}} \right)^k \right. \right.$$

$$\left. \left. - \sum_{k\geq 1} \left(\partial_{x^2,y^2} f_{\mathrm{in}} - xy\delta_{x^2,y^2} f_{\mathrm{in}} \right)^k \right) \right)$$

$$= x^2 \int_y \frac{\delta_{x^2,y^2}(zf) y^2 \delta_{x^2,y^2} f_{\mathrm{in}}}{(1 - \partial_{x^2,y^2} f_{\mathrm{in}})^2 - x^2 y^2 (\delta_{x^2,y^2} f_{\mathrm{in}})^2}$$

where

$$f_{\text{in}} = f_{\text{in}}(\sqrt{z}) = f_{\mathcal{M}_{\text{in}}^{(\text{E})}}(\sqrt{z}, \underline{y}) \tag{7.2.21}$$

which has the form

$$f_{\mathcal{M}_{\text{in}}^{(\text{E})}}(x, \underline{y}) = x^2 f_{\mathcal{M}_{\text{nlE}}}(x, \underline{y}) \tag{7.2.22}$$

from Lemma 7.2.1. Therefore

$$f_{\mathcal{M}_{\text{od}}+\mathcal{M}_B} = x^2 \int_y \frac{\delta_{x^2,y^2}(zf)y^2\delta_{x^2,y^2}(zf)}{(1-\partial_{x^2,y^2}(zf))^2 - x^2y^2(\delta_{x^2,y^2}(zf))^2},$$

by (1.6.9),

$$= \int_y \frac{\delta_{x^2,y^2}(zf)\partial_{x^2,y^2}(z^2 f)}{1 - 2\partial_{x^2,y^2}(zf) - x^2y^2\delta_{x^2,y^2}(zf^2)} \tag{7.2.23}$$

in which the identity

$$x^2y^2\delta^2_{x^2,y^2}(zf) - \partial^2_{x^2,y^2}(zf) = x^2y^2\delta_{x^2,y^2}(zf^2) \tag{7.2.24}$$

has been applied. Moreover, (7.2.24) can be easily derived from the definitions shown in (1.6.7–8) and the identity in (1.6.9).

From the first of (7.2.16), (7.2.18) and (7.2.19) with (1.6.9),

$$f_{\mathcal{M}_{\text{ev}}} = \int_y \Bigg(\partial_{x^2,y^2}(zf) \sum_{k\geq 1} \sum_{l=0}^{\lfloor \frac{k}{2} \rfloor} \binom{k}{2l}$$

$$\times \left(\sum_{\mathcal{M}_{\text{in}}^{(\text{E})}} \underline{y}^{\underline{n}(M)} \sum_{i=0}^{m(M)-1} x^{2i+1} y^{2m(M)-2i-1} \right)^{2l}$$

$$\times \left(\sum_{\mathcal{M}_{\text{in}}^{(\text{E})}} \underline{y}^{\underline{n}(M)} \sum_{i=1}^{m(M)-1} x^{2i} y^{2m(M)-2i} \right)^{k-2l} \Bigg)$$

$$= \int_y \left(\partial_{x^2,y^2}(zf) \sum_{k\geq 1} \sum_{l=0}^{\lfloor \frac{k}{2} \rfloor} \binom{k}{2l} (xy\delta_{x^2,y^2} f_{\text{in}})^{2l} (\partial_{x^2,y^2} f_{\text{in}})^{k-2l} \right)$$

$$= \int_y \frac{\partial_{x^2y^2}(zf)}{2} \left(\sum_{k\geq 1} (xy\delta_{x^2,y^2} f_{\text{in}} + \partial_{x^2y^2} f_{\text{in}})^k \right.$$

$$-\sum_{k\geq 1}(xy\delta_{x^2,y^2}f_{\rm in}-\partial_{x^2y^2}f_{\rm in})^k\Bigg)$$

$$=\int_y\left(\frac{\partial_{x^2y^2}(zf)\big(\partial_{x^2y^2}f_{\rm in}(1-\partial_{x^2y^2}f_{\rm in})+x^2y^2\delta^2_{x^2y^2}f_{\rm in}\big)}{(1-\partial_{x^2y^2}f_{\rm in})^2-x^2y^2\delta^2_{x^2y^2}(zf_{\rm in})}\right)$$

by (7.2.21) and (7.2.24),

$$=\int_y\left(\frac{\partial_{x^2,y^2}(zf)\big(\partial_{x^2,y^2}(zf)+x^2y^2\delta_{x^2,y^2}(zf^2)\big)}{1-2\partial_{x^2,y^2}(zf)-x^2y^2\delta_{x^2,y^2}(zf^2)}\right). \tag{7.2.25}$$

According to (7.2.6), from (7.2.23) and (7.2.25) $f_{\mathcal{M}_A}$ is soon derived.

Then, From (7.2.11) we have

$$f_{\hat{\mathcal{M}}}=xf\int_y(yf_y) \tag{7.2.26}$$

where $f=f(x)$ and $f_y=f(y)$ defined in (7.2.18). From (7.2.13) and (7.2.26),

$$f_{\mathcal{M}_B}=xf\int_y\Bigg(yf_y\sum_{k\geq 1}\sum_{l=0}^{\lfloor\frac{k-1}{2}\rfloor}\binom{k}{2l+1}$$

$$\times\left(\sum_{M\in\mathcal{M}^{(\rm E)}_{\rm in}}y^{\underline{n}(M)}\sum_{i=0}^{m(M)-1}x^{2i+1}y^{2m(M)-2i-1}\right)^{2l+1}$$

$$\times\left(\sum_{M\in\mathcal{M}^{(\rm E)}_{\rm in}}y^{\underline{n}(M)}\sum_{i=1}^{m(M)-1}x^{2i}y^{2m(M)-2i}\right)^{k-2l-1}\Bigg)$$

$$=xf\int_y\Bigg(yf_y\sum_{k\geq 1}\sum_{l=0}^{\lfloor\frac{k-1}{2}\rfloor}\binom{k}{2l+1}(xy\delta_{x^2,y^2}f_{\rm in})^{2l+1}$$

$$\times(\partial_{x^2,y^2}f_{\rm in})^{k-2l-1}\Bigg)$$

$$=x^2f\int_y\frac{y^2f_y\delta_{x^2,y^2}f_{\rm in}}{(1-\partial_{x^2,y^2}f_{\rm in})^2-x^2y^2\delta_{x^2,y^2}f_{\rm in}}$$

by (7.2.21) and (1.6.9),

$$= f\int_y \frac{f_y \partial_{x^2,y^2}(z^2 f)}{(1-\partial_{x^2,y^2}(zf))^2 - x^2y^2\delta_{x^2,y^2}(zf)}$$

by (7.2.24),

$$= f\int_y \frac{f \partial_{x^2,y^2}(z^2 f)}{1-2\partial_{x^2,y^2}(zf) - x^2y^2\delta_{x^2,y^2}(zf^2)} \tag{7.2.27}$$

where in parentheses $f = f_{\sqrt{z}} = f(\sqrt{z})$ as shown in (7.2.18).

Then, from (7.2.23) and (7.2.27), $f_{\mathcal{M}_{\rm od}}$ can be obtained soon even not directly necessary for the functional equation.

Theorem 7.2.1 *The enufunction* $f = f_{\mathcal{M}_{\rm nlE}}(x, \underline{y})$ *defined in* (7.2.1) *and* (7.2.18) *satisfies the following functional equation:*

$$f = 1 + \int_y \frac{\partial_{x^2,y^2}(zf) + x^2y^2\delta_{x^2,y^2}(zf^2)}{1-2\partial_{x^2,y^2}(zf) - x^2y^2\delta_{x^2,y^2}(zf^2)} \tag{7.2.28}$$

where $f = f(\sqrt{z}) = f(\sqrt{z}, \underline{y})$ *under the operations.*

Proof From (7.2.4), (7.2.10), (7.2.15), (7.2.17), (7.2.19), (7.2.23) and (7.2.25), we have

$$\begin{aligned} f = 1 &+ \int_y \partial_{x^2,y^2}(zf) \\ &+ \int_y \frac{\partial_{x^2,y^2}(zf)\big(\partial_{x^2,y^2}(zf) + x^2y^2\delta_{x^2,y^2}(zf^2)\big)}{1-2\partial_{x^2,y^2}(zf) - x^2y^2\delta_{x^2,y^2}(zf^2)} \\ &+ \int_y \frac{\delta_{x^2,y^2}(zf)\partial_{x^2,y^2}(z^2 f)}{1-2\partial_{x^2,y^2}(zf) - x^2y^2\delta_{x^2,y^2}(zf^2)}. \end{aligned}$$

By using the following two identities:

$$\int_y \partial_{x^2y^2}(zf) + \int_y \frac{\partial_{x^2,y^2}(zf)\big(\partial_{x^2,y^2}(zf) + x^2y^2\delta_{x^2,y^2}(zf^2)\big)}{1-2\partial_{x^2,y^2}(zf) - x^2y^2\delta_{x^2,y^2}(zf^2)}$$

$$= \int_y \frac{\partial_{x^2,y^2}(zf)\big(1-\partial_{x^2,y^2}(zf)\big)}{1-2\partial_{x^2,y^2}(zf) - x^2y^2\delta_{x^2,y^2}(zf^2)}$$

and

$$\delta_{x^2,y^2}(zf)\partial_{x^2,y^2}(z^2f) - \partial^2_{x^2,y^2}(zf) = x^2y^2\delta_{x^2,y^2}(zf^2),$$

we find that

$$f = 1 + \int_y \frac{\partial_{x^2,y^2}(zf)\big(1 - \partial_{x^2,y^2}(zf)\big)}{1 - 2\partial_{x^2,y^2}(zf) - x^2y^2\delta_{x^2,y^2}(zf^2)}$$

$$+ \int_y \frac{\delta_{x^2,y^2}(zf)\partial_{x^2,y^2}(z^2f)}{1 - 2\partial_{x^2,y^2}(zf) - x^2y^2\delta_{x^2,y^2}(zf^2)}.$$

Thus, by grouping the last two terms, the theorem is obtained. □

§7.3 General simple maps

A map is said to be *simple* if it has neither loop nor multi-edge. Or in other words, a *simple map* is a loopless map which has no multi-edge. Let $\mathcal{M}_{\text{sim}}$ be the set of all rooted simple planar maps. For a map $M \in \mathcal{M}_{\text{sim}}$, let $m(M)$ be the valency of root-face and $n_i(M)$, the number of nonrooted faces of valency $i, i \geq 1$. Here, one might see that $m(M)$ and $n(M)$ have different meanings from what we met before.

The enufunction of $\mathcal{M}_{\text{sim}}$ with face partition

$$f_{\mathcal{M}_{\text{sim}}}(x, \underline{y}) = \sum_{M \in \mathcal{M}_{\text{sim}}} x^{m(M)} \underline{y}^{\underline{n}(M)} \tag{7.3.1}$$

is considered for finding a functional equation. Let $\mathcal{M}_{\text{sim}}$ be divided into three classes $\mathcal{M}_{\text{sim}_1}$, $\mathcal{M}_{\text{sim}_2}$ and $\mathcal{M}_{\text{sim}_3}$, *i.e.*,

$$\mathcal{M}_{\text{sim}} = \sum_{i=1}^{3} \mathcal{M}_{\text{sim}_i} \tag{7.3.2}$$

where $\mathcal{M}_{\text{sim}_1} = \vartheta$, the vertex map,

$$\mathcal{M}_{\text{sim}_2} = \{M | \forall M \in \mathcal{M}_{\text{sim}}, a = Kr \text{ is a cut-edge}\}.$$

Of course, $\mathcal{M}_{\text{sim}_3}$ consists of those maps in $\mathcal{M}_{\text{sim}}$ with the root-edge on a circuit.

Lemma 7.3.1 *Let* $\mathcal{M}_{\langle \text{sim}\rangle_2} = \{M - a | \forall M \in \mathcal{M}_{\text{sim}_2}\}$, $a = Kr$. *Then*

$$\mathcal{M}_{\langle \text{sim}\rangle_2} = \mathcal{M}_{\text{sim}} \times \mathcal{M}_{\text{sim}} \tag{7.3.3}$$

where $\times$ *represents the Cartesion product of sets.*

Proof For any $M \in \mathcal{M}_{\langle \text{sim}\rangle_2}$, there is a map $M' \in \mathcal{M}_{\text{sim}_2}$ such that $M = M' - a'$ where $a' = Kr'$ is the root-edge of M'. Because a' is a cut-edge, we have $M = M_1 + M_2$ such that both M_1 and M_2 are in $\mathcal{M}_{\text{sim}}$. Thus $M \in \mathcal{M}_{\text{sim}} \times \mathcal{M}_{\text{sim}}$.

Conversely, for any $M_1, M_2 \in \mathcal{M}_{\text{sim}}$ such that

$$M = M_1 + M_2 \in \mathcal{M}_{\text{sim}} \times \mathcal{M}_{\text{sim}},$$

we may always construct a map M' by adding an edge

$$a' = (o_1, o_2) = Kr'$$

as the root-edge where o_1 and o_2 are the root-vertices of M_1 and M_2, respectively. Of course, a' is a cut-edge of M'. By considering $M' \in \mathcal{M}_{\text{sim}}$, we see $M' \in \mathcal{M}_{\text{sim}_2}$. However, $M = M' - a'$. That implies $M \in \mathcal{M}_{\langle \text{sim}\rangle_2}$. □

For a map $M = (\mathcal{X}_{\alpha,\beta}, \mathcal{P})$, let $\Delta_i(M)$ be the map obtained by adding a new edge from the root-vertex to the i-th vertex on the root-face boundary, $i = 0, 1, \cdots$, that is to add an edge $Kr' = (v_r, v_{\mathcal{P}\alpha\beta^i r})$ such that

$$v_{r'} = (r', r, S) \text{ and } v_{\beta r'} = (\alpha\beta r', (\mathcal{P}\alpha\beta)^i r, T_i)$$

are vertices in $\Delta_i(M)$, where S and T_i are such that

$$v_r = (r, S) \text{ and } v_{(\mathcal{P}\alpha\beta)^i r} = ((\mathcal{P}\alpha\beta)^i r, T_i)$$

are in M. The map $\Delta_i(M)$ is called the *i-th increment* while $\nabla_i(M)$ in §7.1 can also be called the *dual* i-th increment of M. Let

$$\tilde{\mathcal{M}} = \sum_{M \in \mathcal{M}_{\text{sim}}} \left\{\Delta_i(M) | \ 0 \leq i \leq m(M)\right\}. \tag{7.3.4}$$

It is easily seen that $\mathcal{M}_{\text{sim}} \subseteq \tilde{\mathcal{M}}$ and that not all maps in $\tilde{\mathcal{M}}$ are simple. Let

$$\mathcal{M}_{\text{L}} = \{M | \forall M \in \tilde{\mathcal{M}}, \text{without multi-edge}$$

$$\text{with only one loop, the root-edge}\};$$

$$\mathcal{M}_{\rm N} = \{M|\forall M \in \tilde{\mathcal{M}}, \text{without loop,}$$

$$\text{with only one multi-edge } a \text{ is in}\}.$$

Lemma 7.3.2 *For* $\mathcal{M}_{\rm sim_3}$ *we have*

$$\mathcal{M}_{\rm sim_3} = \tilde{\mathcal{M}} - \mathcal{M}_{\rm L} - \mathcal{M}_{\rm N}. \tag{7.3.5}$$

Proof For any $M \in \mathcal{M}_{\rm sim_3}$, because there is neither loop nor multi-edge in M, we have $M \notin \mathcal{M}_{\rm L}$ and $M \notin \mathcal{M}_N$. Since the root-edge a of M is not a cut-edge, we may obtain a map $M' \in \mathcal{M}_{\rm sim}$ by deleting a. It is easy to see that there exists an $i, 2 \leq i \leq m(M) - 2$, such that $\Delta_i(M') = M$. That implies $M \in \tilde{\mathcal{M}}$ and hence a member of the set on the right hand side of (7.3.5).

Conversely, for any M in the set on the right, there is a map $M' \in \mathcal{M}_{\rm sim}$ such that $\Delta_i(M') = M$, for an $i, 0 \leq i \leq m(M')$. Because $M \notin \mathcal{M}_{\rm L}$ and $M \notin \mathcal{M}_{\rm N}$, we have $M \in \mathcal{M}_{\rm sim}$. By considering the root-edge a' is not a cut-edge, $M \in \mathcal{M}_{\rm sim_3}$. □

Lemma 7.3.3 *Let* $\mathcal{M}_{\langle {\rm L}\rangle} = \{M - a|\forall M \in \mathcal{M}_{\rm L}\}$ *with* $a = v_r(M)$. *Then*

$$\mathcal{M}_{\langle {\rm L}\rangle} = \mathcal{M}_{\rm sim} \odot \mathcal{M}_{\rm sim} \tag{7.3.6}$$

where $\odot$ *is the* $1v$*-production.*

Proof For any $M \in \mathcal{M}_{\langle {\rm L}\rangle}$, let $M' \in \mathcal{M}_{\rm L}$ such that $M' - a' = M$. Because M' has only one loop which is the root-edge a' without multi-edge, $M = M_1 \dot{+} M_2$ such that $M_1, M_2 \in \mathcal{M}_{\rm sim}$ where M_1 and M_2 are in the inner and outer domains of a' in M'. Thus $M \in \mathcal{M}_{\rm sim} \odot \mathcal{M}_{\rm sim}$.

Conversely, for any $M \in \mathcal{M}_{\rm sim} \odot \mathcal{M}_{\rm sim}$, write $M = M_1 \dot{+} M_2$ such that $M_1, M_2 \in \mathcal{M}_{\rm sim}$. Let M' be the map obtained by adding a loop a' at the common vertex of M_1 and M_2 which are respectively in the inner and the outer domains of a'. Of course, $M' \in \mathcal{M}_{\rm L}$ and $M = M' - a'$. Thus, $M \in \mathcal{M}_{\langle {\rm L}\rangle}$. □

Lemma 7.3.4 *Let* $\mathcal{M}_{\langle {\rm N}\rangle} = \{M - a|\forall M \in \mathcal{M}_{\rm N}\}$ *with* $a = v_r(M)$. *Then*

$$\mathcal{M}_{\langle {\rm N}\rangle} = (\mathcal{M}_{\rm sim} - \vartheta) \oslash (\mathcal{M}_{\rm sim} - \vartheta) \tag{7.3.7}$$

where $\oslash$ *is the* 1e*-production.*

Proof For $M \in \mathcal{M}_{\langle \mathrm{N} \rangle}$ such that $M = M' - a'$. Write

$$v'_r = (r', S_1, l, S_2),$$

and

$$v_{\alpha\beta r'} = (\alpha\beta r', T_2, \alpha\beta l, T_1).$$

Then, $M = M_1 +|M_2$ such that M_1 and M_2 are extended from the root-edge $K\alpha\beta l$ and Kl, with

$$v_{\alpha\beta l} = (\alpha\beta l, T_1) \text{ and } (\alpha\beta l, T_2);$$

$$v_l = (l, S_1) \text{ and } (l, S_2),$$

respectively. Of course, any of M_1 and M_2 is not allowed to be the vertex map ϑ. Moreover, $M_1, M_2 \in \mathcal{M}_{\mathrm{sim}}$. That implies M is a member of the set on the right hand side of (7.3.8).

Conversely, for M in the set on the right, let $M = M_1 +|M_2$ where r_1 and r_2 are the roots of M_1 and M_2, respectively. Assume

$$v_{r_i} = (r_i, S_i) \text{ and } v_{\alpha\beta r_i} = (\alpha\beta r_i, T_i)$$

for $i = 1, 2$. Because $M_i \in \mathcal{M}_{\mathrm{sim}} - \vartheta$, the root-edges $a_i = e_{r_i}, i = 1, 2$, are all links with $r_1 = \alpha\beta r_2$. Then, we may construct a map $M' \in \mathcal{M}_{\mathrm{N}}$ by adding a new edge $a' = Kr'$ such that

$$v_{r'} = (r', T_1, r_2, S_2)$$

and

$$v_{\alpha\beta r'} = (\alpha\beta r', T_2, r_1, S_1)$$

with $Kr_1 = Kr_2$. Since $M = M' - a'$, we have $M \in \mathcal{M}_{\langle N \rangle}$. □

Now, we are going to evaluate the contributions of $\mathcal{M}_{\mathrm{sim}_1}$, $\mathcal{M}_{\mathrm{sim}_2}$, and $\mathcal{M}_{\mathrm{sim}_3}$ to the enufunction $f_{\mathcal{M}_{\mathrm{sim}}}(x, \underline{y})$.

For $\mathcal{M}_{\mathrm{sim}_1}$, because ϑ is without edge, all $m(\vartheta), n_i(\vartheta) = 0, i \geq 1$. That is

$$f_{\mathcal{M}_{\mathrm{sim}_1}} = 1. \tag{7.3.8}$$

From Lemma 7.3.1,

$$f_{\mathcal{M}_{\mathrm{sim}_2}} = x^2 f^2_{\mathcal{M}_{\mathrm{sim}}}(x, \underline{y}). \tag{7.3.9}$$

In order to determine $f_{\mathcal{M}_{\text{sim}_3}}$, we have to consider what $f_{\tilde{\mathcal{M}}}, f_{\mathcal{M}_{\text{L}}}$ and $f_{\mathcal{M}_{\text{N}}}$ look like. According to (7.3.4),

$$f_{\tilde{\mathcal{M}}} = \int_y \sum_{M \in \mathcal{M}_{\text{sim}}} \underline{y}^{n(M)} \sum_{i=0}^{m(M)} x^{i+1} y^{m(M)-i+1}$$

$$= \int_y \sum_{M \in \mathcal{M}_{\text{sim}}} xy \frac{x^{m(M)+1} - y^{m(M)+1}}{x-y} \underline{y}^{n(M)}$$

$$= \int_y xy\delta_{x,y}(zf(z)) \tag{7.3.10}$$

where

$$f = f_{\mathcal{M}_{\text{sim}}}(x) = f_{\mathcal{M}_{\text{sim}}}(x, \underline{y}). \tag{7.3.11}$$

From Lemma 7.2.3,

$$f_{\mathcal{M}_{\text{L}}} = xf(x) \int_y \Big(yf(y)\Big). \tag{7.3.12}$$

From Lemma 7.2.4,

$$f_{\mathcal{M}_{\text{N}}} = (f-1) \int_y \Big(f(y) - 1\Big)$$

$$= (f-1) \Big(\int_y f(y) - 1\Big). \tag{7.3.13}$$

Then based on (7.2.10), (7.3.12) and (7.3.13), from Lemma 7.3.2 we have

$$f_{\mathcal{M}_{\text{sim}_3}} = x \int_y y\delta_{x,y}(zf(z)) - xf(x) \int_y \Big(yf(y)\Big)$$

$$-(f-1)\Big(\int_y (f(y) - 1\Big)$$

$$= f - 1 + \int_y \Big(xy\delta_{x,y}\Big(zf(z)\Big)$$

$$+\Big(1 - (1+xy)f\Big)f(y)\Big). \tag{7.3.14}$$

Theorem 7.3.1 *The enufunction* $f = f_{\mathcal{M}_{\text{sim}}}(x, \underline{y})$ *defined in* (7.3.1) *satisfies the following functional equation:*

$$\Big(\int_y \Big((1+xy)f(y)\Big) - x^2 f\Big) f$$

$$= \int_y \Big(f(y) + xy\delta_{x,y}\big(zf(z)\big)\Big). \tag{7.3.15}$$

Proof From (7.3.8), (7.3.9), (7.3.14) and (7.3.2),

$$\begin{aligned} f &= x^2 f^2 + f + \int_y \Big(xy\delta_{x,y}\big(zf(z)\big)\Big) \\ &\quad + \int_y \Big(\big(1-(1+xy)f\big) f(y)\Big) \\ &= x^2 f^2 + f + \int_y \Big(f(y) + xy\delta_{x,y}\big(zf(z)\big)\Big) \\ &\quad - f\int_y \Big((1+xy) f(y)\Big), \end{aligned}$$

i.e.,

$$f\int_y \Big((1+xy) f(y)\Big) - x^2 f^2 = \int_y \Big(f(y) + xy\delta_{x,y}\big(zf(z)\big)\Big).$$

Therefore, (7.3.15) is obtained. □

For $M \in \mathcal{M}_{\text{sim}}$, if $m(M)$ and $n(M)$ represent the valency of the root-face and the number of nonrooted faces in M respectively, then for the enufunction

$$h(x,y) = \sum_{M \in \mathcal{M}_{\text{sim}}} x^{m(M)} y^{n(M)}, \tag{7.3.16}$$

we may by the decomposition above find the contributions of $\mathcal{M}_{\text{sim}_1}$, $\mathcal{M}_{\text{sim}_2}$ and $\mathcal{M}_{\text{sim}_3}$ to $h = h(x,y)$ as

$$h_{\mathcal{M}_{\text{sim}_1}} = 1,\ h_{\mathcal{M}_{\text{sim}_2}} = x^2 h^2. \tag{7.3.17}$$

By noticing that

$$h_{\tilde{\mathcal{M}}} = \frac{xy}{1-x}(h^* - xh)$$

where $h^* = h(1,y)$,

$$h_{\mathcal{M}_{\text{L}}} = xyh^* h,$$

$$h_{\mathcal{M}_{\text{N}}} = y(h-1)(h^*-1).$$

Hence,

$$\begin{aligned} h_{\mathcal{M}_{\text{sim}_3}} &= h_{\tilde{\mathcal{M}}} - h_{\mathcal{M}_{\text{L}}} - h_{\mathcal{M}_{\text{N}}} \\ &= \left(\frac{xyh^*}{1-x} + y(h^*-1) \right) \\ &\quad - \left(\frac{x^2y}{1-x} + xyh^* + yh^* - y \right) h. \end{aligned} \tag{7.3.18}$$

Theorem 7.3.2 *The function h defined in (7.3.16) satisfies the equation as follows:*

$$\begin{aligned} x^2h^2 &- \left(1 + y(h^*-1) + xyh^* + \frac{x^2y}{1-x}\right)h \\ &+ \frac{xy}{1-x}h^* + yh^* - y + 1 = 0. \end{aligned} \tag{7.3.19}$$

Proof From (7.3.17–18) and (7.3.2),

$$\begin{aligned} h = 1 + x^2h^2 &- \left(\frac{x^2y}{1-x} + xyh^* + yh^* - y \right) h \\ &+ \left(\frac{xyh^*}{1-x} + y(h^*-1) \right), \end{aligned}$$

i.e.,

$$\begin{aligned} x^2h^2 &- \left(\frac{x^2y}{1-x} + xyh^* + yh^* - y + 1 \right) h \\ &+ \frac{xy}{1-x}h^* + y(h^*-1) + 1 = 0 \end{aligned}$$

which is (7.3.19). □

In fact, h is in the case of $f_{\mathcal{M}_{\text{sim}}}$ when $y_i = y, i \geq 1$. On the other hand, let

$$\begin{aligned} \hat{g}_{\mathcal{M}_{\text{sim}}}(x,y) &= \prod_{i\geq 1} \int_y^{-1} f_{\mathcal{M}_{\text{sim}}}(x,\underline{y}) \\ &= \sum_{M\in\mathcal{M}_{\text{sim}}} x^{m(M)}y^{l(M)} \end{aligned}$$

where

$$l(M) = \sum_{i\geq 1} in_i(M) = 2k(M) - m(M)$$

for $k(M)$ being the size of M. Let

$$u = xy^{-1},\ w = y^2.$$

Then

$$\begin{aligned} g_{\mathcal{M}_{\rm sim}}(u,w) &= \hat{g}(u\sqrt{w}, \sqrt{w}) \\ &= \sum_{M\in\mathcal{M}_{\rm sim}} (u\sqrt{w})^{m(M)}(\sqrt{w})^{l(M)} \\ &= \sum_{M\in\mathcal{M}_{\rm sim}} u^{m(M)} w^{\frac{m(M)+l(M)}{2}} \\ &= \sum_{M\in\mathcal{M}_{\rm sim}} u^{m(M)} w^{k(M)} \end{aligned} \tag{7.3.20}$$

which is just the enufunction of $\mathcal{M}_{\rm sim}$ with the valency of the root-face and the size as the two parameters.

By the decomposition lemmas above, we are also allowed to determine the equation satisfied by $g = g_{\mathcal{M}_{\rm sim}}(x,y)$. Moreover, From Lemmas 7.3.1, 7.3.3–4 and (7.3.4), the respective contributions of $\mathcal{M}_{{\rm sim}_2}$, $\mathcal{M}_{\rm L}$, $\mathcal{M}_{\rm N}$ and $\tilde{\mathcal{M}}$ are derived to be

$$g_{\mathcal{M}_{{\rm sim}_2}} = x^2yg^2; \tag{7.3.21}$$

$$g_{\mathcal{M}_{\rm L}} = xyg^*g; \tag{7.3.22}$$

$$g_{\mathcal{M}_{\rm N}} = (g^* - 1)(g - 1) \tag{7.3.23}$$

where $g^* = g(1,y)$ and

$$g_{\tilde{\mathcal{M}}} = xy\frac{tg^* - xy}{1-x}. \tag{7.3.24}$$

Further, based on (7.3.22–24), Lemma 7.3.2 allows us to find

$$\begin{aligned} g_{\mathcal{M}_{{\rm sim}_3}} &= xy\Big(\frac{g^* - xg}{1-x}\Big) - xyg^*g - (g^* - 1)(g - 1) \\ &= \Big(1 - \frac{x^2y}{1-x} - xyg^* - g^*\Big)g \end{aligned}$$

$$+\frac{xyg^*}{1-x}+g^*-1. \tag{7.3.25}$$

Theorem 7.3.3 *The function $g = g(x, y)$ defined in (7.3.20) satisfies the equation as*

$$x^2yg^2-\Big(1-\frac{x^2y}{1-x}+xyg^*-g^*\Big)g$$

$$+\frac{xyg^*}{1-x}+g^*=0. \tag{7.3.26}$$

Proof Since it is easily checked that $g_{\mathcal{M}_{\text{sim}_1}} = 1$ as well, from (7.3.2), (7.3.21) and (7.3.25) the theorem is soon deduced. □

From (7.3.26), we may find the following parametric expressions in a usual way:

$$y=\frac{(\xi-1)(2-\xi)}{\xi^2};\ g^*=\xi^*(2-\xi). \tag{7.3.27}$$

Because it is easily checked that

$$H(y)=\frac{\mathrm{d}g^*}{\mathrm{d}y}=\xi^4,$$

we are allowed to find $\partial_y^{n+1}H(y)$ by employing Corollary 1.5.2, and then the number of distinct rooted simple planar maps of size n as

$$N_{\mathcal{M}_{\text{sim}}}(n)=(n+1)\partial_y^{n+1}H(y)$$

$$=\sum_{i=0}^{n-2}\frac{4(2n+1)!(2n-i-4)!}{i!(n-i-2)!(2n-i+1)!} \tag{7.3.28}$$

for $n \geq 2$.

§7.4 Simple bipartite maps

Only rooted simple bipartite planar maps are considered in the this section for the brevity. Let $\mathcal{B}_{\text{sim}}$ be the set of all those considered with the convention that the vertex map ϑ is in $\mathcal{B}_{\text{sim}}$ for convenience. Then $\mathcal{B}_{\text{sim}}$ is partitioned as

$$\mathcal{B}_{\text{sim}} = \mathcal{B}_0 + \mathcal{B}_1 + \mathcal{B}_2 \tag{7.4.1}$$

where $\mathcal{B}_0$ consists of the single map ϑ, $\mathcal{B}_1 = \{M|\forall M \in \mathcal{B}_{\text{sim}}, M \neq \vartheta, e_r(M)$ is separable $\}$ and hence $\mathcal{B}_2 = \{M|\forall M \in \mathcal{B}_{\text{sim}}, M \neq \vartheta, e_r(M)$ is on a circuit $\}$.

For any $M \in \mathcal{B}_{\text{sim}}$, let $2m(M)$ be the valency of root-face and $\underline{n}(M) = (n_4(M), n_6(M), \cdots)$ where $n_{2j}(M)$ is the number of nonrooted faces of valency $2j$, $j \geq 2$, from the bipartiteness. The purpose here is to determine the enufunction of $\mathcal{B}_{\text{sim}}$ with face partition

$$f_{\mathcal{B}_{\text{sim}}}(x, \underline{y}) = \sum_{M \in \mathcal{B}_{\text{sim}}} x^{2m(M)} \underline{y}^{\underline{n}(M)} \tag{7.4.2}$$

where $\underline{y} = (y_4, y_6, \cdots)$ is a vector of infinite dimension.

Because ϑ has neither edge nor nonrooted face by considering $2m(\vartheta) = 0$, the contribution of $\mathcal{B}_0$ to $f_{\mathcal{B}_{\text{sim}}} = f_{\mathcal{B}_{\text{sim}}}(x, \underline{y})$ is given by

$$f_0 = f_{\mathcal{B}_0}(x, \underline{y}) = 1. \tag{7.4.3}$$

Lemma 7.4.1 *Let $\mathcal{B}_{\langle 1 \rangle} = \{M - a|\forall M \in \mathcal{B}_1\}$ with $a = e_r(M)$. Then we have*

$$\mathcal{B}_{\langle 1 \rangle} = \mathcal{B}_{\text{sim}} \times \mathcal{B}_{\text{sim}}. \tag{7.4.4}$$

Proof For any $M \in \mathcal{B}_{\langle 1 \rangle}$, let $M' \in \mathcal{B}_1$ be the map such that $M = M' - a', a' = e_r(M')$. Because a' is a cut-edge, we have $M = M_1 + M_2 = M_1 \cup M_2$ provided no point is in common. From the heritage of the simplicity and the bipartiteness of M, both M_1 and M_2 are simple and bipartite. By the planarity of M, M_1 and M_2 are in $\mathcal{B}_{\text{sim}}$. This implies that M is a member of the set on the right hand side of (7.4.4).

Conversely, for any $M \in \mathcal{B}_{\text{sim}} \times \mathcal{B}_{\text{sim}}$, since $M = M_1 + M_2$ with $M_1, M_2 \in \mathcal{B}_{\text{sim}}$, we may uniquely construct the map M' by adding an edge $a' = Kr'$ which joins the two root-vertices of M_1 and M_2 such that $M = M' - a'$. However, it is easily checked that $M' \in \mathcal{B}_1$. This implies that $M \in \mathcal{B}_{\langle 1 \rangle}$ as well. □

From Lemma 7.4.1, the contribution of $\mathcal{B}_1$ to $f_{\mathcal{B}_{\text{sim}}}$ is as

$$\begin{aligned} f_1 = f_{\mathcal{B}_1}(x, \underline{y}) &= x^2 \sum_{M \in \mathcal{B}_{(1)}} x^{2m(M)} \underline{y}^{\underline{n}(M)} \\ &= x^2 \left(\sum_{\mathcal{B}_{\text{sim}}} x^{2m(M)} \underline{y}^{\underline{n}(M)} \right)^2 \\ &= x^2 f^2 \end{aligned} \tag{7.4.5}$$

where $f = f_{\mathcal{B}_{\text{sim}}}(x, \underline{y})$.

A map M which is simple except for the root-edge in a 2-multi-edge is said to be *near simple*. For $M = (\mathcal{X}, \mathcal{P}) \in \mathcal{B}_{\text{sim}} - \vartheta$, write $\Delta_{2i+1}(M), i \geq 0$ as the $(2i+1)$-st increment of M, *i.e.*, that obtained by adding an edge Kr' between v_r and $v_{(\mathcal{P}\alpha\beta)^{2i+1}r}$ such that

$$v_{r'} = (r', r, S) \text{ and } v_{\beta r'} = (\alpha\beta r', (\mathcal{P}\alpha\beta)^{2i+1} r, T)$$

where

$$v_r = (r, S) \text{ and } v_{(\mathcal{P}\alpha\beta)^{2i+1}r} = ((\mathcal{P}\alpha\beta)^{2i+1} r, T)$$

are the root-vertex and the vertex incident with $(\mathcal{P}\alpha\beta)^{2i+1}r$ of M respectively. From the bipartiteness of M, it is easily checked that $\Delta_{2i+1}(M), i \geq 0$, are all bipartite since all the faces have even valency.

Lemma 7.4.2 *Let $\mathcal{B}_{\text{N}}$ be the set of all rooted near simple bipartite planar maps and*

$$\tilde{\mathcal{B}} = \sum_{M \in \mathcal{B}_{\text{sim}} - \vartheta} \left\{ \Delta_{2i+1}(M) \mid i = 0, 1, \cdots, m(M) - 1 \right\}. \tag{7.4.6}$$

Then we have

$$\mathcal{B}_2 = \tilde{\mathcal{B}} - \mathcal{B}_{\text{N}}. \tag{7.4.7}$$

Proof For any $M \in \mathcal{B}_2$, since the root-edge $a = e_r(M)$ is not separable, $M' = M - a \in \mathcal{B}_{\text{sim}} - \vartheta$. By the bipartiteness, there exists an integer $i \geq 0$ such that $v_{(\mathcal{P}\alpha\beta)^{2i+1}r'} = v_{\beta r}$ where $r' = \mathcal{P}r$ in M. That implies $M = \Delta_{2i+1}(M')$ and hence M is a member of $\tilde{\mathcal{B}} - \mathcal{B}_{\text{N}}$ since M is simple.

Conversely, for any $M \in \tilde{\mathcal{B}} - \mathcal{B}_{\mathrm{N}}$, since there is a map $M' \in \mathcal{B}_{\mathrm{sim}} - \vartheta$ such that $\Delta_{2i+1}(M') = M$ for some $0 \leq i \leq m(M') - 1$, from the bipartiteness and planarity of M' we see that M is bipartite and planar as well. In view of the simplicity of M, we have $M \in \mathcal{B}_2$. □

For $M = (\mathcal{X}_{\alpha\beta}, \mathcal{P}) \in \mathcal{B}_{\mathrm{N}}$, we may write its root-vertex and the other end of the root-edge as

$$v_r = (r, R, s, S) \text{ and } v_{\beta r} = (\alpha\beta r, L, \alpha\beta s, T), \tag{7.4.8}$$

respectively, where Kr and Ks consist of the multi-edge. Let M_1 and M_2 be the submaps expanded by (r, R) with $(\alpha\beta r, T)$ and (s, S) with $(\alpha\beta s, L)$ as the root-vertex with the other end of root-edge respectively. Of course, r and s are the respective roots of M_1 and M_2. Now we denote

$$M = M_1 +\!: M_2 \tag{7.4.9}$$

where $+:$ is called *2v-addition.* For two sets $\mathcal{M}_1$ and $\mathcal{M}_2$ of maps write

$$\mathcal{M}_1 \times\!: \mathcal{M}_2 = \{M_1 +\!: M_2 | \forall M_1 \in \mathcal{M}_1, \forall M_2 \in \mathcal{M}_2\} \tag{7.4.10}$$

where $\times:$ is called the *2v-production.*

Lemma 7.4.3 *For $\mathcal{B}_{\mathrm{N}}$ we have*

$$\mathcal{B}_{\mathrm{N}} = (\mathcal{B}_{\mathrm{sim}} - \vartheta) \times\!: (\mathcal{B}_{\mathrm{sim}} - \vartheta) \tag{7.4.11}$$

where $\times:$ is the 2v-production.

Proof Because any map $M \in \mathcal{B}_{\mathrm{N}}$ has the form (7.4.9), we soon see that M is a member of the set on the right hand side of (7.4.11).

Conversely, for $M \in (\mathcal{B}_{\mathrm{sim}} - \vartheta) \times\!: (\mathcal{B}_{\mathrm{sim}} - \vartheta)$, let $M = M_1 +\!: M_2, M_1, M_2 \in \mathcal{B}_{\mathrm{sim}} - \vartheta$. From the simplicity of M_1 and M_2, M has only one 2-multi-edge the root-edge is in. Further, by the bipartiteness and the planarity of M_1 and M_2 we have $M \in \mathcal{B}_{\mathrm{N}}$. □

According to Lemma 7.4.3, we find that

$$f_{\mathrm{N}} = \sum_{M \in \mathcal{B}_{\mathrm{N}}} x^{2m(M)} \underline{y}^{\underline{n}(M)}$$

$$= \int_y \left(\sum_{M_1 \in \mathcal{B}_{\text{sim}} - \vartheta} y x^{2m(M_1)} \underline{y}^{\underline{n}(M_1} \right)$$

$$\times \left(\sum_{M_2 \in \mathcal{B}_{\text{sim}} - \vartheta} x^{2m(M_2)} \underline{y}^{\underline{n}(M_2)} \right)$$

$$= (f-1) \int_y \left(f_y - 1 \right) \tag{7.4.12}$$

where $f = f_{\mathcal{B}_{\text{sim}}}(x, \underline{y})$, $f_y = f_{\mathcal{B}_{\text{sim}}}(y, \underline{y})$.

By (7.4.6), we obtain that

$$\tilde{f} = \sum_{M \in \tilde{\mathcal{B}}} x^{2m(M)} \underline{y}^{\underline{n}(M)}$$

$$= \int_y xy \sum_{M \in \mathcal{B}_{\text{sim}} - \vartheta} \left(\sum_{i=0}^{m(M)-1} y^{2i+1} x^{2m(M)-2i-1} \right) \underline{y}^{\underline{n}(M)}$$

$$= \int_y \sum_{M \in \mathcal{B}_{\text{sim}} - \vartheta} \left(\frac{y^2 x^{2m(M)} - y^{2m(M)+2}}{x^2 - y^2} \right) \underline{y}^{\underline{n}(M)}$$

$$= \int_y y^2 \delta_{x^2, y^2} f_{\sqrt{z}} \tag{7.4.13}$$

where δ_{x^2,y^2} is the (x^2, y^2)-difference defined by (1.6.7).

From Lemma 7.4.2, the contribution of $\mathcal{B}_2$ to $f = f_{\mathcal{B}_{\text{sim}}}$ is as

$$f_2 = \sum_{M \in \mathcal{B}_2} x^{2m(M)} \underline{y}^{\underline{n}(M)}$$

$$= \tilde{f} - f_{\text{N}}$$

by (7.4.12–13),

$$= (1-f) \int_y \left(f_y - 1 \right) + \int_y y^2 \delta_{x^2,y^2} f_{\sqrt{z}}$$

$$= f - 1 + \int_y \left(y^2 \delta_{x^2,y^2} f_{\sqrt{z}} \right) + (1-f) \int_y f_y. \tag{7.4.14}$$

Theorem 7.4.1 *The function $f = f_{\mathcal{B}_{\text{sim}}}(x, y)$ defined by (7.4.2) with face partition of $\mathcal{B}_{\text{sim}}$ satisfies the functional equation*

$$x^2 f^2 + (f-1) \int_y f_y = \int_y \left(y^2 \delta_{x^2,y^2} f_{\sqrt{z}} \right). \tag{7.4.15}$$

Proof By (7.4.1) we have

$$f = f_0 + f_1 + f_2$$

by (7.4..3), (7.4.5) and (7.4.14),

$$= 1 + x^2 f^2 + f - 1$$

$$+ \int_y \left(y^2 \delta_{x^3 y^2} f_{\sqrt{z}} \right) + (1 - f) \int_y f_y.$$

With some rearrangement, (7.4.15) is obtained. □

In what follows we discuss the case of y^{2i} instead of y_{2i}, $i \geq 1$, with the substitution $z = xy$. In fact, the enufunction becomes

$$g_{\mathcal{B}_{\text{sim}}}(x, z) = \sum_{M \in \mathcal{B}_{\text{sim}}} x^{2m(M)} z^{n(M)} \tag{7.4.16}$$

where $n(M)$ is the size of M.

Now we evaluate the contributions g_0, g_1 and g_2 of $\mathcal{B}_0, \mathcal{B}_1$ and $\mathcal{B}_2$ to $g = g_{\mathcal{B}_{\text{sim}}}(x, z)$, respectively.

First, it is still easy to see that

$$g_0 = g_{\mathcal{B}_0}(x, z) = 1. \tag{7.4.17}$$

By Lemma 7.4.1 we have

$$g_1 = x^2 z g^2. \tag{7.4.18}$$

Because of Lemma 7.4.3 and Lemma 7.4.2 we find that

$$g_{\text{N}} = g_{\mathcal{B}_{\text{N}}}(x, z) = \sum_{M \in \mathcal{B}_{\text{N}}} x^{2m(M)} z^{n(M)}$$

$$= (g^* - 1)(g - 1) \tag{7.4.19}$$

where $g^* = g_{\mathcal{B}_{\text{sim}}}(1, z)$ and

$$\tilde{g} = g_{\tilde{\mathcal{B}}}(x, z) = \sum_{M \in \tilde{\mathcal{B}}} x^{2m(M)} z^{n(M)}$$

$$= xz \sum_{M \in \mathcal{B}_{\rm sim}} \left(\sum_{i=0}^{m(M)-1} x^{2i+1} \right) z^{n(M)}$$

$$= x^2 z \sum_{M \in \mathcal{B}_{\rm sim}} \frac{1 - x^{2m(M)}}{1 - x^2} z^{n(M)}$$

$$= \frac{x^2 z}{1 - x^2}(g^* - g) \tag{7.4.20}$$

where g^* is given as the same as that in (7.4.19).

Theorem 7.4.2 *The enufunction g of $\mathcal{B}_{\rm sim}$ with the valency of root-face and the size as parameters satisfies the following equation:*

$$x^2 z g^2 - \left(\frac{x^2 z}{1 - x^2} + g^* \right) g + \left(\frac{x^2 z}{1 - x^2} + 1 \right) g^* = 0 \tag{7.4.21}$$

where $g^* = g(1, z)$.

Proof From (7.4.1), we have

$$g = g_0 + g_1 + g_2. \tag{7.4.22}$$

By (7.4.19–20), from Lemma 7.4.2,

$$g_2 = \frac{x^2 z}{1 - x^2}(g^* - 1)(g - 1)$$

$$-g(G^* - 1)(g - 1). \tag{7.4.23}$$

In virtue of (7.4.17), (7.4.18) and (7.4.22), we may soon find (7.4.21) from (7.4.22) within rearrangement. □

The discriminant of (7.4.21) is

$$D(x; g) = (x^2 z + (1 - x^2) g^*)^2 - 4x^2 z (1 - x^2)$$

$$\times (x^2 z + 1 - x^2) g^*. \tag{7.4.24}$$

If $x^2 = \xi$ is treated as power series of z, then from

$$D(\xi; g^*, z) = 0;$$

$$\frac{\partial}{\partial \xi} D(\xi; g^*, z) = 0, \tag{7.2.25}$$

by eliminating the term of ${g^*}^2$ we may find that

$$g^* = \frac{\xi z}{(1-\xi)(1+2\xi(2-\xi)(z-1))}. \tag{7.4.26}$$

By substituting (7.4.26) into the first equation of (7.4.25) we may further find that

$$z = -\frac{(1-\xi)(1+\xi-\xi^2)}{\xi^2(2-\xi)}. \tag{7.4.27}$$

By substituting (7.4.27) into (7.4.26), we have

$$g^* = \frac{1+\xi-\xi^2}{(2-\xi)^2}. \tag{7.4.28}$$

From (7.4.27) we see

$$\xi = 1 + \frac{\xi^2(2-\xi)}{1+\xi-\xi^2} z \tag{7.4.29}$$

and hence based on (7.4.29) and (7.4.28), Corollary 1.5.2 is allowed to extract an explicit form of g^* in a power series of z at $\xi = 1$. However, it does not look straightforward to reach a simpler formula, since the denominator in (7.4.29) involves three terms.

§7.5 Notes

7.5.1 Loopless maps have been discussed by some authors [Liu15, 17], [War3] *et al.* in which one can find the equation (7.1.21). The functional equation (7.1.21), a much simpler form of h, can also be found in [BeW3].

7.5.2 No direct way to solve the functional equation (7.1.19) has been in sight for extracting an explicit solution up to now. However, several applications of the equation were indicated in [Liu53].

7.5.3 The equation (7.2.28) firstly appearing in [Liu55] looks complicated. Is there a more simpler equation for the vertex partition enufunction defined by (7.2.1)? What would happen for the corresponding face partition enufunction?

7.5.4 Based on the decomposition for the equation (7.2.28), equations for enufunctions with a few parameters can also be found. One of them is known to be a cubic equation for which a solution has been obtained in [Liu52]. Is there a quadratic equation for loopless Eulerian maps.

7.5.5 The first result for simple planar maps was in [Liu17] and then [Liu16] and [Liu29]. The face partition equation shown by (7.3.15) can be seen in [Liu34]. About the vertex partition, it does not seem to be suitable for finding an equation by using the decomposition presented here.

7.5.6 Although complication occurs in (7.4.28–29) for finding g^* as a power series of z, the parametric expressions can be easily transformed into those for finding a simple formula of the corresponding enufunction of rooted general bipartite planar maps by showing the relationship between the two kinds of maps. More about bipartite planar maps can be seen in [Liu41, Liu61].

7.5.7 On simple Eulerian maps, any explicit formula has not yet been found for the enufunctions.

Chapter 8

General Maps

§8.1 General planar maps

A map is said to be *general* if both loops and multi-edges are allowed. Of course, the vertex map ϑ is also treated as degenerate. Let $\mathcal{M}_{\text{gep}}$ be the set of all rooted general planar maps. For any $M \in \mathcal{M}_{\text{gep}}$, let $a = e_r(M)$ be the root-edge. Then, $\mathcal{M}_{\text{gep}}$ can be divided into three classes: $\mathcal{M}_{\text{gep}_0}$, $\mathcal{M}_{\text{gep}_1}$, and $\mathcal{M}_{\text{gep}_2}$, *i.e.*,

$$\mathcal{M}_{\text{gep}} = \mathcal{M}_{\text{gep}_0} + \mathcal{M}_{\text{gep}_1} + \mathcal{M}_{\text{gep}_2} \tag{8.1.1}$$

such that $\mathcal{M}_{\text{gep}_0}$ consists of a single map ϑ,

$$\mathcal{M}_{\text{gep}_1} = \{M|\forall M \in \mathcal{M}_{\text{gep}}, a \text{ is a loop}\}.$$

Of course,

$$\mathcal{M}_{\text{gep}_2} = \{M|\forall M \in \mathcal{M}_{\text{gep}}, a \text{ is a link}\}$$

in its own right.

Lemma 8.1.1 *Let* $\mathcal{M}_{\langle\text{gep}\rangle_1} = \{M - a|\forall M \in \mathcal{M}_{\text{gep}_1}\}$. *Then, we have*

$$\mathcal{M}_{\langle\text{gep}\rangle_1} = \mathcal{M}_{\text{gep}} \odot \mathcal{M}_{\text{gep}} \tag{8.1.2}$$

where $\odot$ *is the* 1*-production as defined in* §2.1.

Proof For a map $M \in \mathcal{M}_{\langle\text{gep}\rangle_1}$, because there is a map $\tilde{M} \in \mathcal{M}_{\langle\text{gep}\rangle_1}$, such that $M = \tilde{M} - \tilde{a}, \tilde{a} = e_r(\tilde{M})$, the root-edge of $\tilde{M}$, by considering the root-edge $\tilde{a}$ as a loop we see that $M = M_1 \dot{+} M_2$, provided $M_1 \cap M_2 = o$, the common root-vertex of M and $\tilde{M}$. Since

M_1 and M_2 are allowed to be any maps in $\mathcal{M}_{\rm gep}$ including the vertex map ϑ, this implies that $M \in \mathcal{M}_{\rm gep} \odot \mathcal{M}_{\rm gep}$.

Conversely, for any $M \in \mathcal{M}_{\rm gep} \odot \mathcal{M}_{\rm gep}$, since $M = M_1 \dot{+} M_2$, M_1, $M_2 \in \mathcal{M}_{\rm gep}$, we may always construct a map $\tilde{M}$ by adding a loop $\tilde{a}$ at the common vertex of M_1 and M_2 as the root-edge of $\tilde{M}$ such that M_1 and M_2 are in different domains of the loop. Of course, $\tilde{M}$ is a general map. Because the root-edge of $\tilde{M}$ is a loop added, $\tilde{M} \in \mathcal{M}_{\rm gep_1}$. However, it is easily seen that $M = \tilde{M} - \tilde{a}$. Therefore, $M \in \mathcal{M}_{\langle \rm gep \rangle_1}$.

In consequence, the lemma is proved. □

For $\mathcal{M}_{\rm gep_2}$, because the root-edges are all links we consider the set $\mathcal{M}_{(\rm gep)_2} = \{M \bullet a | \forall M \in \mathcal{M}_{\rm gep_2}\}$, $a = e_r(M)$, the root-edge as usual. The smallest map in $\mathcal{M}_{\rm gep_2}$ is the link map $L = (Kr, (r)(\alpha\beta r))$ and it is seen that $L \bullet a = \vartheta$. Thus, $\vartheta \in \mathcal{M}_{(\rm gep)_2}$. For any $M \in \mathcal{M}_{\rm gep_2}$, because the root-edge of M is not a loop we know that $M \bullet a \in \mathcal{M}_{\rm gep}$. Conversely, for any $M \in \mathcal{M}_{\rm gep}$ we may always construct a map $\tilde{M} \in \mathcal{M}_{\rm gep_2}$ by splitting the root-vertex of $\tilde{M}$ into two vertices with a new edge $\tilde{a}$ as the root-edge connecting them. This implies that $\tilde{M} \bullet \tilde{a} = M \in \mathcal{M}_{(\rm gep)_2}$. Therefore we have

$$\mathcal{M}_{(\rm gep)_2} = \mathcal{M}_{\rm gep}. \tag{8.1.3}$$

Lemma 8.1.2 *For $\mathcal{M}_{\rm gep_2}$ we have*

$$\mathcal{M}_{\rm gep_2} = \sum_{M \in \mathcal{M}_{\rm gep}} \left\{ \nabla_i M |\ 0 \le i \le m(M) \right\} \tag{8.1.4}$$

where $m(M)$ is the valency of the root-vertex of M and ∇_i is the operator defined in §7.1.

Proof For any $M \in \mathcal{M}_{\rm gep_2}$, because the root-edge a is a link, we may assume $a = (o_1, o_2)$ such that

$$o_1 = (r, S) \text{ and } o_2 = (\alpha\beta r, T).$$

Let $\tilde{M}$ be the map obtained by contracting the root-edge a into a vertex $\tilde{o} = (T, S)$ as the root-vertex of $\tilde{M}$. It is easily checked that $\tilde{M} \in \mathcal{M}_{\rm gep}$ from (8.1.3) and that

$$M = \nabla_{|S|} \tilde{M},\ 0 \le |S| \le m(\tilde{M})$$

where $m(\tilde{M}) = |S| + |T|$, and $|Z|, Z = S$ or T, stands for the cardinality of Z. That implies M is a member of the set on the right hand side of (8.1.4).

Conversely, for any M in the set on the right, because there exist a map $\tilde{M} \in \mathcal{M}_{\text{gep}}$ and an integer $i, 0 \le i \le m(\tilde{M})$, such that $M = \nabla_i \tilde{M}$, we may soon find that $M \in \mathcal{M}_{\text{gep}_2}$ by considering that the root-edge of M is always a link and that $M \in \mathcal{M}_{\text{gep}}$ as well. Thus, $M \in \mathcal{M}_{\text{gep}_2}$.

Therefore the lemma follows. □

From the two Lemmas above we are now allowed to determine the contributions of $\mathcal{M}_{\text{gep}_i}, i = 0, 1, 2$, to the enufunction

$$g_{\mathcal{M}_{\text{gep}}}(x, \underline{y}) = \sum_{M \in \mathcal{M}_{\text{gep}}} x^{m(M)} \underline{y}^{\underline{n}(M)} \tag{8.1.5}$$

where $\underline{n}(M) = (n_1(M), n_2(M), \cdots, n_i(M), \cdots)$, $n_i(m)$ is the number of vertices of valency i in M and $m(M)$, the valency of the root-vertex of M.

First, since ϑ has neither non-rooted vertex nor edge we soon see that

$$g_{\mathcal{M}_{\text{gep}_0}} = 1. \tag{8.1.6}$$

Then, by Lemma 8.1.1

$$g_{\mathcal{M}_{\text{gep}_1}} = x^2 g^2 \tag{8.1.7}$$

where $g = g_{\mathcal{M}_{\text{gep}}}(x, \underline{y})$ defined by (8.1.5).

Further, from Lemma 8.1.2

$$\begin{aligned} g_{\mathcal{M}_{\text{gep}_2}} &= \int_y \sum_{M \in \mathcal{M}_{\text{gep}}} \left(\sum_{i=1}^{m(M)+1} x^i y^{m(M)-i+2} \right) \underline{y}^{\underline{n}(M)} \\ &= \int_y \sum_{M \in \mathcal{M}_{\text{gep}}} \frac{y x^{m(M)+2} - x y^{m(M)+2}}{x - y} \underline{y}^{\underline{n}(M)} \\ &= x \int_y \left(y \delta_{x,y}(zg) \right). \end{aligned} \tag{8.1.8}$$

Theorem 8.1.1 *The enufunction g defined by (8.1.5) satisfies the following functional equation:*

$$g = 1 + x^2y^2 + x\int_y \Big(y\delta_{x,y}(zg)\Big). \tag{8.1.9}$$

Proof According to (8.1.1), from (8.1.6–8) the theorem is soon obtained. □

By the decomposition provided above we are also allowed to determine the enufunction

$$f_{\mathcal{M}_{\text{gep}}}(x,y) = \sum_{M\in\mathcal{M}_{\text{gep}}} x^{m(M)}y^{n(M)} \tag{8.1.10}$$

where $m(M)$ and $n(M)$ are the valency of the root-vertex and the number of edges of M.

Since ϑ has no edge,

$$f_{\mathcal{M}_{\text{gep}_0}}(x,y) = 1 \tag{8.1.11}$$

as well.

From Lemma 8.1.1, by considering that the loop contributes 1 to the edge number of a map in $\mathcal{M}_{\text{gep}_1}$,

$$f_{\mathcal{M}_{\text{gep}_1}}(x,y) = x^2yf^2 \tag{8.1.12}$$

where $f = f_{\mathcal{M}_{\text{gep}}}(x,y)$ given in (8.1.10).

Further, from Lemma 8.1.2, by considering the root-edge contributes 1 to each of the valency of the root-vertex and the edge number of a map in $\mathcal{M}_{\text{gep}_2}$,

$$\begin{aligned} f_{\mathcal{M}_{\text{gep}_2}}(x,y) &= xy\sum_{M\in\mathcal{M}_{\text{gep}}} \frac{1}{x}\left(\sum_{i=1}^{m(M)+1} x^i\right)y^{n(M)} \\ &= xy\sum_{M\in\mathcal{M}_{\text{gep}}} \frac{1-x^{m(M)+1}}{1-x}y^{n(M)} \\ &= \frac{xy}{1-x}\Big(h - xf\Big) \end{aligned} \tag{8.1.13}$$

where

$$h = f_{\mathcal{M}_{\text{gep}}}(1, y). \tag{8.1.14}$$

Lemma 8.1.3 *The enufunction f defined in* (8.1.10) *satisfies the equation as follows:*

$$x^2y(1-x)f^2 - (1 - x + x^2y)f$$
$$+xyh + (1-x) = 0 \tag{8.1.15}$$

where h is given in (8.1.14).

Proof From (8.1.1) and (8.1.11–13), by grouping and rearrangement, (8.1.15) is soon found. □

The discriminant of the equation (8.1.15), denoted by $D(x, y)$, is

$$D(x,y) = (x^2y - x + 1)^2 - 4(x-1)x^2y(x - 1 - xyh)$$
$$= 1 - 2x + (1-2y)x^2 + \Big(y^2 + 2y$$
$$-H(y)\Big)x^3 + H(y)x^4 \tag{8.1.16}$$

where

$$H(y) = 4y^2h + y^2 - 4y. \tag{8.1.17}$$

Assume that $D(x, y)$ has the form as

$$D(x,y) = (1-\theta x)^2(1 + ax + bx^2)$$
$$= 1 - (2\theta - a)x + (\theta^2 - 2a\theta + b)x^2$$
$$+\theta(a\theta - 2b)x^3 + \theta^2 bx^4. \tag{8.1.18}$$

Comparing with (8.1.16), we find

$$\theta = 1 + \frac{a}{2}, \tag{8.1.19}$$

and

$$\begin{cases} 1 - 2y = \theta(4 - 3\theta) + 6; \\ y^2 + 2y - h = \theta\Big(2(\theta - 1) - 2b\Big); \\ H(y) = \theta^2 b. \end{cases} \tag{8.1.20}$$

Then, we may obtain an equation for b with θ as the parameter as follows:

$$\frac{1}{4}\left(1-4\theta+3\theta^2-b\right)^2+1-4\theta+3\theta^2-b-\theta^2 b$$

$$=2(\theta-1)\theta^2-2\theta b.$$

By rearrangement, it becomes

$$b^2-(10\theta^2-16\theta+6)b+(90\theta^4-32\theta^3$$

$$+42\theta^2-24\theta+5)=0. \tag{8.1.21}$$

The discriminant of (8.1.21) is

$$(10\theta^2-16\theta-6)^2-4(90\theta^4-32\theta^3+42\theta^2-24\theta+5)$$

$$=64\theta^4-192\theta^3+208\theta^2-96\theta+16$$

$$=(8\theta^2-12\theta+4)^2.$$

Therefore, we have

$$b=(\theta-1)^2, \text{or } 9\theta^2-14\theta+5=(9\theta-5)(\theta-1).$$

The latter has to be chosen in our case. By the last of (8.1.20),

$$H(y)=\theta(9\theta-5)(\theta-1).$$

From the first of (8.1.20) and (8.1.17), we extract the parametric expressions:

$$y=(3\theta-2)(1-\theta);$$

$$h=\frac{4\theta-3}{(3\theta-2)^2}. \tag{8.1.22}$$

This allows us to employ Corollary 1.5.2 for finding

$$h=1+\sum_{n\geq 1}\frac{y^n}{n!}\left(\frac{\mathrm{d}^{n-1}}{\mathrm{d}\theta^{n-1}}\frac{(-1)^{n+1}2(6\theta-5)}{(3\theta-2)^{n+3}}\right)\Bigg|_{\theta=1}$$

$$=1+\sum_{n\geq 1}\frac{y^n}{n!}\frac{(-1)^{n+1}2(-1)^{n+2}}{3^{n+2}(n+2)!}$$

$$\times \frac{\mathrm{d}^{n-1}}{\mathrm{d}\theta^{n-1}} \left((6\theta - 5) \frac{\mathrm{d}^{n+2}}{\mathrm{d}\theta^{n+2}} \frac{1}{3\theta - 2} \right) \Bigg|_{\theta=1}$$

$$= 1 + \sum_{n\geq 1} \frac{y^n}{n!} \left(\frac{2(2n+1)!3^{2n+1}}{3^{n+2}(n+2)!} \right.$$

$$\left. - \frac{2\cdot 6(n-1)(2n)!3^{2n}}{3^{n+2}(n+2)!} \right)$$

$$= 1 + \sum_{n\geq 1} \frac{2\cdot 3^n(2n)!}{n!(n+2)!} y^n. \tag{8.1.23}$$

Theorem 8.1.2 *The number of distinct rooted general planar maps of size n is*

$$H_{\mathcal{M}_{\mathrm{gep}}}(n) = \frac{2\cdot 3^n(2n)!}{n!(n+2)!} \tag{8.1.24}$$

for $n \geq 0$.

Proof Since the number

$$H_{\mathcal{M}_{\mathrm{gep}}}(n) = \partial_y^n h,$$

for (8.1.23) and it is 1 in its own right for $n = 0$, the theorem is directly obtained. □

According to (8.1.22) and Theorem 8.1.2 we are also allowed to find an explicit expression of $f_{\mathcal{M}_{\mathrm{gep}}}(x, y)$. Because more space would otherwise be occupied, it is suggested one refers to [Liu13] in which the coefficient was expressed by a triple summation.

§8.2 Planar c-nets

A rooted planar map whose underlying graph is 3-connected is called a *c-net*. How to determine the number of combinatorially distinct c-nets with a given size is the aim of this section. Of course, the size of a c-net is always greater that three. Let $\mathcal{M}_{\text{c-n}}$ be the set of all c-nets. If the enufunction

$$h_{\mathcal{M}_{\text{c-n}}}(y) = \sum_{M\in\mathcal{M}_{\text{c-n}}} y^{n(M)} \tag{8.2.1}$$

where $n(M)$ is the size of M, then

$$\partial_y^n h = H_{\mathcal{M}_{\text{c-n}}}(n), \tag{8.2.2}$$

where $h = h_{\mathcal{M}_{\text{c-n}}}(y)$ given in (8.2.1), is just the number we expect to know.

A loopless map without a face whose boundary is not a circuit is said to be *biloopless*. From the duality, a planar map is biloopless iff so is its planar dual. A map without a face whose boundary is not a circuit is also said to be *dual loopless*. If a biloopless map has neither multi-edge nor two faces whose common boundary has at least two edges, then it is called *bisimple*. By the duality it can also be verified that a planar map is bisimple iff so is its planar dual. Meanwhile, a dual loopless map without two faces whose common boundary has at least two edges is said to be *dual simple*.

For a set of maps, if any map in the set has its dual in the set as well, then it is said to be *selfdual*. In this view, the sets of all biloopless and bisimple planar maps are both selfdual. From §6.3, we also know that the set of all nonseparable planar maps with the loop map $L_1 = (Kr, (r, \alpha\beta r))$ excluded is also selfdual.

Lemma 8.2.1 *The set of all c-nets is selfdual.*

Proof By contradiction. Suppose M is a c-net but its planar dual M^* is not. Then there exist two vertices u and v in M^* such that

$$u = (x, \mathcal{P}x, \cdots, \mathcal{P}^s x)$$

and

$$v = (y, \mathcal{P}y, \cdots, \mathcal{P}^t y) \tag{8.2.3}$$

with the property that there exist two faces f and g such that

$$f = (x, \mathcal{P}\alpha\beta x, \cdots, \mathcal{P}^j y, \cdots)$$

and

$$g = (y, \mathcal{P}\alpha\beta y, \cdots, \mathcal{P}^i x, \cdots) \tag{8.2.4}$$

for some $i, 1 < i < s$ and some $j, 1 < j < t$.

However, by the duality this does mean that there exist two vertices f and g in M as (8.2.4) written in the form (8.2.3) with the property that there exist two faces u and v as (8.2.3) written in the form (8.2.4). Thus M is not a c-net either. A contradiction to the given condition occurs. □

Because the enufunction of rooted non-separable planar maps has been known in §6.3, we are suggested to find a relationship between nonseparable planar maps and c-nets. Now, consider the set $\mathcal{M}_{\rm ns}(\geq 2)$ of all rooted nonseparable planar maps with at least two edges. We classify $\mathcal{M}_{\rm ns}(\geq 2)$ into four parts: $\mathcal{M}_i, i = 1, 2, 3, 4$. Here, $\mathcal{M}_1$ consists of all the maps in $\mathcal{M}_{\rm ns}(\geq 2)$ with only two edges. In fact, there is only one map $(Kr \cup K\mathcal{P}r, (r, \mathcal{P}r)(\alpha\beta r, \alpha\beta\mathcal{P}r))$ in $\mathcal{M}_1$.

$$\begin{aligned}\mathcal{M}_2 = \{M|\ \forall M \in \mathcal{M}_{\rm ns}(\geq 2),\ M - a = \\ M_1 \dot{+} M_2 \dot{+} \cdots \dot{+} M_k,\ k \geq 2\}\end{aligned}$$

where $a = e_r(M)$, the root-edge of M.

$$\begin{aligned}\mathcal{M}_3 = \{M|\ \forall M \in \mathcal{M}_{\rm ns}(\geq 2),\ M - a = M_1 {}_u{+}_v \\ M_2 {}_u{+}_v \cdots {}_u{+}_v M_k,\ k \geq 2\}\end{aligned}$$

where for two maps M_i and M_{i+1},

$$M_i {}_u{+}_v M_{i+1} = M_i \cup M_{i+1},$$

provided $M_i \cap M_{i+1} = \{u, v\}, i = 1, \cdots, k-1$, $a = (u, v)$ which is the root-edge of M as well. Of course, $\mathcal{M}_4$ consists of all other maps in $\mathcal{M}_{\rm ns}(\geq 2)$. Because any two of $\mathcal{M}_i$, $i = 1, 2, 3, 4$, has no intersection which is easily checked, we have

$$\mathcal{M}_{\rm ns}(\geq 2) = \sum_{i=1}^{4} \mathcal{M}_i. \tag{8.2.5}$$

Lemma 8.2.2 *Let $\mathcal{M}_{\bar{2}}$ be the set of all the maps of $\mathcal{M}_{\rm ns}$ not in $\mathcal{M}_2$ and*

$$h_{\mathcal{M}_{\bar{2}}}(x) = \sum_{M \in \mathcal{M}_{\bar{2}}} x^{n(M)}, \tag{8.2.6}$$

where $n(M)$ is the size of M, the enufunction of $\mathcal{M}_{\bar{2}}$. Of course, the restriction, $n(M) \geq 3$, is always assumed. Then we have the following relation:

$$h_{\mathcal{M}_{\rm ns}}(x) - x = x \frac{x^2 + h_{\mathcal{M}_{\bar{2}}}(x)}{x - x^2 - h_{\mathcal{M}_{\bar{2}}}(x)} \tag{8.2.7}$$

where $h_{\mathcal{M}_{\rm ns}}(x)$ *is known in* §6.3.

Proof Because for any $M \in \mathcal{M}_2, M - a = M_1 \dot{+} \cdots \dot{+} M_k,\ k \geq 2$, we are allowed to assume the inner face incident with the root-edge a of M has the form as

$$(\mathcal{P}r, \cdots, (\mathcal{P}\alpha\beta)^{j_i}\mathcal{P}r, \cdots, (\mathcal{P}\alpha\beta)^m\mathcal{P}r)$$

where $(\mathcal{P}\alpha\beta)^m\mathcal{P}r = \alpha\beta r$,

$$M_i \bigcap M_{i+1} = v_{(\mathcal{P}\alpha\beta)^{j_i}\mathcal{P}r},$$

for $i = 1, 2, \cdots, k-1$. Write the roots of $M_1, M_2, \cdots, M_k$ as $r_1 = \mathcal{P}r$,

$$r_i = (\mathcal{P}\alpha\beta)^{j_{i-1}}\mathcal{P}r,$$

for $i = 2, 3, \cdots, k$ respectively. Then we may construct the maps $\hat{M}_i$ from M_i by adding an edge $K\hat{r}_i = (v_{r_i}, v_{r_{i+1}})$, for $i = 1, 2, \cdots, k-1$, and $(v_{r_k}, v_{\alpha\beta r})$ for $i = k$ such that the root-vertices $o_i = (\hat{r}_i, r_i, S_i)$, where (r_i, S_i) determines the root-vertex of M_i for $i = 1, 2, \cdots, k$. Easily verified that all $\hat{M}_i$ are allowed to be any map in $\mathcal{M}_{\bar{2}}$.

Since for any $M \in \mathcal{M}_2$, $M - a = M_1 \dot{+} M_2 \dot{+} \cdots \dot{+} M_k$, we have

$$n(M - a) = n(M) - 1,\ n(\hat{M}_i) = n(M_i) + 1,$$

for $i = 1, 2, \cdots, k$, and all $\hat{M}_i \in \mathcal{M}_{\bar{2}}$. Thus,

$$h_{\mathcal{M}_2}(x) = x \sum_{k \geq 2} \left(x + \frac{h_{\mathcal{M}_{\bar{2}}}(x)}{x} \right)^k.$$

Moreover,

$$\begin{aligned}
h_{\mathcal{M}_{\rm ns}}(x) - x &= x^2 + h_{\mathcal{M}_{\bar{2}}}(x) + h_{\mathcal{M}_2}(x) \\
&= x \left(x + \frac{h_{\mathcal{M}_{\bar{2}}}(x)}{x} \right) + x \sum_{k \geq 2} \left(x + \frac{h_{\mathcal{M}_{\bar{2}}}(x)}{x} \right)^k \\
&= x \left(x + \frac{h_{\mathcal{M}_{\bar{2}}}(x)}{x} \right) \Big/ \left(1 - x - \frac{h_{\mathcal{M}_{\bar{2}}}(x)}{x} \right) \\
&= \frac{x\left(x^2 + h_{\mathcal{M}_{\bar{2}}}(x)\right)}{x - x^2 - h_{\mathcal{M}_{\bar{2}}}(x)}.
\end{aligned}$$

This is the lemma. □

From the lemma we soon find

$$h_{\mathcal{M}_{\bar{2}}}(x) = \frac{x\Big(h_{\mathcal{M}_{\rm ns}}(x) - x\Big)}{h_{\mathcal{M}_{\rm ns}}(x)} - x^2. \tag{8.2.8}$$

Lemma 8.2.3 *For $\mathcal{M}_2$ we have*

$$h_{\mathcal{M}_2}(x) = \frac{\Big(h_{\mathcal{M}_{\rm ns}}(x) - x\Big)^2}{h_{\mathcal{M}_{\rm ns}}(x)}. \tag{8.2.9}$$

Proof Since

$$h_{\mathcal{M}_2}(x) = h_{\mathcal{M}_{\rm ns}}(x) - x - x^2 - h_{\mathcal{M}_{\bar{2}}}(x),$$

according to (8.2.8) the lemma can be soon found by grouping and rearrangement. □

Lemma 8.2.4 *For $\mathcal{M}_3$ we have*

$$h_{\mathcal{M}_3}(x) = \frac{\Big(h_{\mathcal{M}_{\rm ns}}(x) - x\Big)^2}{h_{\mathcal{M}_{\rm ns}}(x)}. \tag{8.2.10}$$

Proof From the selfduality of $\mathcal{M}_{\rm ns}$, noticing that $\mathcal{M}_3$ is the planar dual of $\mathcal{M}_2$, Lemma 8.2.3 leads the lemma. □

For $M \in \mathcal{M}_4$, let $\{u, v\}$ be a *separator* by which M is divided into a pair U and W such that $U \cup W = M$ while $U \cap W = \{u, v\}$. The pair $\{U, W\}$ is called a *separation* of M. Assume that the root-edge a of M is always in U without loss of generality. Then, a separation is an ordered pair of submaps in M. A separation $\{U, W\}$ is said to be *terminal* if for any separation $\{U', W'\}$, $W' \supseteq W$ implies $W' = W$.

Fact 1 *Any terminal separation $\{U, W\}$ with $U \cap W = \{u, v\}$ has at least a vertex in $\{u, v\}$ which is not incident with the root-edge a of M.*

Proof Otherwise, M is in either $\mathcal{M}_2$ or $\mathcal{M}_3$. This is a contradiction to $M \in \mathcal{M}_4$. □

Fact 2 *Let $\{U_1, W_1\}$ with $U_1 \cap W_1 = \{u_1, v_1\}$ and $\{U_2, W_2\}$ with $U_2 \cap W_2 = \{u_2, v_2\}$ be two distinct terminal separations of $M \in \mathcal{M}_4$. Then $\{u_1, v_1\}$ and $\{u_2, v_2\}$ are distinct.*

Proof Otherwise, assume $u_1 = u_2$ and $v_1 = v_2$. Then by the terminality of $\{U_1, W_1\}$ and $\{U_2, W_2\}$ we have $W_1 = W_2$, and hence $U_1 = U_2$. This is a contradiction to the given condition. □

Now, for any $M \in \mathcal{M}_4$ the size of M is of course at least 4. We may construct a map M_{cor} by identifying each Y of a terminal separation $\{X, Y\}$ with $X \cap Y = \{x, y\}$ as an edge (x, y). Since any cut-vertex of M_{cor} would be converted to a cut-vertex of M and a separator with two vertices of M_{cor} would be that of M, we see that M_{cor} has to be 3-connected. From the planarity, M_{cor} is a c-net with at least 4 edges. The c-net M_{cor} associated with M is called the *core* of M.

Let $\mathcal{M}$ and $\mathcal{N}$ be two sets of maps. For $M \in \mathcal{M}$, let $M \times \mathcal{N}$ be the set of maps each of which is obtained by replacing some edge or edges with maps in $\mathcal{N}$ such that the edge or edges are identified with the root-edge or root-edges of the maps and then by that all the root-edge or root-edges are deleted. Further, write

$$\mathcal{M}^{\mathcal{N}} = \bigcup_{M \in \mathcal{M}} \{M \times \mathcal{N}\}.$$

Lemma 8.2.5 *For $\mathcal{M}_4$ we have*

$$\mathcal{M}_4 = \mathcal{M}_{\text{c-n}}^{\mathcal{M}_{\text{ns}}(\geq 2)}. \tag{8.2.11}$$

Proof Because any c-net in $\mathcal{M}_{\text{c-n}}$ can be seen as a core of a map in $\mathcal{M}_4$, from Facts 1–2 we know that $\mathcal{M}_4$ is a subset of the set on the right hand side of (8.2.11).

Conversely, it is easily shown that any map in the set on the right hand side of (8.2.11) is also a member of $\mathcal{M}_4$ by considering what has been described above. □

From Lemma 8.2.5, by observing the construction of a map in the set on the right hand side of (8.2.11), we have

$$h_{\mathcal{M}_4}(x) = \frac{x h_{\mathcal{M}_{\text{c-n}}}\left(\dfrac{h_{\mathcal{M}_{\text{ns}}}(x) - x}{x}\right)}{\dfrac{h_{\mathcal{M}_{\text{ns}}}(x) - x}{x}}$$

$$= \frac{x^2}{h_{\mathcal{M}_{\text{ns}}}(x) - x} h_{\mathcal{M}_{\text{c-n}}}\left(\frac{h_{\mathcal{M}_{\text{ns}}}(x) - x}{x}\right). \tag{8.2.12}$$

Theorem 8.2.1 *The enufunction $h_{\mathcal{M}_{\text{c-n}}}(y)$ of $\mathcal{M}_{\text{c-n}}$ can be determined by $h_{\mathcal{M}_{\text{ns}}}(y)$ in the following way:*

$$h_{\mathcal{M}_{\text{c-n}}}(y) = y^2 - \frac{2y^3}{1+y} - xy \tag{8.2.13}$$

where

$$y = \frac{1}{x}\Big(h_{\mathcal{M}_{\text{ns}}}(x) - x\Big). \tag{8.2.14}$$

Proof Since $h_{\mathcal{M}_{\text{ns}}}(x) - x = x^2 + h_{\mathcal{M}_2}(x) + h_{\mathcal{M}_3}(x) + h_{\mathcal{M}_4}(x)$, from Lemmas 8.2.3–4 and (8.2.12) we have

$$h_{\mathcal{M}_{\text{ns}}}(x) - x = x^2 + \frac{2\Big(h_{\mathcal{M}_{\text{ns}}}(x) - x\Big)^2}{h_{\mathcal{M}_{\text{ns}}}(x)}$$

$$+ \frac{x^2 h_{\mathcal{M}_{\text{c-n}}}\left(\dfrac{h_{\mathcal{M}_{\text{ns}}} - x}{x}\right)}{h_{\mathcal{M}_{\text{ns}}} - x}. \tag{8.2.15}$$

Dividing the two sides by x, and then by considering (8.2.14), we find that

$$y = x + \frac{2y^2}{y+1} + \frac{h_{\mathcal{M}_{\text{c-n}}}(y)}{y}.$$

By multiplying the two sides by y, (8.2.13) is soon obtained within rearrangement. □

According to (6.4.9) and (8.2.14),

$$x = \xi(1-\xi)^2,$$
$$y = \frac{\xi(1-2\xi)}{(1-\xi)^2}. \tag{8.2.16}$$

Let

$$\eta = \frac{1-2\xi}{1-\xi}. \tag{8.2.17}$$

Then (8.2.16) becomes

$$y = \eta(1-\eta),$$
$$X = -\frac{\eta-1}{(\eta-2)^3} = -x. \tag{8.2.18}$$

It is easily checked that

$$\frac{\mathrm{d}X}{\mathrm{d}y} = \frac{1}{(2-\eta)^4} \tag{8.2.19}$$

and then that X satisfies the following first order ordinary differential equation:

$$\left(4x^2+7x-2\right)\frac{\mathrm{d}X}{\mathrm{d}y} + (6x-15)X = 2 \tag{8.2.20}$$

with the initial condition $X(0) = 0$.

Theorem 8.2.2 *The number of distinct c-nets of size n is given by*

$$N_{\mathcal{M}_{\text{c-n}}}(n) = \Lambda_n + \frac{21}{32}\sum_{i=-1}^{\lfloor\frac{n-1}{2}\rfloor-4}\Big(7(n-3i)^2 - 91(n - 2i) + 300\Big)(i+3)\Big(7(n-2i)-54\Big) \times \frac{(i+2)\Big(2(n-2i)-17\Big)!}{2^{2i+2}(n-2i-9)!(n-2i-3)!} \tag{8.2.21}$$

where

$$\Lambda_n = \begin{cases} -2 + \dfrac{(n-1)^2}{2^{n-2}}, & n = 1(\text{mod } 2); \\[2ex] 2 + \dfrac{40-22n-n^2}{2^{n+1}}, & n = 0(\text{mod } 2) \end{cases} \tag{8.2.22}$$

for $n \geq 6$.

Proof From the equation (8.2.20), we can find

$$X(y) = \frac{1}{2(y+2)^3}\Big(2y^2 - 10y - 1 + \sqrt{(1-4y)^3}\Big)$$

is the solution. Then, the only thing that remains in principle is to expand $\frac{1}{1+y}$, $\frac{1}{(y+2)^3}$ and $\sqrt{(1-4y)^3}$ into power series of y. However, it is with some complication which does not allow us to occupy more space here. The detail is referred to [Liu4]. □

Of course, another way which can be used to extract an explicit expression of $h_{\mathcal{M}_{\text{c-n}}}(y)$ in a power series of y is by employing Corollary 1.5.1 on the basis of the parametric expressions shown in (8.2.18). It does not seem to be simpler than what is just explained above.

§8.3 Convex polyhedra

Because any graph represented by the skeleton of a convex polyhedron in 3-dimensional Euclidean space has to be 3-connected and conversely any 3-connected planar map can always be seen as a convex polyhedron (see [Fid1] or [Liu58]), it is not necessary to distinguish between 3-connected planar maps and convex polyhedra.

We are here especially concerned with cubic convex polyhedra, *i.e.*, 3-connected cubic planar maps. Let $\mathcal{C}$ be the set of all rooted 3-connected cubic planar maps. Our aim is to determine the enufunction of $\mathcal{C}$ with the size as the parameter, *i.e.*,

$$f_{\mathcal{C}}(x) = \sum_{C \in \mathcal{C}} x^{m(C)} \tag{8.3.1}$$

where $m(C)$ is the size of C.

For $M \in \mathcal{C}_{\text{ns}}$, the set of all rooted cubic planar nonseparable maps, if $M = M_1 \cup M_2$ such that $M_1 \cap M_2 = \{u, v\}$ and both M_1 and M_2 are submaps of M, then we may always assume that one of M_1 and M_2 has all the vertices of valency 3 except for exactly two of valency 1 and the other, one with the two exceptional vertices of valency 2. The former is called a 1-*trap* and the latter a 2-*trap*. Or more precisely, a 1-trap is a rooted planar map such that all of its vertices are of valency 3 except for only two of valency 1. Of course, a 2-trap is defined by that as for 1-trap with 1 replaced by 2. For convenience, we may always assume that the 1-trap never contains the root-edge of M.

A 1-trap has no proper submap which is also a 1-trap is called a *maxtrap*. The two univalent vertices in a 1-trap is called its *ends*, and others, *inner vertices*.

Lemma 8.3.1 *Any two maxtraps have no inner vertex in common.*

Proof By contradiction. Assume T_1 and T_2 are two maxtraps of a map $M \in \mathcal{C}_{\mathrm{ns}}$ with an inner vertex in common. From the definition of a trap, the common vertex should be an end of T_1 and T_2. From the maximality, T_1 has to equal T_2. This is a contradiction to the assumption. □

Let $\mathcal{C}_{\mathrm{I}}$, $\mathcal{C}_{\mathrm{II}}$ and $\mathcal{C}_{\mathrm{III}}$ be the subsets of $\mathcal{C}_{\mathrm{ns}}$ which consist of all such maps M that $M \bullet a, a = e_r(M)$, has a cut-vertex, $M - a$ has a cut-vertex, and neither $M \bullet a$ nor $M - a$ has a cut-vertex, respectively. Of course, it is easily seen that

$$\mathcal{C}_{\mathrm{ns}} = \mathcal{C}_{\mathrm{I}} + \mathcal{C}_{\mathrm{II}} + \mathcal{C}_{\mathrm{III}}. \tag{8.3.2}$$

For $M \in \mathcal{C}_{\mathrm{III}}$, let M' be the map obtained by replacing each maxtrap of M with an edge. By the cubicness, M' has at least 6 edges. Because any separating pair $\{x, y\}$ of vertices in M' is also a separating pair in M, M' is not allowed to have any separating pair and hence M' is 3-connected, *i.e.*, $M' \in \mathcal{C}$.

Lemma 8.3.2 *Let $\mathcal{L}$ be the set of all the rooted planar maps in each of which all vertices are 3-valent, except only for two univalent vertices such that cut-vertices, if any, are all on a path between the two exceptional vertices. Then we have*

$$\mathcal{C}_{\mathrm{III}} = \mathcal{C}^{\mathcal{L}} \tag{8.3.3}$$

where the meaning of a set with a set power is as shown in §8.2.

Proof From the discussion above, $\mathcal{C}_{\mathrm{III}}$ is a subset of the set on the right hand side of (8.3.3).

Conversely, because for any M belonging to the set on the right has the form $M \in N \times \mathcal{L}$ for $N \in \mathcal{C}$, according to the 3-connectedness of N, M is in $\mathcal{C}_{\mathrm{ns}}$ but in neither $\mathcal{C}_{\mathrm{I}}$ nor $\mathcal{C}_{\mathrm{II}}$. That implies $M \in \mathcal{C}_{\mathrm{III}}$. Thus, the set on the right is a subset of $\mathcal{C}_{\mathrm{III}}$. □

For $M \in \mathcal{L}$ in which the link map L is included as the degenerate case, let v_r and $v_x, x = (\mathcal{P}\alpha\beta)^l r, l > 1$, be the two univalent vertices. Write $\delta(M) = M'$ obtained by replacing $Kr \cup Kx$ with $Kr' = \{\alpha\beta r, \beta r, \beta x, \alpha\beta x\}$ where r' is the root-edge of M'.

Lemma 8.3.3 *Let $\delta(\mathcal{L} - L) = \{\delta(M) | \forall M \in \mathcal{L} - L\}$. Then we have*

$$\delta(\mathcal{L} - L) = \mathcal{C}_{\mathrm{ns}} \tag{8.3.4}$$

where $\mathcal{L}$ is given in Lemma 8.3.2.

Proof For any $M \in \delta(\mathcal{L} - L)$ it is easily checked that M is a rooted planar cubic map. By the condition on maps in $\mathcal{L}$, it can be seen that M is nonseparable. Thus $\delta(\mathcal{L} - L) \subseteq \mathcal{C}_{\mathrm{ns}}$.

Conversely, for any $M \in \mathcal{C}_{\mathrm{ns}}$, from the nonseparability, we may uniquely construct a map M' by replacing Kr with $Kr' \cup Kx$ such that $Kr' = \{r', \alpha r', \alpha r, r\}$ and $Kx = \{x, \alpha x, \beta r, \alpha\beta r\}$ and by introducing two new vertices (r') and (x) which are of course univalent. From the nonseparability of M, all cut vertices of M' have to be on a path joining (r') and (x), if any. This implies $M' \in \mathcal{L}$. However, it is seen that $M = \delta(M')$ and hence $M \in \delta(\mathcal{L} - L)$. Thus $\mathcal{C}_{\mathrm{ns}} \subseteq \delta(\mathcal{L} - L)$. □

From this lemma, we are allowed to write

$$\mathcal{L} - L = \delta^{-1}(\mathcal{C}_{\mathrm{ns}}). \tag{8.3.5}$$

Let $f_{\mathcal{C}_{\mathrm{III}}}$, $f_{\mathcal{C}_{\mathrm{ns}}}$ and $f_{\mathcal{L}}$ be the enufunctions of $\mathcal{C}_{\mathrm{III}}$, $\mathcal{C}_{\mathrm{ns}}$ and $\mathcal{L}$ with the size as the parameter respectively. Then from Lemmas 8.3.2–3, we find that

$$f_{\mathcal{C}_{\mathrm{III}}}(x) = f_{\mathcal{C}}(x + x f_{\mathcal{C}_{\mathrm{ns}}}). \tag{8.3.6}$$

Lemma 8.3.4 *Let* $\mathcal{C}_{\langle \mathrm{II}\rangle} = \{M - a | \forall \in \mathcal{C}_{\mathrm{II}}\}$, $a = e_r(M)$. *Then, we have*

$$\mathcal{C}_{\langle \mathrm{II}\rangle} = \mathcal{R} \times\!\cdot \mathcal{L} \times\!\cdot \mathcal{R} \tag{8.3.7}$$

where $\mathcal{R}$ *consists of all the rooted planar nonseparable maps with all vertices of valency* 3 *except only for two of valency* 2 *on the root-face boundary each and the operation* $\times\cdot$ *is as shown in* §6.1.

Proof It is easily verified that for any $M \in \mathcal{C}_{\mathrm{II}}$,

$$M = M_1 +\!\cdot M_2 +\!\cdot M_3$$

such that $M_1, M_3 \in \mathcal{R}$ and $M_2 \in \mathcal{L}$. This implies $\mathcal{C}_{\langle \mathrm{II}\rangle}$ is a subset of the set on the right hand side of (8.3.7).

Conversely, for any

$$M = M_1 +\!\cdot M_2 +\!\cdot M_3 \in \mathcal{R} \times\!\cdot \mathcal{L} \times\!\cdot \mathcal{R},$$

we may construct a map M' by adding an edge between the root-vertex of M_1 and the nonrooted 2-valent vertex of M_3 in the inner domain of the outerface boundary of M. Because it can be seen that $M' \in \mathcal{C}_{\mathrm{II}}$ and $M = M' - a', a' = e_r(M')$, we find that $M \in \mathcal{C}_{\langle \mathrm{II}\rangle}$ and hence the set on the right is also a subset of $\mathcal{C}_{\langle \mathrm{II}\rangle}$. □

For $M \in \mathcal{R}$, let $\lambda(M)$ be the map obtained by adding an edge from the root-vertex of M to the nonrooted 2-valent vertex in the inner domain of the outerface boundary as the root-edge of $\lambda(M)$. Write $\lambda(\mathcal{R}) = \{\lambda(M) | \forall M \in \mathcal{R}\}$. It can be verified that

$$\lambda(\mathcal{R}) = \mathcal{C}_{\bar{\mathrm{II}}} \tag{8.3.8}$$

where $\mathcal{C}_{\bar{\mathrm{II}}} = \mathcal{C}_{\mathrm{I}} + \mathcal{C}_{\mathrm{III}}$ and further

$$\mathcal{C}_{\bar{\mathrm{II}}} = \mathcal{C}_{\mathrm{ns}} - \mathcal{C}_{\mathrm{II}}. \tag{8.3.9}$$

Let $f_{\mathcal{C}_{\mathrm{II}}} = f_{\mathcal{C}_{\mathrm{II}}}(x)$, $f_{\mathcal{R}}(x)$ and $f_{\mathcal{C}_{\bar{\mathrm{II}}}}$ be the enufunctions of $\mathcal{C}_{\mathrm{II}}, \mathcal{R}$ and $\mathcal{C}_{\bar{\mathrm{II}}}$ with the size as the parameter, respectively. Then from Lemma 8.3.4, (8.3.8), and (8.3.5) we find that

$$f_{\mathcal{C}_{\mathrm{II}}} = x\big(f_{\mathcal{R}}(x)\big)^2 f_{\mathcal{L}}(x)$$

$$= x\left(\frac{1}{x}f_{\mathcal{C}_{\text{II}}}\right)^2(x + xf_{\mathcal{C}_{\text{ns}}})$$

$$= (f_{\mathcal{C}_{\text{ns}}} - f_{\mathcal{C}_{\text{II}}})^2(1 + f_{\mathcal{C}_{\text{ns}}}). \tag{8.3.10}$$

Hence we see that only the following solution

$$f_{\mathcal{C}_{\text{II}}}(x) = \frac{f_{\mathcal{C}_{\text{ns}}}^2}{1 + f_{\mathcal{C}_{\text{ns}}}} \tag{8.3.11}$$

is available.

Lemma 8.3.5 *Let $\mathcal{C}_{(\text{I})} = \{M \bullet a | \forall M \in \mathcal{C}_{\text{I}}\}$. Then, we have*

$$\mathcal{C}_{(\text{I})} = (O + \mathcal{C}_2)^{\times \cdot 2} \tag{8.3.12}$$

where O is the loop map and $\mathcal{C}_2$ is the set of all the near cubic non-separable maps with the root-vertex of valency 2.

Proof For $M \in \mathcal{C}_{(\text{I})}$, there exists a map $M' \in \mathcal{C}_{\text{I}}$ such that $M = M' \bullet a'$ has a cut-vertex which must be the root-vertex. From the cubicness of M', $M = M_1 + \cdot M_2$ has each of M_1 and M_2 as either O or a map in $\mathcal{C}_2$. This implies M is a member of the set on the right hand side of (8.3.12).

Conversely, for $M = M_1 + \cdot M_2, M_i \in O + \mathcal{C}_2, i = 1, 2$, we may always obtain a map M' by splitting the root-vertex o' into o_1 and o_2 with a new edge (o_1, o_2) such that both o_1 and o_2 are 3-valent in M'. It is easy to see that $M = M' \bullet a'$ where $a' = (o_1, o_2)$. Since $M' \in \mathcal{C}_{\text{I}}$, we have $M \in \mathcal{C}_{(\text{I})}$. □

Let $f_{\mathcal{C}_{\text{I}}}(x)$ and $f_{\mathcal{C}_2}(x)$ be the respective enufunctions of $\mathcal{C}_{\text{I}}$ and $\mathcal{C}_2$ with the size as the parameter. By considering that

$$f_{\mathcal{C}_2}(x) = xf_{\mathcal{C}_{\text{ns}}}(x),$$

from Lemma 8.3.5 we have

$$f_{\mathcal{C}_{\text{I}}}(x) = x(x + xf_{\mathcal{C}_{\text{ns}}})^2. \tag{8.3.13}$$

Theorem 8.3.1 *Enufunctions $f_{\mathcal{C}}(x)$ and $f_{\mathcal{C}_{\rm ns}} = f_{\mathcal{C}_{\rm ns}}(x)$ defined above satisfies the following relation:*

$$f_{\mathcal{C}}(x + x f_{\mathcal{C}_{\rm ns}}) + x(1 + f_{\mathcal{C}_{\rm ns}})^2 = \frac{f_{\mathcal{C}_{\rm ns}}}{1 + f_{\mathcal{C}_{\rm ns}}}. \tag{8.3.14}$$

Proof Since $f_{\mathcal{C}_{\rm ns}} = f_{\mathcal{C}_{\rm I}} + f_{\mathcal{C}_{\rm II}} + f_{\mathcal{C}_{\rm III}}$ from (8.3.2), by substituting (8.3.6), (8.3.11), and (8.3.13) into this form, we may soon obtain (8.3.14) after rearrangement. □

From (8.3.14), in order to determine $f_{\mathcal{C}}(x)$ we have to investigate $f_{\mathcal{C}_{\rm ns}}(x)$ first by looking for what will happen on the set $\mathcal{N}$ of rooted nonseparable near cubic planar maps with the convention that the loop map O is not in $\mathcal{N}$. Let

$$f_{\mathcal{N}}(x,y) = \sum_{N\in\mathcal{N}} x^{m(N)} y^{n(N)} \tag{8.3.15}$$

where $m(N)$ and $n(N)$ are the root-vertex valency and the size of N.

Because $\mathcal{N}$ can be decomposed into two classes $\mathcal{N}_1$ and $\mathcal{N}_2$ such that

$$\mathcal{N}_1 = \{N|\forall N \in \mathcal{N}, N \bullet a \text{ has no cut-vertex}\}$$

and

$$\mathcal{N}_2 = \{N|\forall N \in \mathcal{N}, N \bullet a \text{ has a cut-vertex}\},$$

we have

$$\mathcal{N} = \mathcal{N}_1 + \mathcal{N}_2. \tag{8.3.16}$$

For $\mathcal{N}_1$, let $\mathcal{N}_{(1)} = \{N \bullet a|\forall N \in \mathcal{N}_1\}$. Then it is easily checked that

$$\mathcal{N}_{(1)} = \mathcal{N} - \mathcal{C}_2 \tag{8.3.17}$$

where $\mathcal{C}_2$ is as shown in (8.3.12) above. For $\mathcal{N}_2$, let $\mathcal{N}_{(2)} = \{N \bullet a|\forall N \in \mathcal{N}_2\}$. Then, we can see that

$$\mathcal{N}_{(2)} = (O + \mathcal{N})^{\times 2}. \tag{8.3.18}$$

Let $f_{\mathcal{N}_1}$ and $f_{\mathcal{N}_2}$ be the corresponding enufunctions of $\mathcal{N}_1$ and $\mathcal{N}_2$ with the two parameters respectively. From (8.3.17), we have

$$f_{\mathcal{N}_1} = \frac{xy}{x^2}\left(f - x^2 f_2\right) \tag{8.3.19}$$

where $f = f_{\mathcal{N}}(x, y)$ and f_2 is the coefficient of x^2 in f. From (8.3.18), we may find

$$f_{\mathcal{N}_2} = \frac{xy}{x^2}\left(x^2 y + f\right)^2. \tag{8.3.20}$$

Theorem 8.3.2 *The enufunction $f = f_{\mathcal{N}}(x, y)$ as defined in (8.3.15) with $f_2 = yf^*$ as in (8.3.19) satisfies the following equation:*

$$yf^2 + (2x^2y^2 + y - x)f + x^4y^3 - x^2y^2f^* = 0 \tag{8.3.21}$$

where

$$f^* = f_{\mathcal{C}_{\rm ns}}(y). \tag{8.3.22}$$

Proof According to (8.3.16), (8.3.19), (8.3.20) and (8.3.22), we have

$$f = \frac{y}{x}\left(f - x^2 y f^*\right) + \frac{y}{x}\left(x^2 y + f\right)^2.$$

By multiplying the two sides by x and then by rearrangement, (8.3.21) is soon obtained. □

Since the discriminant of the equation (8.3.21) is given by

$$\begin{aligned} D(x, y) = {} & y^2 + x^2 + 4x^2y^3(1 + f^*) \\ & - f x^3 y^2 - 2xy, \end{aligned} \tag{8.3.23}$$

from $x = \xi$ as a power series of y such that

$$D(\xi, y) = 0; \ \frac{\mathrm{d}}{\mathrm{d}\xi} D(\xi, y) = 0,$$

we may find that

$$\begin{aligned} y &= \frac{\xi}{1 + 2\xi^3}; \\ 1 + f^* &= \frac{1 + 2\xi^3}{2}\left(3 - \xi^2(1 + 2\xi^3)\right). \end{aligned} \tag{8.3.24}$$

Then, according to (8.3.14) and (8.3.22),

$$z = y\left(1 + f_{\mathcal{C}_{\rm ns}}\right);$$

$$f_{\mathcal{C}} = \frac{z-y}{z} - z. \tag{8.3.25}$$

In consequence, based on (8.3.24) and (8.3.25) $f_{\mathcal{C}} = f_{\mathcal{C}}(z)$ can be determined by employing Corollary 1.5.1 or by trying to find a differential equation in the way similar to that in §8.2.

§8.4 Quadrangulations via c-nets

A *quadrangulaton* is a map such that all faces are *quadrangles*, in other words, circuits of length four, or in short, 4-circuits. In particular, quadrangulations we are concerned here shall always be assumed to have no circuit of length 2. If a quadrangulation has all 4-circuits as face boundaries, then it is said to be *proper*.

For any quadrangulation $M = (\mathcal{X}_{\alpha,\beta}(X), \mathcal{P})$, write

$$M^* = (\mathcal{X}^*_{\alpha^*,\beta^*}(X^*), \mathcal{P}^*)$$

as the dual where $\mathcal{X}^*_{\alpha^*,\beta^*} = \mathcal{X}_{\beta,\alpha}$, and $\mathcal{P}^* = \mathcal{P}\alpha\beta$ as described in §1.1. Then, we may construct another map $\tilde{M} = (\tilde{\mathcal{X}}_{\tilde{\alpha},\tilde{\beta}}, \tilde{\mathcal{P}})$ such that

$$\tilde{\mathcal{X}}_{\tilde{\alpha},\tilde{\beta}} = \bigcup_{\tilde{x}\in X\cup X^*} K\tilde{x};\ \tilde{\mathcal{P}} = \mathcal{P}\mathcal{P}^* \tag{8.4.1}$$

where

$$\tilde{K}_{\tilde{x}} = \begin{cases} \{x, \alpha x, \alpha^*\mathcal{P}^*x^*, \mathcal{P}^*x^*\},\ \tilde{x} = x \in X; \\ \\ \{x^*, \alpha^*x^*, \alpha\mathcal{P}x, \mathcal{P}x\},\ \tilde{x} = x^* \in X^* \end{cases} \tag{8.4.2}$$

in which x and x^* are the corresponding elements determined by the duality. In fact, the difference between $\tilde{K}x$ and $\tilde{K}x^*$ are in the exchange of the superscripts: star and no star.

Since for any $x \in X$, it is easily checked that

$$\begin{aligned}&(x, \mathcal{P}^{*2}x^*, \mathcal{P}\alpha\beta\mathcal{P}^{-1}x, \mathcal{P}^*\alpha^*\beta^*\mathcal{P}^*x^*)\\ &\quad = (x, \tilde{\mathcal{P}}\tilde{\alpha}\tilde{\beta}x, (\tilde{\mathcal{P}}\tilde{\alpha}\tilde{\beta})^2x, (\tilde{\mathcal{P}}\tilde{\alpha}\tilde{\beta})^3x)\end{aligned} \tag{8.4.3}$$

and for any $x^* \in X^*$ as well, the map $\tilde{M}$ is a quadrangulation which is called the *associate quadrangulation* of the dual pair (M, M^*).

For planar quadrangulation M, if M is not proper, then there is a quadrangle which does not form a face boundary. Such a quadrangle is called a *quad*. A quad C is said to be *maximal* if there does not exist another quad C' such that C' is in the inner domain of C on M. From the maximality and the planarity, any two maximal quads have no point in common with their inner domains.

Lemma 8.4.1 *Any two maximal quads have at most two vertices in common, which are the two ends of the common edge.*

Proof If two maximal quads have at least three vertices in common, then they have to have at least two adjacent edges in common. From the quadrangularity this leads to a contradiction to the maximality.

By the same reason, two maximal quads do not have two common vertices which are not the two ends of a common edge. □

Let $\mathcal{M}_{\mathrm{pq}}$ be the set of all rooted proper quadrangulations. For $M \in \mathcal{M}_{\mathrm{pq}}$, let F be the set of all faces of M. Denote the set of all the subsets with i elements of F by $\binom{F}{i}$. For $f^{(i)} = \{f_1, \cdots, f_i\} \in \binom{F}{i}$, let $M(f^{(i)} + M^{(i)})$,

$$M^{(i)} = (M_1, \cdots, M_i) \in \prod_{j=1}^{i} \mathcal{M}_j,$$

be the map obtained by adding M_j into f_i with the outerface boundary of M_j and the boundary of f_i being identified on M. Then $M(f^{(i)} + M^{(i)})$ is called the *face addition* of M with $f^{(i)}$ and $M^{(i)}$. Write

$$M\Big(f^{(i)} \times \prod_{j=1}^{i} \mathcal{M}_i\Big) = \Big\{M(f^{(i)} + M^{(i)})\Big| \;\; \forall M^{(i)} \in \prod_{j=1}^{i} \mathcal{M}_j\Big\}, \tag{8.4.4}$$

which is called the *face production* of M with $f^{(i)}$ and $\prod_{j=1}^{i} \mathcal{M}_j$. And further,

$$\mathcal{M}^{F\times\prod_{j=1}^{|F|}\mathcal{M}_j} = \bigcup_{M\in\mathcal{M}} \bigcup_{\substack{f^{(i)}\in\binom{F}{i}\\ i\geq 0}} M\Big(f^{(i)} \times \prod_{j=1}^{i} \mathcal{M}_j\Big), \tag{8.4.5}$$

which is called the *face power* of $\mathcal{M}$ with $\prod_{j=1}^{|F|}\mathcal{M}_j$.

If $\mathcal{M}_1=\mathcal{M}_2=\cdots=\mathcal{M}_{|F|}=\mathcal{N}$, then we briefly write

$$M\Big(f^{(i)}\times\prod_{j=1}^{i}\mathcal{M}_j\Big)=M(f^{(i)}\times\mathcal{N}^i) \tag{8.4.6}$$

and

$$\mathcal{M}^{F\times\prod_{j=1}^{|F|}\mathcal{M}_j}=\mathcal{M}^{F\times\mathcal{N}^{|F|}}. \tag{8.4.7}$$

Lemma 8.4.2 *Let $\mathcal{Q}$ be the set of all rooted planar quadrangulations. Then we have*

$$\mathcal{Q}=\mathcal{M}_{\mathrm{pq}}^{F\times\mathcal{Q}^{|F|}}. \tag{8.4.8}$$

Proof For any map $Q\in\mathcal{Q}$, assume Q has maximal quads M_1, $M_2,\cdots,M_i$ all together. Then from Lemma 8.4.1, we may find a map M by replacing all the maps $M_j, 1\le j\le i$, with faces $f_j, 1\le j\le i$. It is easily checked that $M\subset\mathcal{M}_{\mathrm{pq}}$ and that

$$Q=M(f^{(i)}+M^{(i)})$$

where $f^{(i)}=(f_1,\cdots,f_i)$ and $M^{(i)}=(M_1,\cdots,M_i)$. This implies that Q is a member of the set on the right hand side of (8.4.8).

Conversely, since both $\mathcal{Q}$ and $\mathcal{M}_{\mathrm{pq}}$ are planar quadrangulations, it is obvious that any map Q in the set on the right hand side of (8.4.8) is a planar quadrangulation properly rooted. This is to say that Q is a member of $\mathcal{Q}$.

In consequence, the lemma follows. □

For a quadrangulation $M=(\mathcal{X}_{\alpha,\beta},\mathcal{P})$, we may construct a pair of maps denoted by $\tilde{M}=(\tilde{\mathcal{X}}_{\tilde{\alpha},\tilde{\beta}},\tilde{\mathcal{P}})$ and $\hat{M}=(\hat{\mathcal{X}}_{\hat{\alpha},\hat{\beta}},\hat{\mathcal{P}})$ which are obtained by expanding from

$$\begin{aligned}\tilde{K}x&=\{x,\alpha\beta(\mathcal{P}\alpha\beta)^3\alpha x,\alpha\beta(\mathcal{P}\alpha\beta)\alpha x,\alpha(\mathcal{P}\alpha\beta)^2\alpha x\}\\&=\{x,\alpha\mathcal{P}x,\alpha\beta(\mathcal{P}\alpha\beta)\alpha x,\alpha(\mathcal{P}\alpha\beta)^2\alpha x\}\end{aligned} \tag{8.4.9}$$

with the orbit $\mathrm{orb}_{\tilde{\mathcal{P}}}x=(x,\mathcal{P}x,\mathcal{P}^2x,\cdots)$ denoted by $\tilde{v}_x$ as a vertex as well and

$$\hat{K}\beta x=\{\beta x,\alpha\mathcal{P}\beta x,\alpha(\mathcal{P}\alpha\beta)^3\alpha x,\alpha\beta(\mathcal{P}\alpha\beta)^2\alpha x\} \tag{8.4.10}$$

with $\mathrm{orb}_{\hat{\mathcal{P}}}(\beta x) = (\beta x, \mathcal{P}\beta x, \mathcal{P}^2\beta x, \cdots)$ denoted by $\hat{v}_{\beta x}$ respectively for a face

$$\Big(\alpha x, (\mathcal{P}\alpha\beta)\alpha x, (\mathcal{P}\alpha\beta)^2\alpha x, (\mathcal{P}\alpha\beta)^3\alpha x\Big)$$

chosen. From Theorem 1.1.5, it is seen that the pair of maps $\tilde{M}$ and $\hat{M}$ are uniquely defined in this way.

Lemma 8.4.3 *For a quadrangulation M, let $\tilde{M}$ and $\hat{M}$ be the maps determined by* (8.4.9) *and* (8.4.10) *shown above respectively. Then $\tilde{M}$ and $\hat{M}$ are dual one to another.*

Proof From (8.4.9–10), we may assume that $\tilde{x}$ in $\tilde{M}$ is corresponding to x in M and that $\hat{x}$ in $\hat{M}$ is to βx in M without loss of generality. Since

$$\tilde{\mathcal{P}}\tilde{\alpha}\tilde{\beta}\tilde{x} = \tilde{\mathcal{P}}\tilde{\alpha}\tilde{\beta}x = \mathcal{P}(\alpha(\mathcal{P}\alpha\beta)^2\alpha x)$$

from (8.4.9) and

$$\hat{\mathcal{P}}\hat{x} = \mathcal{P}\beta x = (\mathcal{P}\alpha\beta)\alpha x,$$

by considering

$$(\mathcal{P}\alpha\beta)\alpha x = \beta\Big(\mathcal{P}\alpha(\mathcal{P}\alpha\beta)^2\alpha x\Big),$$

we soon see that $\tilde{M}$ and $\hat{M}$ are duals of each other. □

For a planar map M, let $\{v, u\}$ be a *separator*, *i.e.*, $M = M_1 \cup M_2$ provided $M_1 \cap M_2 = \{u, v\}$. Both M_1 and M_2 are submaps of M. For convenience, we may assume that

$$u = (s_u, S_u, s'_u, t_u, T_u, t'_u)$$

and

$$v = (s_v, S_v, s'_v, t_v, T_v, t'_v) \tag{8.4.11}$$

such that $\langle s_u, S_u, s'_u\rangle$ and $\langle s_v, S_v, s'_v\rangle$ are in M_1 and that $\langle t_u, T_u, t'_u\rangle$ and $\langle t_v, T_v, t'_v\rangle$ are in M_2 without loss of generality. From the separability of $\{u, v\}$, there are two faces f and g of M such that

$$f = (s_u, F_u, \alpha\beta s'_v, t_v, G, \alpha\beta t'_u)$$

and

$$g = (t_u, G_u, \alpha\beta t'_v, s_v, F_v, \alpha\beta s'_u) \tag{8.4.12}$$

with $\langle s_u, F_u, \alpha\beta s'_v\rangle$ and $\langle s_v, F_v, \alpha\beta s'_u\rangle$ in M_1 and the other two parts in M_2. Then, the associate quadrangulation $\tilde{M}$ of the dual pair (M, M^*) has a quad which is incident with u, v f and g.

Lemma 8.4.4 *A planar map M is a c-net iff the associate quadrangulation $\tilde{M}$ of (M, M^*) has no quad.*

Proof The sufficiency is obvious since, otherwise, if M is not a c-net, *i.e.*, M has a separator which consists of two vertices, it causes that $\tilde{M}$ should have a quad from what was mentioned above, and hence a contradiction to the given condition appears.

To prove the necessity by contradiction. For any planar quadrangulation $\tilde{M}$, let M and M^* be the dual pair corresponding to $\tilde{M}$. If $\tilde{M}$ has a quad which is assumed to be with the four vertices u, v, w and x in correspondence with the direction of the 4-circuit, from the planarity and the construction of $\tilde{M}$, $\{u, w\}$ and $\{v, x\}$ are a separator of one and the other between M and M^*. This implies the necessity.

In consequence, the lemma follows. □

A map which has all of its faces quadrangles with probably only one exception is called a *near quadrangulation.* Of course, the exception face is chosen to be the rooted one as usual. Let $\mathcal{M}_{\rm nq}$ be the set of all rooted planar near quadrangulations and $\mathcal{M}_{\rm pnq} \subseteq \mathcal{M}_{\rm nq}$ consists of all those proper quadrangulations. Because all the boundaries of inner faces construct a basis of the cycle space, it is easy to see that the root-face has a valency of an even number for any map in $\mathcal{M}_{\rm nq}$. Write

$$q_{\mathcal{M}_{\rm nq}}(x, y) = \sum_{M \in \mathcal{M}_{\rm nq}} x^{m(M)} y^{n(M)} \tag{8.4.13}$$

and

$$q_{\mathcal{M}_{\rm pnq}}(x, y) = \sum_{M \in \mathcal{M}_{\rm pnq}} x^{m(M)} y^{n(M)} \tag{8.4.14}$$

where $n(M)$ and $2m(M)$ are the number of inner vertices and the valency of root-face of M respectively.

Let $q^* = q^*(y)$ be the coefficient of x^2 in $q_{\mathcal{M}_{\rm nq}}(x, y)$, *i.e.*, q^* is the enufunction of quadrangulations in $\mathcal{M}_{\rm nq}$ with the only one parameter, the number of inner vertices. Then , from Lemma 8.4.2 we have

$$q_{\mathcal{M}_{\mathrm{nq}}}(x,y) = \sum_{M\in\mathcal{M}_{\mathrm{pnq}}} x^{m(M)}y^{n(M)}q^{*n(M)+m(M)-1}$$

$$= q^{*^{-1}} \sum_{M\in\mathcal{M}_{\mathrm{pnq}}} (xq^*)^{m(M)}(yq^*)^{n(M)}$$

$$= q^{*^{-1}} q_{\mathcal{M}_{\mathrm{pnq}}}(xq^*, yq^*)$$

in which $n(M)+m(M)-1$ is the number of nonrooted faces of M from the near quadrangularity and the Euler formula for planar maps. This implies the following relation:

$$q^* q_{\mathcal{M}_{\mathrm{nq}}}(x,y) = q_{\mathcal{M}_{\mathrm{pnq}}}(xq^*, yq^*). \tag{8.4.15}$$

Further, let $p^* = p^*(y)$ be the enufunction of proper quadrangulations in $\mathcal{M}_{\mathrm{pnq}}$ with the number of inner vertices as the only parameter. In what follows, we investigate the relationship between q^* and p^*.

A quadrangulation which has a pair of edges with a common end and the other two ends both on the outerface boundary is said to be *diagonal.* The pair of edges is called a diagonal of the quadrangulation.

Let $\mathcal{M}_{\mathrm{q}}$ and $\mathcal{M}_{\mathrm{pq}}$ be the sets of all rooted planar and proper planar quadrangulations respectively. Then each of them can be divided into diagonal and nondiagonal ones. Write p^*_{D} as the enufunction of diagonal proper quadrangulations including the map which is a quadrangle itself and p^*_{N}, that of all others in $\mathcal{M}_{\mathrm{q}}$. Because it can be shown that there are only two diagonal proper quadrangulations which are those with the root $r = x_1$ and $\alpha\beta x_3$ in

$$Q = \Big(\sum_{i=1}^{6} Kx_i, \mathcal{P}\Big)$$

where

$$\mathcal{P} = (x_1, x_2, x_3)(\alpha\beta x_3, x_4)(\alpha\beta x_4, x_5, x_6)$$
$$(\alpha\beta x_6, \alpha\beta x_1)(\alpha\beta x_2, \alpha\beta x_5),$$

we soon see that

$$p^* = 1 + 2y + p^*_{\mathrm{N}}. \tag{8.4.16}$$

Lemma 8.4.5 *For p^* and q^*, we have the following relation:*

$$p^* + \frac{1}{q^*} = 1 + 2y + \frac{1 - yq^*}{1 + yq^*}. \tag{8.4.17}$$

Proof Let q^*_{D} and q^*_{N} be the contributions of diagonal and not to q respectively. By a similar discussion to Lemma 8.4.2, we have

$$q^*_{\mathrm{N}} = q^* p^*_{\mathrm{N}}(yq^*) + 1 \tag{8.4.18}$$

and hence

$$q^* = q^*_{\mathrm{D}} + q^* p^*_{\mathrm{N}}(yq^*) + 1. \tag{8.4.19}$$

For diagonal quadrangulations, there are two distinct possibilities: the root-vertex is incident with the diagonal, *i.e.*, the pair of edges to make a quadrangulation diagonal, and not. Let q^*_{R} and $q^*_{\bar{\mathrm{R}}}$ be the respective corresponding contributions to q^*_{D}. From the symmetry,

$$q^*_{\mathrm{R}} = q^*_{\bar{\mathrm{R}}} = \tfrac{1}{2} q^*_{\mathrm{D}}. \tag{8.4.20}$$

Because any diagonal quadrangulation M with the root-vertex incident to a diagonal has the form of identifying the two edges $e_{r_1}, e_{\mathcal{P}\alpha\beta r_1}$ of M_1 respectively with $e_{r_2}, e_{\mathcal{P}\alpha\beta r_2}$ of M_2 for M_1 being a nondiagonal or diagonal not incident to the root-vertex quadrangulation and M_2 being any map in $\mathcal{M}_{\mathrm{q}}$ such that the root of M is $r(M) = (\mathcal{P}_1\alpha\beta)^2 r_1$, we may find that

$$q^*_{\mathrm{R}} = y(q^*_{\mathrm{N}} + q^*_{\bar{\mathrm{R}}})q^*. \tag{8.4.21}$$

By substituting (8.4.20) into (8.4.21),

$$q^*_{\mathrm{D}} = \frac{2yq^*_{\mathrm{N}}q^*}{1 - yq^*} \tag{8.4.22}$$

can be soon obtained. Then by substituting (8.4.18) into (8.4.22),

$$q^*_{\mathrm{D}} = \frac{2yq^{*2} p^*_{\mathrm{N}}(yq^*) + 2yq^*}{1 - yq^*}. \tag{8.4.23}$$

From (8.4.23) and (8.4.19), we have

$$p^*_{\mathrm{N}}(yq^*) = \frac{1 - yq^*}{1 + yq^*} - \frac{1}{q^*}. \tag{8.4.24}$$

Finally, by (8.4.20) the lemma is derived. □

According to the duality, from the expressions in (5.4.15–16), we may find the corresponding parametric expression for determining the enufunction of rooted general planar quadrangulations. Then, by considering that all general quadrangulations can be obtained by replacing edges on maps in $\mathcal{M}_{\rm q}$ with maps in $\mathcal{M}_{\rm nq}$ with one additional edge forming a multiedge to the root-edge as the root-face boundary of each, we may further extract the parametric expressions for determining q^*. Finally, according to Lemma 8.4.5 we may derive the parametric expressions for p^*.

By checking the differential equation (8.2.20), we may obtain that

$$N_{\mathcal{M}_{\rm c\text{-}n}}(m) = Q_{\rm pq}(m-2) \tag{8.4.25}$$

where $Q_{\rm pq}(m-2)$ is the number of rooted proper planar quadrangulations with $m-2$ inner vertices and $N_{\mathcal{M}_{\rm c\text{-}n}}(m)$ is the number of c-nets of size m for $m \geq 6$.

Theorem 8.4.1 *The number of rooted proper planar quadrangulations with n inner vertices is*

$$\begin{aligned} N_{\mathcal{M}_{\rm pq}}(n) = \Delta_n + \frac{21}{32} \sum_{i=-1}^{\lfloor \frac{n+1}{2} \rfloor} \Big(7(n-3i)^2 - 119n \\ +266i + 146\Big)(i+3)\Big(7(n-2i) - 40\Big) \\ \times \frac{(i+2)(2(n-2i)-13)!}{2^{2i+2}(n-2i-7)!(n-2i-1)!} \end{aligned} \tag{8.4.26}$$

where

$$\Delta_n = \begin{cases} -2 + \dfrac{(n+1)^2}{2^n}, & n = 1 \,(\mathrm{mod}\ 2); \\ 2 - \dfrac{n^2+24n+8}{2^{n+3}}, & n = 0 \,(\mathrm{mod}\ 2) \end{cases} \tag{8.4.27}$$

for $n \geq 4$.

Proof From (8.4.25) and (8.2.21–22) the theorem is obtained. □

It is, in fact, shown that the correspondence between c-nets and proper quadrangulations as described above is a bijection.

§8.5 General maps on the surfaces

Let $\mathcal{M}$ be the set of all rooted maps on orientable surfaces of all possible genera. For $M \in \mathcal{M}$, write $M = (\mathcal{X}, \mathcal{P})$ such that

$$v_x = (x, \mathcal{P}x, \mathcal{P}^2x, \cdots, \mathcal{P}^jx, \cdots) \tag{8.5.1}$$

is the orbit of $\mathcal{P}$ at $x \in \mathcal{X}$, or in other words, the vertex of M.

The root of $M-a, a = e_r(M)$, the root-edge, is chosen to be $\mathcal{P}\alpha\beta r$, r is the root of M when $\mathcal{P}\alpha\beta r \neq r$; $\mathcal{P}r$, otherwise if $M - a$ is still a map.

Three classes of $\mathcal{M}$ are considered as

$$\mathcal{M} = \mathcal{M}_{\mathrm{I}} + \mathcal{M}_{\mathrm{II}} + \mathcal{M}_{\mathrm{III}} \tag{8.5.2}$$

where $\mathcal{M}_{\mathrm{I}}$ consists of the vertex map ϑ,

$$\mathcal{M}_{\mathrm{II}} = \{M |\ \forall M \in \mathcal{M} - \vartheta, \text{ the underlying graph } G(M-a) \text{ of } M-a \text{ is disconnected}\}$$

and hence

$$\mathcal{M}_{\mathrm{III}} = \{M |\ \forall M \in \mathcal{M} - \vartheta,\ G(M-a) \text{ is connected}\}.$$

The enufunction concerned with only one parameter here is as

$$f_{\mathcal{M}}(x) = \sum_{M \in \mathcal{M}} x^{m(M)} \tag{8.5.3}$$

where $m(M)$ is the size of M.

Lemma 8.5.1 *Let $\mathcal{M}_{\langle \mathrm{II} \rangle} = \{M - a | \forall M \in \mathcal{M}_{\mathrm{II}}\}$ where $a = e_r(M)$ as well. Then we have*

$$\left|\mathcal{M}_{\langle \mathrm{II} \rangle}\right| = \left|\mathcal{M} \times \mathcal{M}\right|. \tag{8.5.4}$$

Proof For any $M \in \mathcal{M}_{\langle \mathrm{II} \rangle}$, there exists a map $M' \in \mathcal{M}_{\mathrm{II}}$ such that $M' - a' = M_1 + M_2$ where $a' = e_r(M')$. Because both of M_1 and M_2 are allowed to be any map in $\mathcal{M}$, we have $(M_1, M_2) \in \mathcal{M} \times \mathcal{M}$. This implies $|\mathcal{M}_{\langle \mathrm{II} \rangle}| \leq |\mathcal{M} \times \mathcal{M}|$.

Conversely, for any $(M_1, M_2) \in \mathcal{M} \times \mathcal{M}$ we may uniquely construct a map $M' = (\mathcal{X}', \mathcal{P}')$ by adding a new edge Kr' joining the root-vertices of M_1 and M_2 such that $\mathcal{X}' = \mathcal{X} \cup Kr'$ and

$$v_{r'} = (r', v_{r_1}), \ v_{\beta r'} = (v_{r_2}, \alpha\beta r')$$

determine $\mathcal{P}'$ with all other orbits being the same as those of $\mathcal{P}$. It is easily seen that

$$M' - a' = M_1 + M_2, \ a' = Kr'.$$

Since $M' \in \mathcal{M}_{\mathrm{II}}, M = M_1 + M_2 = M' - a \in \mathcal{M}_{\langle \mathrm{II} \rangle}$. This implies $|\mathcal{M} \times \mathcal{M}| \leq |\mathcal{M}_{\langle \mathrm{II} \rangle}|$. □

For $M = (\mathcal{X}, \mathcal{P}) \in \mathcal{M}_{\mathrm{III}}$, because $M - a$ is still a map, the root of $M - a$ has two possibilities: $r(M - a) = \mathcal{P}\alpha\beta r(M)$ and $\mathcal{P}r(M)$. Let $\tilde{M} = (\tilde{\mathcal{X}}, \tilde{\mathcal{P}}) = M - a$. Then the latter corresponds to exactly one map $M = (\mathcal{M}, \mathcal{P})$ obtained by adding an edge $a = Kr$ from $\tilde{M}$ in the way that $\mathcal{X} = \tilde{\mathcal{X}} \cup Kr$ with $v_r = (\alpha\beta r, r, v_{\tilde{r}})$ where $\tilde{r} = r(\tilde{M})$, the root of $\tilde{M}$. In fact, $\tilde{a} = e_{\tilde{r}}$ is a loop. However, the former has the following $\frac{1}{2}|\tilde{\mathcal{X}}|$ ways to obtain M. For convenience, let $v_x = (x, S), x \in \tilde{\mathcal{X}}$. From the orientability of $\tilde{M}$, the group Ψ_H generated by $H = \{\alpha\beta, \tilde{\mathcal{P}}\}$ has exactly two orbits on $\tilde{\mathcal{X}}$, as shown in §1.1. Write $\tilde{\mathcal{X}}_1$ as one of the two orbits. Then for any $x \in \tilde{\mathcal{X}}_1$, we may construct a map M by adding a new edge $a = e_r(M) = Kr$ joining v_x and $v_{\tilde{r}}$ such that

$$v_x = (x, r, S) \text{ and } v_r = (\alpha\beta r, \tilde{r}, \tilde{R})$$

where $v_{\tilde{r}} = (\tilde{r}, \tilde{R})$ in $\tilde{M}$. Of course, $v_x = v_{\tilde{r}}$ is allowed as a special case. Since $|\tilde{\mathcal{X}}_1| = \frac{1}{2}|\tilde{\mathcal{X}}|$, the discussion above shows that there is $2m(\tilde{M})+1$ distinct maps M such that $\tilde{M} = M - a$ for $\tilde{M} \in \mathcal{M}_{\langle \mathrm{III} \rangle}$.

Lemma 8.5.2 *Let* $\mathcal{M}_{\langle \mathrm{III} \rangle} = \{M - a | \forall M \in \mathcal{M}_{\mathrm{III}}\}$ *where* $a = e_r(M)$. *Then, we have*

$$\mathcal{M}_{\langle \mathrm{III} \rangle} = \mathcal{M}. \tag{8.5.5}$$

Proof Because for any $M \in \mathcal{M}_{\mathrm{III}}, M - a$ is also a map, we soon see that $\mathcal{M}_{\langle\mathrm{III}\rangle} \subseteq \mathcal{M}$.

Conversely, for any $\tilde{M} \in \mathcal{M}$ we may construct a map M in any way described above such that $\tilde{M} = M - a$ and $M \in \mathcal{M}_{\mathrm{III}}$. This implies $M \in \mathcal{M}_{\langle\mathrm{III}\rangle}$. Thus, $\mathcal{M} \subseteq \mathcal{M}_{\langle\mathrm{III}\rangle}$. □

Denote by $\mathcal{H}(\tilde{M})$ for $\tilde{M} \in \mathcal{M}$ the set of all the $2m(\tilde{M}) + 1$ maps $M \in \mathcal{M}_{\mathrm{III}}$.

Lemma 8.5.3 *For $\mathcal{M}_{\mathrm{III}}$, we have*

$$\mathcal{M}_{\mathrm{III}} = \sum_{M \in \mathcal{M}} \mathcal{H}(M). \tag{8.5.6}$$

Proof For any $M \in \mathcal{M}_{\mathrm{III}}$, let $\tilde{M} = M - a$. From Lemma 8.5.2, we see that $M \in \mathcal{H}(\tilde{M}), \tilde{M} \in \mathcal{M}$. Thus, M is a member of the set on the right hand side of (8.5.6) and hence $\mathcal{M}_{\mathrm{III}}$ is a subset.

Conversely for any $M \subset \mathcal{H}(\tilde{M})$, $\tilde{M} \in \mathcal{M}$, because $M - a$, $a = e_r$, the root-edge of M, is still a map, we see that $M \in \mathcal{M}_{\mathrm{III}}$. This implies that the set on the right is also a subset of $\mathcal{M}_{\mathrm{III}}$. □

From what is discussed above, the formula (8.5.6) provides a 1-to-1 correspondence between the sets on the two sides.

Because ϑ has no edge, the contribution of $\mathcal{M}_{\mathrm{I}}$ to $f_{\mathcal{M}}(x)$ is

$$f_{\mathcal{M}_{\mathrm{I}}}(x) = 1. \tag{8.5.7}$$

According to Lemma 8.5.1, the contribution of $\mathcal{M}_{\mathrm{II}}$ to $f_{\mathcal{M}}(x)$ is

$$f_{\mathcal{M}_{\mathrm{II}}}(x) = x f_{\mathcal{M}}^2(x). \tag{8.5.8}$$

Then, based on Lemma 8.5.3, we have the contribution of $\mathcal{M}_{\mathrm{III}}$ to $f_{\mathcal{M}}(x)$ is

$$\begin{aligned} f_{\mathcal{M}_{\mathrm{III}}}(x) &= x \sum_{M \in \mathcal{M}} \Big(2m(M) + 1\Big) x^{m(M)} \\ &= x f_{\mathcal{M}}(x) + 2x \sum_{M \in \mathcal{M}} m(M) x^{m(M)} \end{aligned}$$

$$= xf_{\mathcal{M}}(x) + 2x^2 \frac{\mathrm{d}}{\mathrm{d}x} f_{\mathcal{M}}(x). \tag{8.5.9}$$

Theorem 8.5.1 *The enufunction $f = f_{\mathcal{M}}(x)$ of rooted maps on orientable surfaces with the size as the parameter satisfies the following differential equation:*

$$2x^2 \frac{\mathrm{d}f}{\mathrm{d}x} = -1 + (1-x)f - xf^2 \tag{8.5.10}$$

such that $f_0 = f(0) = 1$.

Proof Based on (8.5.2), by considering (8.5.7–9) the theorem can soon be obtained. □

According to Theorem 8.5.1, we are allowed to determine f by writing f as a power series of x and then by comparing the coefficients on the two sides of (8.5.10) step by step, or by solving the differential equation (8.5.10) directly. The latter looks complicated.

As for the enumeration of nonorientable maps, or maps on nonorientable surfaces, it appears with more complications than that for orientable case. However, for the simplicity, we here only discuss nonorientable maps with a single vertex.

Let $\mathcal{U}$ be the set of all rooted nonorientable maps with one vertex. Three classes $\mathcal{U}_{\mathrm{I}}, \mathcal{U}_{\mathrm{II}}$ and $\mathcal{U}_{\mathrm{III}}$ of $\mathcal{U}$ are partitioned into. Here, $\mathcal{U}_{\mathrm{I}}$ consists of a single map which is the loop map $L_2 = (Kr, (r, \beta r))$ on the projective plane as shown in §1.1.

$$\mathcal{U}_{\mathrm{II}} = \{M | \forall M \in \mathcal{U} - L_2, M - a \text{ is orientable}\}.$$

And hence,

$$\mathcal{U}_{\mathrm{III}} = \{M | \forall M \in \mathcal{U} - L_2, M - a \text{ is nonorientable}\}.$$

Of course, $a = e_r(M)$ is the root-edge of M as usual.

The enufunction considered here is

$$g_{\mathcal{U}}(x) = \sum_{M \in \mathcal{U}} x^{m(M)} \tag{8.5.11}$$

where $m(M)$ is the size of M.

It is easily seen that the contribution of $\mathcal{U}_{\rm I}$ to $g_{\mathcal{U}}(x)$ is

$$g_{\mathcal{U}_{\rm I}}(x) = x \tag{8.5.12}$$

since L_2 has only one edge.

For $M = (\mathcal{X}_{\alpha,\beta}, \mathcal{P}) \in \mathcal{U}_{\rm II}$, since $M - a$ is orientable, the root-vertex of M has to be in the form as

$$(r, \beta r, R).$$

The root of $M - a$ is chosen to be $\mathcal{P}\alpha r$.

Lemma 8.5.4 *Let* $\mathcal{U}_{\langle \rm II\rangle} = \{M - a|\forall M \in \mathcal{U}_{\rm II},\ a = e_r(M)$. *Then, we have*

$$\mathcal{U}_{\langle \rm II\rangle} = \mathcal{D} \tag{8.5.13}$$

where $\mathcal{D}$ *is the set of orientable maps with a single vertex excluding the vertex map.*

Proof For any $M \in \mathcal{U}_{\langle \rm II\rangle}$ there exists a map $U \in \mathcal{U}_{\rm II}$ such that $M = U - a$, $a = e_r(U)$, which is an orientable one with a single vertex as well and hence $M \in \mathcal{D}$.

Conversely, for any $M \in \mathcal{D}$, we may always construct a map $\tilde{M}$ by adding an edge $K\tilde{r}$, a loop precisely, such that the root-vertex of $\tilde{M}$ is in the form as

$$(\tilde{r}, \beta\tilde{r}, R, \alpha r)$$

where $(\alpha r, R)$ is the conjugate orbit of the root vertex of M. Because M has only one vertex, $\tilde{M}$ has only one vertex as well. By considering that $\tilde{M}$ has the same number of faces as M does and one edge greater than M, from Euler formula $\tilde{M}$ is nonorientable and hence $\tilde{M} \in \mathcal{U}_{\rm II}$. Since $\tilde{M} - a = M$, we see that $M \in \mathcal{U}_{\langle \rm II\rangle}$. □

Let $h = f_{\mathcal{D}}(x)$ be the enufunction of $\mathcal{D}$ with the size as the parameter. Then from Lemma 8.5.4, the contribution of $\mathcal{U}_{\rm II}$ to $g = g_{\mathcal{U}}(x)$ is

$$g_{\mathcal{U}_{\rm II}}(x) = xh. \tag{8.5.14}$$

Lemma 8.5.5 *Let* $\mathcal{U}_{\langle \rm III\rangle} = \{M - a|\forall M \in \mathcal{U}_{\rm III}\}$. *Then we have*

$$\mathcal{U}_{\langle \rm III\rangle} = \mathcal{U}. \tag{8.5.15}$$

Proof For $M \in \mathcal{U}_{\langle \mathrm{III} \rangle}$, there exists a map $\tilde{M} \in \mathcal{U}_{\mathrm{III}}$ such that $M = \tilde{M} - \tilde{a}$ where $\tilde{a} = e_r(\tilde{M})$. Thus M is nonorientable. Because $\tilde{M}$ has only one vertex, so does M. This implies $M \in \mathcal{U}$.

Conversely, for $M \in \mathcal{U}$, we can construct a map $\tilde{M}$ by adding an edge $K\tilde{r}$ such that the root-vertex of $\tilde{M}$ as

$$(\tilde{r}, \alpha\beta\tilde{r}, r, R)$$

where (r, R) is the root-vertex of M. It is easy to see that $\tilde{M} - \tilde{a} = M$. By the discussion in §1.1, we also know that $\tilde{M}$ is nonorientable, and hence $\tilde{M} \in \mathcal{U}_{\mathrm{III}}$. This implies $M \in \mathcal{U}_{\langle \mathrm{III} \rangle}$. □

For any $M = (\mathcal{X}, \mathcal{P}) \in \mathcal{U}$, we may find $|\mathcal{X}| + 1$ maps $\tilde{M}$ in $\mathcal{U}_{\mathrm{III}}$ such that $\tilde{M} - \tilde{a} = M$, $\tilde{a} = e_r(\tilde{M})$, in the following way. Let x be any member of $\mathcal{X}$. Then, $\tilde{M}$ can be uniquely constructed as the root-vertex with the form

$$(\tilde{r}, R, \alpha\beta\tilde{r}, r, T)$$

where (r, T, R) is the root-vertex of M. There are

$$|\mathcal{X}| = 4m(M)$$

maps obtained in this way. From Axiom 3 in §1.1, all of them are distinct. It can also be shown that they are all in $\mathcal{U}_{\mathrm{III}}$. Another one is that with the root-vertex as

$$(\tilde{r}, r, R, \alpha\beta\tilde{r})$$

where (r, R) is the root-vertex of M.

Lemma 8.5.6 *For $\mathcal{U}_{\mathrm{III}}$, we have*

$$\mathcal{U}_{\mathrm{III}} = \sum_{M \in \mathcal{U}} \tilde{\mathcal{H}}(M) \tag{8.5.16}$$

where $\tilde{\mathcal{H}}(M)$ is the set of the $4m(M) + 1$ maps obtained from M in the way described above.

Proof For any $M \in \mathcal{U}_{\mathrm{III}}$, from Lemma 8.5.5 $\tilde{M} = M - a \in \mathcal{U}$. Since $M \in \tilde{\mathcal{H}}(\tilde{M})$, we see that M is a member of the set on the right hand side of (8.5.16).

Conversely, for M in the set on the right, there exists a map $\tilde{M} \in \mathcal{U}$ such that $M \in \tilde{\mathcal{H}}(\tilde{M})$. From the procedure of constructing M from $\tilde{M}$, by the nonorientability of $\tilde{M}$, we find that $M \in \mathcal{U}_{\rm III}$. □

According to Lemma 8.5.6, the contribution of $\mathcal{U}_{\rm III}$ to $g = g_{\mathcal{U}}(x)$ is

$$g_{\mathcal{U}_{\rm III}}(x) = x \sum_{m \in \mathcal{U}} \Big(4m(M) + 1\Big) x^{m(M)}$$

$$= xg + 4x^2 \frac{\mathrm{d}g}{\mathrm{d}x}$$

$$= x\left(g + 4x\frac{\mathrm{d}g}{\mathrm{d}x}\right). \tag{8.5.17}$$

Theorem 8.5.2 *The enufunction $g = g_{\mathcal{U}}(x)$ of $\mathcal{U}$ defined by (8.5.11) satisfies the following differential equation:*

$$4x^2\frac{\mathrm{d}g}{\mathrm{d}x} = (1-x)g - x(1+h) \tag{8.5.18}$$

where $h = f_{\mathcal{D}}(x)$ is the enufunction of $\mathcal{D}$ with the size as the parameter such that $g_0 = g(0) = 0$.

Proof Since $\mathcal{U}$ is partitioned into $\mathcal{U}_{\rm I}$, $\mathcal{U}_{\rm II}$ and $\mathcal{U}_{\rm III}$, from (8.5.12), (8.5.14) and (8.5.17), the theorem is soon derived. □

Now the only thing remaining for g is to determine h. By considering $\mathcal{D}$ as a subset of $\mathcal{U}$, we see that $\mathcal{D}$ is also a subset of $\mathcal{U}_{\rm III}$. It is easily checked that $\mathcal{D}$ is itself satisfied by Lemma 8.5.3. So we have

$$h = x \sum_{M \in \mathcal{D}+\vartheta} \Big(2m(M) + 1\Big) x^{m(M)}$$

$$= x(h+1) + 2x^2\frac{\mathrm{d}h}{\mathrm{d}x}. \tag{8.5.19}$$

Because there is only one map of size 1 in $\mathcal{D}$, we soon find that h satisfies the following differential equation:

$$2x^2\frac{\mathrm{d}h}{\mathrm{d}x} = (1-x)h - x;$$

$$\left.\frac{\mathrm{d}h}{\mathrm{d}x}\right|_{x=0} = 1. \tag{8.5.20}$$

From the equation, we soon find that

$$h = \sum_{i\geq 1} \frac{(2i-1)!}{2^{i-1}(i-1)!} x^i. \tag{8.5.21}$$

§8.6 Notes

8.6.1 Although the decomposition used in §8.1 is different from that in [Liu13], the equation satisfied by the enufunctions here is in the same form as there. This tells us that the numbers of general planar maps respectively with the root-face valency m and the root-vertex valency m of size n are the same. This is just guaranteed by the planar duality in this case. General planar maps with binary weights on edges can be employed to represent knots and hence the methods described can be used to enumerate knots [Liu59].

8.6.2 The number shown in (8.1.24) can also be seen in Tutte's original article[Tut7] where the equation was given by different method from here. Tutte also provided the parametric expressions for determining the number of general maps for the order m and the dual order n given [Tut13]. However, no simpler formula has been found yet. On this aspect, one can also find some other methods and results in [Arq1–3], [Cor1–2], [CoM1], [CDV1], [CoR1] and [CoV1].

8.6.3 It would be possible in a way similar to that discussed in §7.1 to determine the enufunction of the set of all biloopless maps where the number of the maps with the order m and the dual order n is a symmetric function of m and n.

8.6.4 For bisimple maps, it would be possible to find a functional equation satisfied by the enufunction of the maps with vertex partition and then to determine the number with the order m and the dual order n as two parameters as a symmetric function of m and n as well in a similar manner to §7.3.

8.6.5 It is still an open problem to find a compact formula for the number of cubic convex polyhedra.

8.6.6 In [Bro1], Brown showed the relationship between nonseparable maps and quadrangulations on the disk. Then, in [MuSc1], Mullin and Schellenberg provided the relationship between quadrangulations and c-nets on the sphere. What happens on more general surfaces is still a problem.

8.6.7 General rooted maps on orientable surface were first considered by Walsh and Lehman[WaL1–3]. However, general nonorientable maps have not yet been investigated before.

8.6.8 On the number of nonisomorphic planar maps can be seen in [Lis1-5] and [Wor1]. However, such a problem on surfaces has not been observed yet.

8.6.9 Other approaches to the enumeration of general maps on surfaces (orientable and non- orientable) can be seen in [JaV1–3] and [Jac1–4].

Chapter 9

Chrosum Equations

§9.1 Tree equations

Let $\mathcal{T}$ be the set of all rooted plane trees. Because trees can only be embedded on the plane among all kinds of surfaces, orientable and nonorientable, all tree maps are plane trees.

For $T \in \mathcal{T}$, let $m(T)$ and $s(T)$ be the valencies of the root-vertex and the root-face of T, and let

$$\underline{n}(T) = (n_1(T), n_2(T), \cdots)$$

where $n_i(T)$ is the number of nonrooted vertices with valency $i, i \geq 1$ in T. Write

$$\mathcal{T}(m; \underline{n}) = \{T | \forall T \in \mathcal{T}, m(T) = m, \underline{n}(T) = \underline{n}\} \tag{9.1.1}$$

for $m, \underline{n} = (n_1, n_2, \cdots)$ of all nonnegative integers. Then we have

$$\mathcal{T} = \sum_{m,\underline{n} \geq 0} \mathcal{T}(m, \underline{n}). \tag{9.1.2}$$

Since it is easily seen that if $m = 0$, then $\underline{n} = \underline{0}$, *i.e.*, all $n_i = 0, i \geq 1$. This coordinates the vertex map ϑ. We are allowed to write

$$\mathcal{T} = \vartheta + \mathcal{T}_{\mathrm{I}} \tag{9.1.3}$$

where

$$\mathcal{T}_{\mathrm{I}} = \sum_{m>0,\underline{n} \geq 0} \mathcal{T}(m, \underline{n}). \tag{9.1.4}$$

Lemma 9.1.1 *Let $\mathcal{T}\langle m\rangle = \{T - a | \forall T \in \mathcal{T}(m)\}$ for $m > 0$ where*

$$\mathcal{T}(m) = \sum_{\underline{n}\geq 0} \mathcal{T}(m, \underline{n}) \tag{9.1.5}$$

and $a = e_r(T)$ for $T \in \mathcal{T}$. Then, we have

$$\mathcal{T}\langle m\rangle = \mathcal{T}(m-1) \times \mathcal{T}. \tag{9.1.6}$$

Proof For any $T \in \mathcal{T}\langle m\rangle$, there is another tree $T' \in \mathcal{T} - \vartheta$ such that $T' - a' = T_1 + T_2$,

$$m(T_1) = m(T') - 1 = m - 1$$

and $m(T_2) \geq 0$. Because both T_1 and T_2 are tree maps, we soon see that $T \in \mathcal{T}(m-1) \times \mathcal{T}$.

Conversely, for any $T \in \mathcal{T}(m-1) \times \mathcal{T}$, let $T = T_1 + T_2$ such that $T_1 \in \mathcal{T}(m-1)$ and $T_2 \in \mathcal{T}$. By adding an edge a' which joins the root-vertices of T_1 and T_2, we may obtain another tree map T'. Of course, $T' \neq \vartheta$. Since

$$m(T') = m(T_1) + 1 = m - 1 + 1 = m,$$

we have $T' \in \mathcal{T}(m)$. By considering that $T = T' - a', T \in \mathcal{T}\langle m\rangle$. $\square$

Lemma 9.1.2 *Let $\mathcal{T}[m] = \{T \bullet a | \forall T \in \mathcal{T}(m)\}$ for $m > 0$ where $a = e_r(T)$, the root-edge of T. Then we have*

$$\mathcal{T}[m] = \mathcal{T}(m-1) \odot \mathcal{T} \tag{9.1.7}$$

where $\odot$ appeared in §2.1.

Proof For $T \in \mathcal{T}[m]$, there is a map $T' \in \mathcal{T}(m)$ such that

$$T = T' \bullet a',\ a' = e_r(T').$$

Write the root-vertex of T as (r, S, R). Then, the root-vertex of T' should be (r', R) whose order is m, the valency of the root-vertex of T'. However, (r, S) and (R) can be extended into submaps which are both tree maps denoted by T_1 and T_2, of T. It is easy to see that

$T = T_1 \odot T_2$. Because the root-vertex of T_1 is (R), its valency is one less than that of $(r'R)$ and hence $T_1 \in \mathcal{T}(m-1)$. On the other hand, since the valency of the root-vertex of T_2 is allowed to be any nonnegative integer, by considering T_2 being a tree map we soon see that $T_2 \in \mathcal{T}$. Moreover, T_1 and T_2 have exactly one common vertex which is the root-vertex, *i.e.*, $T = T_1 \dot{+} T_2$. Thus, $T \in \mathcal{T}(m-1) \odot \mathcal{T}$.

Conversely, for $T \in \mathcal{T}(m-1) \odot \mathcal{T}$, since $T = T_1 \dot{+} T_2$ such that $m(T_1) = m-1, T_1 \in \mathcal{T}(m-1)$ and $T_2 \in \mathcal{T}$, we may write the root-vertices of T_1 and T_2 are

$$(R) \text{ and } (r, S)$$

while the root-vertex of T is, of course, (r, S, R). Let T' be the map obtained by introducing an edge Kr' such that the root-vertex and the nonrooted vertex incident with Kr' are (r', R) and $(\alpha\beta r', r, S)$ respectively. It is easy to check that $T' \in \mathcal{T}$ and the valency of its root-vertex is one greater than that of T_1, *i.e.*,

$$m(T') = m(T_1) + 1 = m - 1 + 1 = m.$$

Hence, $T' \in \mathcal{T}(m)$. By considering that $T = T' \bullet a'$, $a' = Kr'$, we see that $T \in \mathcal{T}[m]$. □

Based on Lemma 9.1.1–2, we are allowed to evaluate what kind of equations are satisfied by the chrosum function

$$f(P, \mathcal{T}) = \sum_{T \in \mathcal{T}} P(T, \lambda) x^{m(T)} z^{s(T)} \underline{y}^{\underline{n}(T)} \tag{9.1.8}$$

where $P(T,\lambda), m(T), s(T)$ and $\underline{n}(T)$ are the chromatic polynomial, the valency of root-vertex, the valency of root-face and the vector as

$$\underline{n}(T) = (n_1(T), n_2(T), \cdots)$$

with $n_i(T)$ being the number of nonrooted vertices of valency $i, i \geq 0$, of T respectively.

Because the vertex map ϑ has all of $m(\vartheta), s(\vartheta)$ and $\underline{n}(\vartheta)$ zero, the contribution of ϑ to $f(P, \mathcal{T})$ is

$$f(P, \vartheta) = \lambda \tag{9.1.9}$$

which is just the chromatic polynomial $P(\vartheta, \lambda)$ of ϑ.

From the property of *chromials*, or chromatic polynomials, the contribution of $\mathcal{T}_{\mathrm{I}}$ can be written as

$$f(P, \mathcal{T}_{\mathrm{I}}) = \sum_{T \in \mathcal{T}_{\mathrm{I}}} P(T - a, \lambda) x^{m(T)} z^{s(T)} \underline{y}^{\underline{n}(T)}$$

$$- \sum_{T \in \mathcal{T}_{\mathrm{I}}} P(T \bullet a, \lambda) x^{m(T)} z^{s(T)} \underline{y}^{\underline{n}(T)} \tag{9.1.10}$$

in which the first and the last summations are denoted by f_{A} and f_{B} respectively.

According to (9.1.6) in Lemma 9.1.1, we have

$$f_{\mathrm{A}} = xz^2 \sum_{T \in \mathcal{T}} P(T, \lambda) x^{m(T)} z^{s(T)} \underline{y}^{\underline{n}(T)}$$

$$\times \int_y y \sum_{T \in \mathcal{T}} P(T, \lambda) x^{m(T)} z^{s(T)} \underline{y}^{\underline{n}(T)}$$

$$= xz^2 \left(f(P, \mathcal{T}) \int_y y f_y(P, \mathcal{T}) \right) \tag{9.1.11}$$

where $f_y(P, \mathcal{T}) = f(P, \mathcal{T})|_{x=y}$.

From Lemma 9.1.2, by considering that

$$P(T_1 \dot{+} T_2, \lambda) = \frac{1}{\lambda} P(T_1, \lambda) P(T_2, \lambda),$$

we find

$$f_{\mathrm{B}} = \frac{xz^2}{\lambda} \sum_{T \in \mathcal{T}} P(T, \lambda) x^{m(T)} z^{s(T)} \underline{y}^{\underline{n}(T)}$$

$$\times \sum_{T \in \mathcal{T}} P(T, \lambda) z^{s(T)} \underline{y}^{\underline{n}(T)}$$

$$= \frac{xz^2}{\lambda} f(P, \mathcal{T}) \int_y y f_y(P, \mathcal{T}). \tag{9.1.12}$$

Theorem 9.1.1 *Let $f = f(P, \mathcal{T})$ and $f_y = f_y(P, \mathcal{T})$ be defined in (9.1.11). Then we have the following functional equation:*

$$\left(1 - xz^2 \left(1 - \frac{1}{\lambda}\right) \int_y y f_y \right) f = \lambda. \tag{9.1.13}$$

Proof According to (9.1.11–12), we have

$$f(P,\mathcal{T}_{\rm I}) = f_{\rm A} - f_{\rm B}$$
$$= \frac{xz^2}{\lambda}\Big(1-\frac{1}{\lambda}\Big)f\int_y yf_y.$$

By (9.1.3) and (9.1.9), (9.1.13) is then obtained. □

Although (9.1.13) looks simpler, the appearance of a quadratic term shows not to be easy enough for extracting a simple solution directly. However, we may first restrict ourselves to discussing the chrosum function only with λ and the size as parameters. That is the case for $x = z = 1$ and $y_i = \sqrt{y}^i, i \geq 1$, of $f(P,\mathcal{T})$. Write

$$g = \sum_{T\in\mathcal{T}} P(T,\lambda)y^{n(T)} \tag{9.1.14}$$

where $n(T)$ is the size of T. Then, by the procedure employed above we can find that g satisfies the equation as follows:

$$\left(1-y\Big(1-\frac{1}{\lambda}\Big)g\right)g = \lambda. \tag{9.1.15}$$

This is an ordinary quadratic equation. The solution should be

$$g = \frac{1-\sqrt{1-4y(1-\frac{1}{\lambda})\lambda}}{2y(1-\frac{1}{\lambda})} \tag{9.1.16}$$

by considering that g is with all the coefficients nonnegative integers in the form of a power series of y.

Based on (9.1.16), by expanding the square root function into power series of y and then by reduction, we may find that

$$g = \lambda + \sum_{n\geq 1}\frac{\lambda(\lambda-1)^n(2n)!}{n!(n+1)!}x^n. \tag{9.1.17}$$

Because $C_{n+1} = \frac{(2n)!}{n!(n+1)!}$ (which is called the *Catalan number* in combinatorics) is just the number of rooted tree maps of size n, (9.1.7) shows that the chrosum of rooted tree maps of size n is

$$P(n,\lambda)C_{n+1}$$

where

$$P(n, \lambda) = \lambda(\lambda - 1)^n$$

is the chromial of any tree of size n. This also shows that all trees are uniquely 2-colorable. Further, we can infer the general solution of (9.1.13).

§9.2 Outerplanar equations

Because any map with a loop is not colorable and multi-edge is meaningless in map coloration, all maps here are always assumed to be simple. Within reductions, we are allowed to consider 2-connected maps, *i.e.*, nonseparable simple maps.

Let $\mathcal{M}$ be the set of all rooted outerplanar maps, nonseparable and simple of course. Moreover, the link map L is assumed to be in $\mathcal{M}$.

Now, $\mathcal{M}$ is partitioned into two parts:

$$\mathcal{M} = \mathcal{M}_0 + \mathcal{M}_{\mathrm{I}} \tag{9.2.1}$$

where $\mathcal{M}_0 = L$, *i.e.*, the set of only one map L, the link map.

For $M \in \mathcal{M}$, denote by $m(M)$ the valency of the root-vertex of M. Write

$$\mathcal{M}(m) = \{M | \forall M \in \mathcal{M}, m(M) = m\}. \tag{9.2.2}$$

Then we have

$$\mathcal{M}_{\mathrm{I}} = \sum_{m \geq 2} \mathcal{M}(m) \tag{9.2.3}$$

while $\mathcal{M}(1) = \mathcal{M}_0$.

For two maps $M_1 = (\mathcal{X}_1, \mathcal{P}_2)$ and $M_2 = (\mathcal{X}_2, \mathcal{P}_2)$ with the respective roots $r_1 = r(M_1)$ and $r_2 = r(M_2)$, let $M = (\mathcal{X}, \mathcal{P})$ be such a map that

$$\mathcal{X} = \mathcal{X}_1 + \mathcal{X}_2$$

and that $\mathcal{P}$ is determined by identifying the vertices $v_{\beta r_1} = (\alpha\beta r_1, T_1)$ and $v_{r_2} = (r_2, S_2)$ in the form as

$$v_{\beta r} = (\alpha\beta r_1, r_2, S_2, T_1)$$

with the root $r = r_1$. The operation is called HT_{r}*-addition* denoted by $+_{\mathrm{r}}$. Then we may write

$$M = M_1 +_{\mathrm{r}} M_2$$

in this case. Similarly for the *HT_r-production* denoted by $\times_r$ between two sets of maps.

Because the HT_r-addition satisfies neither the commutative law nor associative law for $M_i = (\mathcal{X}_i, \mathcal{P}_i)$, $i = 1, 2, \cdots, k$, we define

$$M = M_1 \dot{+}_r M_2 \dot{+}_r \cdots \dot{+}_r M_k$$

to be $(\mathcal{X}, \mathcal{P})$ such that

$$\mathcal{X} = \sum_{i=1}^{k} \mathcal{X}_i$$

and $\mathcal{P}$ is determined by composing new vertices

$$v_i = (\alpha\beta r_i, r_{i+1}, S_{i+1}, T_i)$$

from $v_{\beta r_i}$ and $v_{r_{i+1}}$ for $i = 1, 2, \cdots, k-1$ with the root $r = r_1$. This operation for $k \geq 2$ is in particular called the *chain HT_r-addition.* Correspondingly, the *Chain HT_r-production* for sets of maps.

Lemma 9.2.1 *Let $\mathcal{M}_{\langle I\rangle}(m) = \{M - a | \forall M \in \mathcal{M}_I(m)\}$ where $a = e_r(M)$. Then we have*

$$\mathcal{M}_{\langle I\rangle}(m) = \sum_{k\geq 1} \mathcal{M}(m-1) \times_r \mathcal{M}^{\times_r k} \tag{9.2.4}$$

for $m \geq 2$ where $\times_r$ is the HT_r-production.

Proof For $M = (\mathcal{X}, \mathcal{P}) \in \mathcal{M}_{\langle I\rangle}(m)$, let $M' \in \mathcal{M}(m)$ such that

$$M = M' - a', \; a' = e_r(M') = Kr'.$$

Assume that the nonrooted face incident to a' in M' is

$$\left(\alpha\beta r', (\mathcal{P}\alpha\beta)\alpha\beta r', (\mathcal{P}\alpha\beta)^2\alpha\beta r', \cdots\right).$$

Then, there exist integers $k \geq 1$ (by the simplicity and outerplanarity) and $j_1, j_2, \cdots, j_k \geq 2$ such that

$$v_{(\mathcal{P}\alpha\beta)^{j_1}\alpha\beta r'}, v_{(\mathcal{P}\alpha\beta)^{j_2}\alpha\beta r'}, \cdots, v_{(\mathcal{P}\alpha\beta)^{j_k}\alpha\beta r'}$$

are all the cut-vertices in $M = M' - a'$ from the nonseparability of M'. Thus, by the outerplanarity we have

$$M = M_0 \dot{+}_r M_1 \dot{+}_r \cdots \dot{+}_r M_k$$

such that $r(M_0) = (\mathcal{P}\alpha\beta)\alpha\beta r'$,

$$r(M_i) = (\mathcal{P}\alpha\beta)^{j_i}\alpha\beta r'$$

for $i = 1, 2, \cdots, k$. By the outerplanarity, it is only possible that $j_i = i + 1$ for $i = 1, 2, \cdots, k$. This means that the inner face incident to the root-edge of M' should be

$$(\alpha\beta r', (\mathcal{P}\alpha\beta)\alpha\beta r', \cdots, (\mathcal{P}\alpha\beta)^{k+1}\alpha\beta r')$$

and hence of valency $k+2$. The root of M is chosen to be $\mathcal{P}r'$. In this case, $+\cdot$ is represented by $+\cdot_{\mathrm{r}}$ and $\times\cdot$, by $\times\cdot_{\mathrm{r}}$ respectively. Then, the valency of the root-vertex in M, which is the same as that in M_0, is one less than that in M'. Because all of $M_i, i \geq 0$, are in $\mathcal{M}$, we have that M is a member of the set on the right hand side of (9.2.4).

Conversely, for any M in the set on the right, since there are $M_0, M_1, \cdots, M_k \in \mathcal{M}$ for some $k \geq 0$ in the form as

$$M = M_0 +\cdot_{\mathrm{r}} M_1 +\cdot_{\mathrm{r}} \cdots +\cdot_{\mathrm{r}} M_k,$$

we may uniquely find the map M' by adding an edge $a' = Kr'$ on M such that the root-vertex and the nonrooted vertex incident to the root-edge Kr' is respectively

$$(r', r_0, R) \text{ and } (\alpha\beta r', S, \alpha\beta r_k)$$

where r_i is the root of M_i for $i = 0$ and k, and that (r_0, R) and $(S, \alpha\beta r_k)$ are respectively the root-vertex in M_0 and the nonrooted end of the root-edge in M_k. Easy to check that $M' \in \mathcal{M}_{\mathrm{I}}(m)$ if $m(M_0) = m - 1$ and $M = M' - a'$. Thus, $M \in \mathcal{M}_{\langle \mathrm{I} \rangle}(m)$. □

For $M \in \mathcal{M}_{\mathrm{I}}$, let M be with the root-vertex and the nonrooted vertex incident to the root-edge $a = Kr$ as

$$v_r = (r, R) \text{ and } v_{\alpha\beta r} = (\alpha\beta r, s, S),$$

respectively. Write $\mathcal{M}_{(\mathrm{I})} = \{M \bullet a | \forall M \in \mathcal{M}_{\mathrm{I}}\}$. Then, since $M \bullet a$ has two possibilities: $M \bullet a$ is still simple or not, *i.e.*, $M \bullet a$ has multi-edge, $\mathcal{M}_{(\mathrm{I})}$ is partitioned into two classes in the form as

$$\mathcal{M}_{(\mathrm{I})} = \mathcal{M}_{\mathrm{S}} + \mathcal{M}_{\mathrm{N}} \tag{9.2.5}$$

where $\mathcal{M}_S$ is of maps all simple and $\mathcal{M}_N$, of all the others.

For coloring, a multi-edge in any $M \in \mathcal{M}_N$ is allowed to be with the convention that the edges in a multi-edge are identified.

Let M' be the map such that $M = M' \bullet a'$. Write

$$v_{r'} = (r, s, R) \text{ and } v_{\alpha\beta r'} = (\alpha\beta r', t, S, \alpha\beta l).$$

Then, we have that

$$\left(\alpha\beta r', \mathcal{P}\alpha\beta r', (\mathcal{P}\alpha\beta)^2 r'\right) = (\alpha\beta r', s, l)$$

is the nonrooted face incident to a'. Define the operation $+\!\!|$ called *ed*1*-addition* on maps by

$$M = M_1 +\!\!| \; M_2 \tag{9.2.6}$$

where M_1 and M_2 are the submaps of M expanded from

$$\{v_{r_1} = (l, T),\ v_{\alpha\beta r_1} = (\alpha\beta l, t, S)\}$$

and

$$\{v_{r_2} = (s, R),\ v_{\alpha\beta r_2} = (\alpha\beta s, W)\}$$

as the end pairs of the root-edges, respectively. Here, W and T appear in the vertex $(l, T, W, \alpha\beta s)$ of M'. Of course, M has $s = \alpha\beta l$ in the convention. Likewise, the corresponding operation $\times\!|$ on sets of maps is called the *ed*1*-production.*

Lemma 9.2.2 *For $\mathcal{M}_N$, we have*

$$\mathcal{M}_N = \mathcal{M} \times\!| \, \mathcal{M} \tag{9.2.7}$$

*where $\times\!|$ is the ed*1*-production corresponding to the ed*1*-addition as defined above.*

Proof By the outerplanarity, for any $M' \in \mathcal{M}_I$ such that $M' \bullet a' = M \in \mathcal{M}_N$, the nonrooted triangular face has $\{v_{\alpha\beta l}, v_l\}$ and $\{v_s, v_{\alpha\beta s}\}$ both separating pairs of M', where $v_{\alpha\beta l} = v_{\alpha\beta r'}, v_s = v_{r'}, v_{\alpha\beta s} = v_l$. Then any $M \in \mathcal{M}_N$ has the form shown in (9.2.6) and hence, M is a member of the set on the right hand of (9.2.7).

Conversely, for any map M in the set on the right, because M has the form shown as (9.2.6) we may uniquely find $M' \in \mathcal{M}_I$ such that $M = M' \bullet a'$ by reversing the procedure mentioned above. Thus, $M \in \mathcal{M}_{\mathrm{N}}$. □

Lemma 9.2.3 *For $\mathcal{M}_{\mathrm{S}}$, we have*

$$\mathcal{M}_{\mathrm{S}} = (\mathcal{M} - L) \times\!| \mathcal{M} \tag{9.2.8}$$

where L is the link map.

Proof For $M \in \mathcal{M}_{\mathrm{S}}$, let $M' \in \mathcal{M}_{\mathrm{I}}$ be such that $M = M' \bullet a'$. Since M is simple, M' has the nonrooted face f incident with the root-edge a' not triangle. Write the root-vertex of M' as

$$(r', s, R_2) \tag{9.2.9}$$

and the nonrooted end of the root-edge $a' = Kr'$ as

$$(r, R_1, \alpha\beta r'). \tag{9.2.10}$$

Then the root-vertex of M should be

$$(r, R_1, s, R_2). \tag{9.2.11}$$

Thus M has the form (9.2.6) such that M_1 and M_2 are the submaps of M expanded from (s, r, R_1) and (s, R_2) as the respective root-vertices. Because the face f in M' is not a triangle, M_1 is not allowed to be L. Hence M is a member of the set on the right hand side of (9.2.8).

Conversely, for any map M in the set on the right, we may uniquely construct the map M' by expanding from

$$(r', s, R_2) \text{ and } (\alpha\beta r', r, R_1)$$

as respectively the rooted end and the nonrooted end of the new root-edge $a' = Kr'$ on the basis of M whose root-vertex is as shown in (9.2.11). Because $M' \in \mathcal{M}_{\mathrm{I}}$ with the nonrooted face f incident with a' not triangle, we have that $M = M' \bullet a' \in \mathcal{M}_{\mathrm{S}}$. □

The chrosum function here is considered to be with face partition as

$$g(P, \mathcal{M}) = \sum_{M \in \mathcal{M}} P(M, \lambda) x^{m(M)} z^{s(M)} \underline{y}^{\underline{n}(M)} \tag{9.2.12}$$

where $m(M)$ and $s(M)$ are the valencies of the root-vertex and root-face, respectively, $\underline{n}(M) = (n_3(M), n_4(M), \cdots)$, $n_i(M)$ is the number of nonrooted faces of valency $i, i \geq 3$.

By easily checking, we soon find that the contribution of $\mathcal{M}_0$ to $g(P, \mathcal{M})$ is

$$g_0 = \lambda(\lambda - 1)xz^2. \tag{9.2.13}$$

In view of the formula

$$P(M, \lambda) = P(M - a, \lambda) - P(M \bullet a, \lambda)$$

for chromatic polynomials, the contribution of $\mathcal{M}_{\mathrm{I}}$ to $g(P, \mathcal{M})$ is

$$g_{\mathrm{I}} = \sum_{M \in \mathcal{M}_{\mathrm{I}}} P(M - a, \lambda)x^{m(M)}z^{s(M)}\underline{y}^{\underline{n}(M)}$$
$$- \sum_{M \in \mathcal{M}_{\mathrm{I}}} P(M \bullet a, \lambda)x^{m(M)}z^{s(M)}\underline{y}^{\underline{n}(M)}. \tag{9.2.14}$$

The first and the last summations in (9.2.14) are denoted by g_A and g_B respectively. From (9.2.5),

$$g_B = \left(\sum_{\substack{M \in \mathcal{M}_{\mathrm{I}} \\ M \bullet a \in \mathcal{M}_{\mathrm{S}}}} + \sum_{\substack{M \in \mathcal{M}_{\mathrm{I}} \\ M \bullet a \in \mathcal{M}_{\mathrm{N}}}} \right) P(M, \lambda)x^{m(M)}z^{s(M)}\underline{y}^{\underline{n}(M)} \tag{9.2.15}$$

in which the first and the last summations are denoted by g_{S} and g_{N}.

Form Lemma 9.2.1,

$$g_A = \frac{xz}{\lambda} \sum_{M \in \mathcal{M}} P(M, \lambda)x^{m(M)}z^{s(M)}\underline{y}^{\underline{n}(M)}$$
$$\times \sum_{k \geq 1} \frac{y_{k+1}}{\lambda^{k-1}} \left(\sum_{M \in \mathcal{M}} P(M, \lambda)z^{s(M)-1}\underline{y}^{\underline{n}(M)} \right)^k$$
$$= xg(P, \mathcal{M}) \int_y \left(y^2 \sum_{k \geq 1} \left(\frac{yh}{\lambda z} \right)^k \right)$$

by considering that $h = g(P, \mathcal{M})|_{x=1}$,

$$= xg(P, \mathcal{M}) \int_y \frac{y^3 h}{\lambda z - yh}. \tag{9.2.16}$$

On account of Lemma 9.2.3 and by considering that the part in $\mathcal{M} - L$ has no contribution to the root-vertex of the corresponding maps in $\mathcal{M}_{\mathrm{I}}$, from

$$P(M_1 +\!| \, M_2, \lambda) = \frac{1}{\lambda(\lambda-1)} P(M_1, \lambda) P(M_2, \lambda),$$

we have that

$$\begin{aligned} g_{\mathrm{S}} &= \frac{xz}{\lambda(\lambda-1)} \left(\sum_{M \in \mathcal{M}} P(M,\lambda) x^{m(M)} z^{s(M)-1} \underline{y}^{\underline{n}(M)} \right) \\ &\quad \times \left(\sum_{i \geq 3} \frac{y_{i+1}}{y_i} \sum_{M_i \in \mathcal{M}-L} P(M_i,\lambda) z^{s(M_i)-1} \underline{y}^{\underline{n}(M_i)} \right) \\ &= \frac{x}{\lambda(\lambda-1)} g(P,\mathcal{M}) \sum_{i \geq 3} \frac{y_{i+1}}{y_i} h_i \end{aligned} \qquad (9.2.17)$$

where $M_i \in \mathcal{M} - L$ is a map with the valency i of the nonrooted face incident to the root-edge of $M_i, i \geq 3$, and

$$h_i = \sum_{M_i \in \mathcal{M}-L} P(M_i,\lambda) z^{m(M_i)} \underline{y}^{\underline{n}(M_i)} \qquad (9.2.18)$$

for $i \geq 3$. Because h_i is the chrosum function of $\mathcal{M} - L$ with the valency i of the nonrooted face incident to the root-edge according to the face partition and the valency of root-face without considering the valency of root-vertex, we soon see that

$$h = \lambda(\lambda-1)z^2 + \sum_{i \geq 3} h_i. \qquad (9.2.19)$$

Similarly, by Lemma 9.2.2,

$$\begin{aligned} g_{\mathrm{N}} &= \frac{xz}{\lambda(\lambda-1)} y_3 \left(\sum_{M \in \mathcal{M}} P(M,\lambda) x^{m(M)} z^{s(M)-1} \underline{y}^{\underline{n}(M)} \right) \\ &\quad \times \left(\sum_{M \in \mathcal{M}} P(M,\lambda) z^{s(M)-1} \underline{y}^{\underline{n}(M)} \right) \\ &= \frac{xy_3}{\lambda(\lambda-1)} g(P,\mathcal{M}) h. \end{aligned} \qquad (9.2.20)$$

Theorem 9.2.1 *Let $g = g(P, \mathcal{M})$. Then, g satisfies the following functional equation:*

$$\left(1 + \frac{x}{\lambda(\lambda-1)z}(H - y_3 h)\right) g$$

$$= x\left(\lambda(\lambda-1)z^2 + g\int_y \frac{y^2}{z - yh}\right) \quad (9.2.21)$$

where $h = g|_{x=1}$ and

$$H = \sum_{i\geq 3} \frac{y_{i+1}}{y_i} h_i \quad (9.2.22)$$

for h_i defined in (9.2.18), $i \geq 3$.

Proof From (9.2.1), (9.2.14) and (9.2.5), we see that

$$g = g_0 + g_I$$

$$= g_0 + g_A - g_B$$

$$= g_0 + g_A - (g_S + g_N).$$

By substituting (9.2.13), (9.2.16), (9.2.17) and (9.2.20), the theorem is soon obtained by grouping and rearrangement. □

For simplicity, we investigate the case when $x = 1$, $z = \sqrt{y}$ and $y_i = \sqrt{y}^i, i \geq 3$. That is the chrosum function

$$f = \sum_{M \in \mathcal{M}} P(M, \lambda) y^{n(M)} \quad (9.2.23)$$

with only the size $n(M)$ as the parameter except λ.

According to what has been done, we may find that

$$f = \lambda(\lambda-1) + y\frac{f^2}{1-f}$$

$$-\frac{y}{\lambda(\lambda-1)}\left(f\left(f - \lambda(\lambda-1)y\right) + f^2\right). \quad (9.2.24)$$

Theorem 9.2.2 *The chrosum function f defined by* (9.2.23) *satisfies the following equation:*

$$\frac{2}{\lambda(\lambda-1)}\left(1+y-\frac{2}{\lambda-1}\right)f^2 - \left(\lambda-\lambda(\lambda-1)y-y\right)f = \lambda^2(\lambda-1)y+\lambda y. \qquad (9.2.25)$$

Proof A direct result of (9.2.24). □

The equation (9.2.25) can in principle be solved by Lagrangian inversion as shown in §1.5.

§9.3 General equations

Let $\mathcal{M}$ be the set of rooted nonseparable maps excluding the loop map because loops are meaningless for map coloring. For $M \in \mathcal{M}$, $M - a$ and $M \bullet a, a = e_r(M)$, are the maps obtained by deleting and contracting the root-edge a, respectively. We have to pay attention that the link map which consists of a single edge, a link, is assumed to be in $\mathcal{M}$. Write

$$\mathcal{M} = \mathcal{M}_0 + \mathcal{M}_{\mathrm{I}} \qquad (9.3.1)$$

where $\mathcal{M}_0$ consists of a single map, the link map L and $\mathcal{M}_{\mathrm{I}}$, the set of all others.

The chrosum function for $\mathcal{M}$ is considered to be

$$f(P,\mathcal{M}) = \sum_{M\in\mathcal{M}} P(M,\lambda)x^{p(M)}y^{q(M)}z^{m(M)}t^{s(M)} \qquad (9.3.2)$$

where $p(M)$, $q(M)$, $m(M)$ and $s(M)$ are the number of nonrooted vertices, the number of nonrooted faces, the valency of root-face and the valency of root-vertex in M respectively.

For a map $M \in \mathcal{M}_{\mathrm{I}}$, let $m_C(M-a)$ be the contribution of the root-face boundary of $M-a$ to $r(M)$ where $a = e_r(M)$. Assume that $v_r(M)$ and $v_{\alpha\beta r}(M)$ have the respective forms as

$$(r, \mathcal{P}r, T) \text{ and } (\alpha\beta r, \mathcal{P}\alpha\beta r, S).$$

Then for $M-a$ we have

$$v_r(M-a) = (\mathcal{P}r, T) \text{ and } v_{\alpha\beta r}(M-a) = (\mathcal{P}\alpha\beta r, S)$$

where $r(M-a) = \mathcal{P}r$. Since there is an integer $k \geq 0$ such that the root-face of $M-a$ has the form as

$$\Big(\mathcal{P}r, \cdots, (\mathcal{P}\alpha\beta)^{j_1}\mathcal{P}r, \cdots, (\mathcal{P}\alpha\beta)^{j_k}\mathcal{P}r, \cdots, \mathcal{P}\alpha\beta r,$$

$$\cdots, (\mathcal{P}\alpha\beta)^{l_k}\mathcal{P}r, \cdots, (\mathcal{P}\alpha\beta)^{l_1}\mathcal{P}r, \cdots\Big)$$

where $v_{(\mathcal{P}\alpha\beta)^{j_i}\mathcal{P}r} = v_{(\mathcal{P}\alpha\beta)^{l_i}\mathcal{P}r}$, $i = 1, 2, \cdots, k$, and $j_1 < \cdots < j_k < l_k < \cdots < l_1$. Here, k is assumed to be maximal for this property. Then, we see that

$$M - a = M_0 \hat{+} M_1 \hat{+} \cdots \hat{+} M_k \tag{9.3.3}$$

where $\hat{+}$ is the chain $1v$-addition as shown in §6.1 and the outer-face boundaries of $M_0, M_1, \cdots, M_k$ are

$$\Big(\mathcal{P}r, \cdots, (\mathcal{P}\alpha\beta)^{j_1-1}\mathcal{P}r, (\mathcal{P}\alpha\beta)^{l_1}\mathcal{P}r, \cdots\Big),$$

$$\Big((\mathcal{P}\alpha\beta)^{j_1}\mathcal{P}r, \cdots, (\mathcal{P}\alpha\beta)^{l_2}\mathcal{P}r, \cdots, (\mathcal{P}\alpha\beta)^{l_1-1}\mathcal{P}r\Big),$$

$$\cdots\cdots,$$

$$\Big((\mathcal{P}\alpha\beta)^{j_k}\mathcal{P}r, \cdots, \mathcal{P}\alpha\beta r, \cdots, (\mathcal{P}\alpha\beta)^{l_k-1}\mathcal{P}r\Big)$$

respectively. In this case, the lengths of the segments

$$\langle(\mathcal{P}\alpha\beta)^{l_1}\mathcal{P}r, \cdots, (\mathcal{P}\alpha\beta)^{-1}\mathcal{P}r\rangle,$$

$$\langle(\mathcal{P}\alpha\beta)^{l_2}\mathcal{P}r, \cdots, (\mathcal{P}\alpha\beta)^{l_1-1}\mathcal{P}r\rangle,$$

$$\cdots\cdots,$$

$$\langle\mathcal{P}\alpha\beta r, \cdots, (\mathcal{P}\alpha\beta)^{l_k-1}\mathcal{P}r\rangle$$

are respectively $m_{C_0} = m_C(M_0), m_{C_1} = m_C(M_1), \cdots, m_{C_k} = m_C(M_k)$. These segments are said to be *marked*.

Lemma 9.3.1 *Let $\mathcal{M}_{\langle \mathrm{I}\rangle}(s) = \{M - a | \forall M \in \mathcal{M}_{\mathrm{I}}, s(M) = s\}$ with $a = e_r(M)$, $s \geq 2$. Then we have*

$$\mathcal{M}_{\langle \mathrm{I}\rangle}(s) = \sum_{k\geq 0} \mathcal{M}_+(s-1) \hat{\times} \mathcal{M}_+^{\hat{\times}k} \tag{9.3.4}$$

where $\hat{\times}$ *is the chain* 1v-*production shown in* §6.1,

$$\mathcal{M}_+(s-1) = \{M|\forall M \in \mathcal{M}_+, s(M) = s-1\}$$

and

$$\mathcal{M}_+ = \sum_{M \in \mathcal{M}} \{M(1), \cdots, M(m(M)-1)\} \tag{9.3.5}$$

such that $M(i)$ *is isomorphic to* M *with* $m_C(M(i)) = i$, *for* $i = 1, 2, \cdots, m(M)-1$.

Proof For any $M \in \mathcal{M}_{\langle \mathrm{I} \rangle}(s)$, M has the form shown by (9.3.3). From the definition, M_0 has to be with $s(M_0) = s-1$ and all M_i, $0 \le i \le k$, are allowed to be any map of $\mathcal{M}_+$. This implies M is a member of the set on the right hand side of (9.3.4) as well.

Conversely from any M, assumed to be of the form (9.3.3), in the set on the right hand side of (9.3.4), we may uniquely obtain a map M' by adding a new edge Kr' between the root-vertex (R) and the noncut-end (S) of the marked segment of M_k to form vertices (r', R) and $(\alpha\beta r', S)$ respectively. Because M_0 is with $s(M_0) = s-1$, it is easily seen that $M \in \mathcal{M}_\mathrm{I}$ and $s(M) = s$. However, $M = M' - Kr'$. This implies $M \in \mathcal{M}_{\langle \mathrm{I} \rangle}(s)$. □

On the other hand, because a map with a loop is meaningless for coloring, it is allowed to discuss

$$\tilde{\mathcal{M}}_\mathrm{I} = \{M|\forall M \in \mathcal{M}_\mathrm{I}, M \bullet a \text{ is without loop}\}$$

only. For $M \in \mathcal{M}_\mathrm{I}$, there exists an integer k such that

$$M \bullet a = M_0 \dot{+} M_1 \dot{+} \cdots \dot{+} M_k \tag{9.3.6}$$

where $\dot{+}$ is the inner 1v-addition shown in §7.1 and all $M_i, i = 0, 1, \cdots, k$, are maps in $\mathcal{M}$ with the attention that $M \bullet a$ is allowed to have multi-edges here. It is easily seen that $m(M \bullet a) = m(M_0) = m(M) - 1$.

In fact, we may assume that M has its root-vertex and the non-rooted vertex incident to the root-edge to be with the respective forms

$$(r, s_k, S_k, s_{k-1}, S_{k-1}, \cdots, s_0, S_0) \tag{9.3.7}$$

and

$$(\alpha\beta r, t_0, T_0, t_1, T_1, \cdots, t_k, T_k) \tag{9.3.8}$$

such that for any s_i, there exists an integer $l > 0, (\mathcal{P}\alpha\beta)^l s_i = t_{i+1}, i = 0, 1, \cdots, k$, $t_{k+1} = t_0$. Thus, M_i is the submap of $M \bullet a$ expanded by the root-vertex

$$(t_i, T_i, s_i, S_i) \tag{9.3.9}$$

where t_i is the root of $M_i, i = 0, 1, \cdots, k$. The segment $\langle s_i, S_i\rangle$ is also said to be *marked*. The length of the marked segment is denoted by $s_C(M_i), i = 0, 1, \cdots k$.

Lemma 9.3.2 *Let* $\mathcal{M}_{(\mathrm{I})}(m) = \{M \bullet a | \forall M \in \tilde{\mathcal{M}}_{\mathrm{I}}, m(M) = m\}$ *with* $a = e_r(M)$. *Then we have*

$$\mathcal{M}_{(\mathrm{I})}(m) = \sum_{k \geq 0} \overset{+}{\mathcal{M}}(m-1) \times\cdot \overset{+}{\mathcal{M}}^{\times\cdot k}, \tag{9.3.10}$$

where $\times\cdot$ *is the inner* 1*v-production described in* §7.1,

$$\overset{+}{\mathcal{M}}(m-1) = \{M | \forall M \in \overset{+}{\mathcal{M}}, m(M) = m-1\} \tag{9.3.11}$$

and

$$\overset{+}{\mathcal{M}} = \sum_{M \in \mathcal{M}} \{M\langle 1\rangle, M\langle 2\rangle, \cdots, M\langle s(M)-1\rangle\} \tag{9.3.12}$$

such that $M\langle i\rangle$ *is isomorphic to* M *with* $s_c(M\langle i\rangle) = i, i = 0, 1, 2, \cdots, s(M) - 1$.

Proof For $M \in \mathcal{M}_{(\mathrm{I})}(m)$, there is a map $\tilde{M} \in \tilde{\mathcal{M}}_{\mathrm{I}}$, $m(\tilde{M}) = m$, such that $M = \tilde{M} \bullet \tilde{a}$, where $\tilde{a} = e_r(\tilde{M})$. Because M has the form of (9.3.6), from what was discussed above we see that M is a member of the set on the right hand side of (9.3.10).

Conversely, for M, $m(M) = m - 1$, being a member of the set on the right, since M has the form shown in (9.3.6), we may uniquely construct a map $\tilde{M}$ by splitting the root-vertex of M into two with an edge $K\tilde{r}$ such that (9.3.7) and (9.3.8) with $\tilde{r}$ instead of r are the root-vertex and the nonrooted vertex incident to the root-edge of $\tilde{M}$. It is easily seen that $\tilde{M} \in \tilde{\mathcal{M}}_{\mathrm{I}}$ and $M = \tilde{M} \bullet \tilde{a}$ with $m(\tilde{M}) = m(\tilde{M} \bullet \tilde{a}) + 1 = m$. This implies that $M \in \mathcal{M}_{(\mathrm{I})}(m)$. □

Now, we evaluate the contributions of $\mathcal{M}_0$ and $\mathcal{M}_{\mathrm{I}}$ to $f(P, \mathcal{M})$. For $\mathcal{M}_0$, it is easily checked that

$$f(P, \mathcal{M}_0) = \lambda(\lambda - 1)xz^2t \tag{9.3.13}$$

by considering $P(L,\lambda) = \lambda(\lambda-1)$ and simply counting $p(L) = 1$, $q(L) = 0$, $m(L) = 2$, and $s(L) = 1$.

By the relationship among $P(M,\lambda)$, $P(M-a,\lambda)$ and $P(M\cdot a,\lambda)$, we see that

$$f(P,\mathcal{M}_{\mathrm{I}}) = \sum_{M\in\mathcal{M}_{\mathrm{I}}} P(M-a,\lambda)x^{p(M)}y^{q(M)}z^{m(M)}t^{s(M)}$$

$$-\sum_{M\in\mathcal{M}_{\mathrm{I}}} P(M\bullet a,\lambda)x^{p(M)}y^{q(M)}z^{m(M)}t^{s(M)}. \qquad (9.3.14)$$

In (9.3.14), the first and the last summations are denoted by f_{A} and f_{B}, respectively.

From Lemma 9.3.1, by the formula

$$P(M_0\hat{+}\cdots\hat{+}M_k,\lambda) = \lambda^{-k}P(M_0,\lambda)\cdots P(M_k,\lambda),$$

we find that

$$f_{\mathrm{A}} = yzt\left(\sum_{M\in\mathcal{M}_+} P(M,\lambda)x^{p(M)}y^{q(M)}z^{m_C(M)}t^{s(M)}\right)$$

$$\times\sum_{k\geq 0}\frac{1}{\lambda^k}\left(\sum_{M\in\mathcal{M}_+} P(M,\lambda)x^{p(M)}y^{q(M)}z^{m_C(M)}\right)^k$$

$$= yzt\left(\sum_{M\in\mathcal{M}} P(M,\lambda)x^{p(M)}y^{q(M)}\sum_{i=1}^{m(M)-1} z^i t^{s(M)}\right)$$

$$\times\frac{1}{\lambda}\sum_{k\geq 0}\left(\sum_{M\in\mathcal{M}} P(M,\lambda)x^{p(M)}y^{q(M)}\sum_{i=1}^{m(M)-1} z^i\right)^k$$

$$= yzt\left(\sum_{M\in\mathcal{M}} P(M,\lambda)x^{p(M)}y^{q(M)}\frac{z^{m(M)}-z}{z-1}t^{s(M)}\right)$$

$$\times\sum_{k\geq 0}\left(\frac{1}{\lambda}\sum_{M\in\mathcal{M}} P(M,\lambda)x^{p(M)}y^{q(M)}\frac{z^{m(M)}-z}{z-1}\right)^k$$

$$= yzt\partial_z f\sum_{k\geq 0}\left(\frac{1}{\lambda}\partial_z f_{t=1}\right)^k$$

$$= \frac{yzt\partial_z f}{1 - \frac{1}{\lambda}\partial_z f_{t=1}} \tag{9.3.15}$$

where for $g(z) = f(z)$, or $f_{t=1}(z)$,

$$\partial_z g = \frac{g(z) - zg(1)}{z - 1} \tag{9.3.16}$$

which is just the $\langle z, 1\rangle$-difference as in (1.6.8).

From Lemma 9.3.2, and by

$$P(M_0 \dotplus \cdots \dotplus M_k) = \lambda^{-k} P(M_0, \lambda) \cdots P(M_k, \lambda)$$

as well, we find that

$$\begin{aligned} f_B &= xzt \left(\sum_{M \in \mathcal{M}} P(M,\lambda) x^{p(M)} y^{q(M)} r^{m(M)} t^{s_C(M)} \right) \\ &\quad \times \sum_{k \geq 0} \frac{1}{\lambda^k} \left(\sum_{M \in \mathcal{M}} P(M,\lambda) x^{p(M)} y^{q(M)} t^{s_C(M)} \right)^k \\ &= xzt \left(\sum_{M \in \mathcal{M}} P(M,\lambda) x^{p(M)} y^{q(M)} r^{m(M)} \frac{t^{s(M)} - t}{t - 1} \right) \\ &\quad \times \sum_{k \geq 0} \left(\frac{1}{\lambda} \sum_{M \in \mathcal{M}} P(M,\lambda) x^{p(M)} y^{q(M)} \frac{t^{s(M)} - t}{t - 1} \right)^k \\ &= \frac{xzt\partial_t f}{1 - \frac{1}{\lambda}\partial_t f_{z=1}}. \end{aligned} \tag{9.3.17}$$

Theorem 9.3.1 *The chrosum function $f = f(P, \mathcal{M})$ defined by (9.3.2) satisfies the following equation:*

$$\begin{aligned} &\Big(1 - \frac{1}{\lambda}\partial_z f^+\Big)\Big(1 - \frac{1}{\lambda}\partial_t f^*\Big)\Big(f - \lambda(\lambda - 1)xz^2 t\Big) \\ &\qquad = zt\Big(y\Big(1 - \frac{1}{\lambda}\partial_t f^*\Big)\partial_z f \\ &\qquad\qquad - x\Big(1 - \frac{1}{\lambda}\partial_z f^+\Big)\partial_t f\Big) \end{aligned} \tag{9.3.18}$$

where $f^* = f_{z=1}$ and $f^+ = f_{t=1}$.

Proof From (9.3.1), we have

$$f = f(P, \mathcal{M}_0) + f(P, \mathcal{M}_\mathrm{I})$$

by (9.3.13) and (9.3.14),

$$= \lambda(\lambda - 1)xz^2 + f_\mathrm{A} - f_\mathrm{B}$$

by (9.3.15) and (9.3.17),

$$= \lambda(\lambda - 1)xz^2t - \frac{xzt\partial_t f}{1 - \frac{1}{\lambda}\partial_t f^*}$$

$$+ \frac{yzt\partial_z f}{1 - \frac{1}{\lambda}\partial_z f^+}.$$

Multiplying by $(1 - \frac{1}{\lambda}\partial_t f^+)(1 - \frac{1}{\lambda}\partial_t f^*)$ the two sides and then grouping, (9.3.18) is soon obtained. □

The well definedness of the equation given by this theorem will be shown in §9.5.

§9.4 Triangulation equations

Let $\mathcal{T}_\mathrm{Tri}$ be the set of rooted planar maps which are nonseparable with all nonrooted face triangles. Of course, $\mathcal{T}_\mathrm{Tri}$ is a subset of $\mathcal{M}$ which appears in §9.3. The link map is noticed to be in $\mathcal{T}_\mathrm{Tri}$ as a degenerate case. For any $T \in \mathcal{T}_\mathrm{Tri}$ let $p(T), q(T), m(T)$ and $s(T)$ be the number of nonrooted vertices, the number of nonrooted faces, the valency of the root-face and the valency of the root-vertex respectively. From Euler formula, we see that

$$q(T) = 2p(T) - m(T) \tag{9.4.1}$$

for any $T \in \mathcal{T}_\mathrm{Tri}$. Then the chrosum function defined as (9.3.2) becomes

$$f(p, \mathcal{T}_\mathrm{Tri}) = \sum_{T \in \mathcal{T}_\mathrm{Tri}} p(T, \lambda)x^{\frac{q(T)+m(T)}{2}} y^{q(T)} z^{m(T)} t^{s(T)}.$$

By substituting y and z for $\sqrt{x}y$ and $\sqrt{x}z$ respectively, we have

$$h(P, \mathcal{T}_{\mathrm{Tri}}) = \sum_{T\in\mathcal{T}_{\mathrm{Tri}}} p(T,\lambda) y^{q(T)} z^{m(T)} t^{s(T)}. \tag{9.4.2}$$

The purpose of this section is to derive an equation satisfied by $h = h(P, \mathcal{T}_{\mathrm{Tri}})$ according to the general procedure shown in §9.3.

The procedure of decomposition of $\mathcal{T}_{\mathrm{Tri}}$ is similar to what has been described in §9.3. Write

$$\mathcal{T}_{\mathrm{Tri}} = \mathcal{T}_{\mathrm{Tri}_0} + \mathcal{T}_{\mathrm{Tri}_{\mathrm{I}}} \tag{9.4.3}$$

where $\mathcal{T}_{\mathrm{Tri}_0}$ consists of only the link map and $\mathcal{T}_{\mathrm{Tri}_{\mathrm{I}}}$, of all others.

Lemma 9.4.1 *Let $\mathcal{T}_{\langle \mathrm{I}\rangle} = \{T - a | \forall T \in \mathcal{T}_{\mathrm{Tri}_{\mathrm{I}}}\}$ with $a = e_r(T)$, the root-edge of T. Then we have*

$$\mathcal{T}_{\langle \mathrm{I}\rangle} = \mathcal{T}^{\Delta} + \mathcal{T}_{\mathrm{Tri}}^{\times 2} \tag{9.4.4}$$

where $\mathcal{T}^{\Delta}$ is the subset of all near triangulations with the root-face boundary not a 2-gon in $\mathcal{T}_{\mathrm{Tri}}$.

Proof For any $T \in \mathcal{T}_{\langle \mathrm{I}\rangle}$, let $\hat{T} \in \mathcal{T}_{\mathrm{I}}$ such that

$$\hat{T} - \hat{a} = T,\ \hat{a} = e_r(\hat{T}).$$

There are two possibilities: $T \in \mathcal{T}_{\mathrm{Tri}}$ or not. Because no inner face of a 2-gon appears in any map of $\mathcal{T}_{\mathrm{Tri}}$, the former implies that T is not allowed to be those with a 2-gon as the root-face boundary . Hence $T \in \mathcal{T}^{\Delta}$. The latter implies that T has at least one cut-vertex. From the triangularity, T is only allowed to have exactly one cut-vertex and hence $T = T_1 \dot{+} T_2$ shown as in §6.1. Because T_1 and T_2 are allowed to be any map of $\mathcal{T}_{\mathrm{Tri}}$. Thus $T \in \mathcal{T}_{\mathrm{Tri}}^{\times\cdot 2}$. In all, T is a member of the set on the right hand side of (9.4.4).

For any map T of the set on the right, we have $T \in \mathcal{T}^{\Delta}$ or $\mathcal{T}_{\mathrm{Tri}}^{\times\cdot 2}$. In the former case, because the boundary of the root-face is not a 2-gon, we can construct a map $\hat{T}$ by adding an edge $K\hat{r}$ with its two ends as

$$(\hat{r}, r, R) \text{ and } (\alpha\beta(\mathcal{P}\alpha\beta r), \alpha\beta\hat{r}, S)$$

where

$$v_r = (r, R) \text{ and } v_{\alpha\beta\mathcal{P}\alpha\beta r} = (\alpha\beta\mathcal{P}\alpha\beta r, S)$$

are the vertices in T. Because $\hat{T} \in \mathcal{T}_{\mathrm{Tri}}$ but not the link map, we see that $\hat{T} \in \mathcal{T}_{\mathrm{Tri}_\mathrm{I}}$. In the latter case, we may uniquely construct $\hat{T}$ in the same way such that $\hat{T} \in \mathcal{T}_{\mathrm{Tri}_\mathrm{I}}$ as well. Since $T = \hat{T} - \hat{a}, \hat{a} = K\hat{r}$, this implies $T \in \mathcal{T}_{\langle \mathrm{I} \rangle}$. □

In view of the meaningfulness of coloring on maps, we still only consider the set

$$\tilde{\mathcal{T}}_{\mathrm{Tri}_\mathrm{I}} = \{T | \forall T \in \mathcal{T}_{\mathrm{Tri}_\mathrm{I}}, T \bullet a \in \mathcal{T}_{\mathrm{Tri}}\} \tag{9.4.5}$$

instead of $\mathcal{T}_{\mathrm{Tri}_\mathrm{I}}$ for

$$\mathcal{T}_{(\mathrm{I})} = \{T \bullet a | \forall T \in \tilde{\mathcal{T}}_{\mathrm{Tri}_\mathrm{I}}\} \tag{9.4.6}$$

where $T \bullet a$ is the map obtained by contracting the root-edge a on T with the replacement of a 2-gon of an inner face boundary by a link.

For $T \in \mathcal{T}_{\mathrm{Tri}}$, write

$$\Delta(T) = \{T_i | i = 2, 3, \cdots, s(T) + 1\} \tag{9.4.7}$$

where $T_i, 2 \le i \le s(T) + 1$, is the map obtained by splitting the edge $e_{\mathcal{P}^{i-2}r}$ into two: Ky and Kz and adding a new edge Kx such that

$$(r, \cdots, \mathcal{P}^{i-3}r, y, x), (\alpha\beta x, z, \mathcal{P}^{i-1}r, R),$$

$$(\alpha\beta z, \alpha\beta y, S)$$

are three vertices instead of the two vertices:

$$(r, \cdots, \mathcal{P}^{i-3}r, \mathcal{P}^{i-2}r, \mathcal{P}^{i-1}r, R),$$

$$(\alpha\beta\mathcal{P}^{i-2}r, S)$$

on T. Of course, r remains to be the root of T_i. It is easily seen that $s(T_i) = i$, $2 \le i \le s(T) + 1$.

Lemma 9.4.2 *For $\tilde{\mathcal{T}}_{\mathrm{Tri}_\mathrm{I}}$, we have*

$$\tilde{\mathcal{T}}_{\mathrm{Tri}_\mathrm{I}} = \sum_{T \in \mathcal{T}_{\mathrm{Tri}}} \Delta(T). \tag{9.4.8}$$

Proof First, prove

$$\tilde{\mathcal{T}}_{\mathrm{Tri_I}} = \sum_{T \in \mathcal{T}_{(\mathrm{I})}} \Delta(T). \tag{9.4.9}$$

For any $T \in \tilde{\mathcal{T}}_{\mathrm{Tri_I}}$, because $T \bullet a \in \mathcal{T}_{(\mathrm{I})}$, we see that $T \in \Delta(T \bullet a)$. Thus T is a member of the set on the right hand side of (9.4.9). Conversely, for T being a member of the set on the right, there exists a map $T' \in \mathcal{T}_{(\mathrm{I})}$ such that $T \in \Delta(T')$. Since $T' = T \bullet a$, we have T is a member of $\tilde{\mathcal{T}}_{\mathrm{Tri_I}}$ by considering (9.4.6).

Second, prove

$$\mathcal{T}_{(\mathrm{I})} = \mathcal{T}_{\mathrm{Tri}}. \tag{9.4.10}$$

For $T \in \mathcal{T}_{(\mathrm{I})}$, from (9.4.5–6) it is seen that $T \in \mathcal{T}_{\mathrm{Tri}}$. Conversely, for any $T \in \mathcal{T}_{\mathrm{Tri}}$, because $\Delta(T) \subseteq \mathcal{T}_{\mathrm{Tri_I}}$ and for any

$$T' \in \Delta(T),\ T' \bullet a = T \in \mathcal{T}_{\mathrm{Tri}},$$

we have $T' \in \tilde{\mathcal{T}}_{\mathrm{Tri_I}}$. This implies $T \in \mathcal{T}_{(\mathrm{I})}$.

From (9.4.9–10), (9.4.8) is soon obtained. □

In what follows, we evaluate the contributions h_0 and h_{I} of $\mathcal{T}_{\mathrm{Tri_0}}$ and $\mathcal{T}_{\mathrm{Tri_I}}$ to $h(P, \mathcal{T}_{\mathrm{Tri}})$ respectively.

Because for the link map L, $q(L) = 0, m(L) = 2, s(L) = 1$ and $P(L, \lambda) = \lambda(\lambda - 1)$, we have

$$h_0 = \lambda(\lambda - 1)z^2 t. \tag{9.4.11}$$

Write

$$h_A = \sum_{T \in \mathcal{T}_{\mathrm{Tri_I}}} P(T - a) y^{q(T)} z^{m(T)} t^{s(T)} \tag{9.4.12}$$

and

$$h_B = \sum_{T \in \mathcal{T}_{\mathrm{Tri_I}}} P(T \bullet a) y^{q(T)} z^{m(T)} t^{s(T)}. \tag{9.4.13}$$

Then we find

$$h_{\mathrm{I}} = h_A - h_B \tag{9.4.14}$$

from the chromial formula $P(T) = P(T - a) - P(T \bullet a)$ for $T \in \mathcal{T}_{\mathrm{Tri_I}}$.

By Lemma 9.4.1 and the formula

$$P(T_1 \dotplus T_2) = \frac{1}{\lambda} P(T_1) P(T_2),$$

we have that

$$\begin{aligned} h_A &= yz^{-1}t\Big(\sum_{T \in \mathcal{T}^{\triangle}} P(T,\lambda) y^{q(T)} z^{m(T)} t^{s(T)} \\ &\quad + \frac{1}{\lambda}\left(\sum_{T \in \mathcal{T}_{\mathrm{Tri}}} P(T,\lambda) y^{q(T)} z^{m(T)} t^{s(T)} \right) \\ &\quad \times \left(\sum_{T \in \mathcal{T}_{\mathrm{Tri}}} P(T,\lambda) y^{q(T)} z^{m(T)} \right) \Big) \\ &= yz^{-1}t\Big(h - z^2 h^{\theta} + \frac{1}{\lambda} h h^{+} \Big) \end{aligned} \tag{9.4.15}$$

where h^{θ} is the coefficient of z^2 in $h = h(P, \mathcal{T}_{\mathrm{Tri}})$ and $h^{+} = h|_{t=1}$.

By Lemma 9.4.2,

$$\begin{aligned} h_B &= \sum_{T \in \tilde{\mathcal{T}}_{\mathrm{Tri}_{\mathrm{I}}}} P(T,\lambda) y^{q(T)} z^{m(T)} t^{s(T)} \\ &= yz \sum_{T \in \mathcal{T}_{\mathrm{Tri}}} P(T,\lambda) y^{q(T)} z^{m(T)} \left(\sum_{j=2}^{s(T)+1} t^j \right) \\ &= yz \sum_{T \in \mathcal{T}_{\mathrm{Tri}}} P(T,\lambda) y^{q(T)} z^{m(T)} \frac{t^2 - t^{m(T)+2}}{1-t} \\ &= yzt^2 \delta_t h \end{aligned} \tag{9.4.16}$$

where

$$\delta_t h = \frac{h^{+} - h}{1-t} \tag{9.4.17}$$

is the $(1,t)$-difference of h as in (1.6.7).

Theorem 9.4.1 *The chrosum function $h = h(P, \mathcal{T}_{\mathrm{Tri}})$ defined by (9.4.2) satisfies the following equation:*

$$\left(yt\left(1 + \frac{1}{\lambda}\right) - z \right) h - yz^2 t^2 \delta_t h$$

$$+z^2t\Big(\lambda(\lambda-1)z - yh^\theta\Big) = 0 \tag{9.4.18}$$

where $h^\theta = \partial_z^2 h$ *as defined by* (1.5.1), $h^+ = h|_{t=1}$ *and* $\delta_t h$ *is the* $(1,t)$-*difference as in* (1.6.7).

Proof Since $h = h_0 + h_{\rm I}$, from (9.4.11), (9.4.14) and (9.4.15–16),

$$\begin{aligned} h &= \lambda(\lambda-1)z^2t + h_A - h_B \\ &= \lambda(\lambda-1)z^2t + yz^{-1}t\Big(h - z^2h^\theta \\ &\quad + \frac{1}{\lambda}hh^+\Big) - yzt^2\delta_t h. \end{aligned}$$

By multiplying the two sides by z, (9.4.18) is soon found within grouping and rearrangement. □

§9.5 Well definedness

The main purpose in this section is to show the well-definedness of the equation (9.3.18). In fact, the method used here is so general that all equations appearing in this book can be shown to be well-defined in this way.

Let us write

$$\begin{aligned} F &= F(f, \partial_z f, \partial_t f, \partial_z f^+, \partial_t f^*) \\ &= \Big(\lambda^{-1}f - (\lambda-1)xz^2t\Big)(1-\lambda^{-1}\partial_z f^+)\partial_t f^* \\ &\quad - \lambda^{-1}yzt\partial_z f\partial_t f^+ \end{aligned} \tag{9.5.1}$$

From (9.3.18), we have

$$\begin{aligned} f &= \lambda(\lambda-1)xz^2t + \Big(\lambda^{-1}f - (\lambda-1)xz^2t\Big)\partial_z f^+ \\ &\quad + zt(y\partial_z f - x\partial_t f) + \lambda^{-1}xzt\partial_t f\partial_z f^+ + F. \end{aligned} \tag{9.5.2}$$

Expand f into a power series of x and y as

$$f = \sum_{p\geq 1}\sum_{q\geq 0} F_{p,q}(z,t;\lambda)x^py^q. \tag{9.5.3}$$

Because $p+q$ is the size of a planar map which has $p+1$ vertices and $q+1$ faces, $F_{p,q}(z,t;\lambda)$ is a polynomial of z,t and λ. According to the linearity of th operator ∂, we have

$$\partial_z f = \sum_{p\geq 1}\sum_{q\geq 0} \partial_z F_{p,q}(z,t;\lambda)x^p y^q; \tag{9.5.4}$$

$$\partial_t f = \sum_{p\geq 1}\sum_{q\geq 0} \partial_t F_{p,q}(z,t;\lambda)x^p y^q \tag{9.5.5}$$

where

$$\partial_z F_{p,q} = \sum_{m\geq 2}\sum_{s\geq 2} F_{p,q,m,s}(\lambda)(z+\cdots+z^{m-1})t^s; \tag{9.5.6}$$

$$\partial_t F_{p,q} = \sum_{m\geq 2}\sum_{s\geq 2} F_{p,q,m,s}(\lambda)z^m(t+\cdots+t^{s-1}) \tag{9.5.7}$$

for $F_{p,q} = F_{p,q}(z,t;\lambda)$ and $F_{p,q,m,s}(\lambda)$, the chrosum as defined in §1.4.

Lemma 9.5.1 *For $p \geq 1$ and $q = 0$, we have*

$$F_{p,0} = \begin{cases} \lambda(1-\lambda)z^2 t, & p = 1; \\ \\ 0, & otherwise. \end{cases} \tag{9.5.8}$$

Proof Because it can be evaluated that on the right hand side of (9.5.2), all the terms have powers greater than or equal to two of x except only for the first term, this leads the value $\lambda(\lambda-1)z^2t$ of (9.5.8) for $p = 1$.

For $p \geq 2$, by considering that F does not have a term not involving y, from (9.5.2) we have

$$\begin{aligned} F_{p,0} = \lambda^{-1} &\sum_{0\leq l\leq p-2} F_{l+1,0}\partial_z F_{p-l-1,0}(z,1;\lambda) \\ &-(\lambda-1)z^2t\partial_z F_{p-1,0}(z,1;\lambda) - zt\partial_t F_{p-1,0} \\ &+\lambda^{-1}zt\sum_{0\leq l\leq p-3} \partial_z F_{l+1,0}(z,1;\lambda)\partial_t F_{p-l-2,0}. \end{aligned} \tag{9.5.9}$$

Employing (9.5.6–7), from (9.5.9) and the case $p = 1$ we find

$$F_{2,0}(z,t;\lambda) = \lambda^{-1}F_{1,0}\partial_z F_{1,0}(z,1;\lambda) - (\lambda-1)z^2t$$

$$\times\partial_z F_{1,0}(z,1;\lambda) - zt\partial_t F_{1,0}$$
$$= \lambda^{-1}\Big(\lambda(\lambda-1)z^2t\lambda(\lambda-1)zt\Big)$$
$$-(\lambda-1)z^2t\lambda(\lambda-1)zt = 0.$$

Further, if for $k < p$ all $F_{k,0}$ have been determined by (9.5.8), then from $P > 2$,

$$F_{p,0} = \lambda^{-1} \sum_{0\le l\le p-2} F_{l+1,0}\partial_z F_{p-l-1,0}(z,1;\lambda)$$
$$+\lambda^{-1}zt \sum_{0\le l\le p-3} \partial_z F_{l+1,0}(z,1;\lambda)\partial_t F_{p-l-2,0}$$
$$= \lambda^{-1}\Big(F_{1,0}\partial_z F_{p-1,0}(z,1;\lambda) + F_{p-1,0}\partial_z F_{1,0}(z,1;\lambda)\Big)$$
$$+\lambda^{-1}zt\Big(\partial_z F_{1,0}(z,1;\lambda)\partial_t F_{p-2,0}$$
$$+\partial_z F_{p-2,0}(z,1;\lambda)\partial_t F_{1,0}\Big).$$

By the hypothesis and that $\partial_t F_{1,0} = 0$, we soon find that

$$F_{p,0} = 0, \text{ for } p \geq 2.$$

Hence the lemma is true. □

This coincides with the fact that only one nonseparable planar map is with only one face. That is the link map.

Lemma 9.5.2 *For $q = 1$, we have*

$$F_{p,1}(z,t;\lambda) = \alpha_{p+1}(\lambda)z^{p+1}t^2 \tag{9.5.10}$$

where

$$\alpha_{p+1}(\lambda) = (\lambda-2)\alpha_p(\lambda) + (\lambda-1)\alpha_{p-1}(\lambda) \tag{9.5.11}$$

with the initial conditions:

$$\alpha_2(\lambda) = \lambda(\lambda-1), \alpha_3(\lambda) = \lambda(\lambda-1)(\lambda-2) \tag{9.5.12}$$

for $p \geq 1$.

Proof From (9.5.2), by evaluating the term of xy, we find that

$$F_{1,1}(z,t;\lambda) = zt\partial_z(\lambda(\lambda-1)z^2t) = \lambda(\lambda-1)z^2t^2$$

and

$$\begin{aligned} F_{2,t}(z,t;\lambda) = \lambda^{-1}\big(&F_{1,0}\partial_z F_{1,1}(z,1,\lambda) \\ &+F_{1,1}\partial_z F_{1,0}(z,1;\lambda)\big) \\ &-(\lambda-1)z^2t\partial_z F_{1,1}(z,1;\lambda) \\ &-zt\partial_t F_{1,1} \\ = \lambda^{-1}\big(&\lambda(\lambda-1)z^2t\lambda(\lambda-1)z \\ &+\lambda(\lambda-1)z^2t^2\cdot\lambda(\lambda-1)z\big) \\ &-(\lambda-1)z^2t\cdot\lambda(\lambda-1)z \\ &-zt\lambda(\lambda-1)z^2t \\ = \lambda&(\lambda-1)^2z^3t^2 - \lambda(\lambda-1)z^3t^2 \\ = \lambda&(\lambda-1)(\lambda-2)z^3t^2. \end{aligned}$$

This enables us to know that (9.5.12) is true, or hence the cases of $p=1$ and 2 in (9.5.10).

By induction on p, assume that (9.5.10) is true for any $k < p$, to prove (9.5.10) is true for $p \geq 3$. In fact, from (9.5.2) we may see that

$$\begin{aligned} F_{p,1} = \lambda^{-1}\big(&F_{1,0}\partial_z F_{p-1,1}(z,1;\lambda) + F_{p-1,1}\partial_z F_{1,0}(z,1;\lambda)\big) \\ &-(\lambda-1)z^2t\partial_z F_{p-1,1}(z,1;\lambda) - zt\partial_t F_{p-1,1} \\ &+\lambda^{-1}zt\partial_t F_{1,0}(z,1;\lambda)\partial_t F_{p-2,1} \end{aligned} \tag{9.5.13}$$

to which F described in (9.5.1) has no contribution. According to (9.5.13) and the induction hypothesis, we find that

$$\begin{aligned} F_{p,1} = \lambda^{-1}\Big(&\lambda(\lambda-1)z^2t\alpha_p(\lambda)\sum_{i-1}^{p-1} z^i \\ &+\alpha_p(\lambda)z^pt^2\lambda(\lambda-1)z\Big) \end{aligned}$$

$$-zt\alpha_p(\lambda)z^pt - (\lambda-1)z^2t\alpha_p(\lambda)\sum_{i=1}^{p-1} z^i$$
$$+\lambda^{-1}zt(\lambda-1)z\alpha_{p-1}(\lambda)z^{p-1}t$$
$$= \Big((\lambda-1)\big(\alpha_p(\lambda)+\alpha_{p-1}(\lambda)\big)-\alpha_p(\lambda)\Big)z^{p+1}t^2$$
$$= \alpha_{p+1}(\lambda)z^{p+1}t^2.$$

The lemma is proved. □

This Lemma coincide with the fact that all the nonseparable planar maps, except the loop map, with two faces are the *circuit maps*

$$C_n = \Big(\bigcup_{i=1}^{n} Kx_i, (x_1, \alpha\beta x_n)(x_2, \alpha\beta x_1)$$
$$\cdots (x_{n-1}, \alpha\beta x_{n-2})(x_n, \alpha\beta x_{n-1})\Big)$$

for $n \geq 2$.

Lemma 9.5.3 *For $q \geq 0$ we have*

$$F_{1,q}(z,t;\lambda) = \begin{cases} \lambda(\lambda-1)z^2t, \ q=0; \\ \\ \lambda(\lambda-1)z^2t^{q+1}, \ q \geq 1. \end{cases} \tag{9.5.14}$$

Proof The case of $q = 0$ is given by Lemma 9.5.1. Because only the term $zty\partial_z f$ in (9.5.2) contributes to $F_{1,q}(z,t;\lambda)$ for $q \geq 1$, we have that

$$F_{1,q} = zt\partial_z F_{1,q-1}. \tag{9.5.15}$$

From $F_{1,0}$ which is just known, according to (9.4.15) we soon see that

$$F_{1,1} = zt\partial_z(\lambda(\lambda-1)z^2t)$$
$$= zt\lambda(\lambda-1)zt$$
$$= \lambda(\lambda-1)z^2t^2.$$

This is the case of $q = 1$ in (9.5.14).

In general, by the induction hypothesis on q and from (9.5.15), we find that

$$\begin{aligned} F_{1,q} &= zt\partial_z(\lambda(\lambda-1)z^2t^q) \\ &= zt\lambda(\lambda-1)zt^q \\ &= \lambda(\lambda-1)z^2t^{q+1}. \end{aligned}$$

Therefore, the lemma is obtained. □

For convenience, let us denote by

$$F(F_{i,j} : (i,j) < (p,q)) \tag{9.5.16}$$

a function which is only dependent on the terms involving $F_{i,j}$ such that $i \leq p$ and $j \leq q$ with $i+j < p+q$.

For $p \geq 2$ and $q \geq 1$, from (9.5.2–7) we have

$$\begin{aligned} F_{p,q} = \lambda^{-1} & \sum_{1\leq i\leq p-1} \sum_{0\leq j\leq q} F_{i,j}\partial_z F_{p-i,q-j}(z,t;\lambda) \\ & -(\lambda-1)z^2t\partial_z F_{p-1,q}(z,1;\lambda) \\ & +zt\partial_z F_{p,q-1} + \lambda^{-1}zt\partial_z F_{p,q-1} \\ & +\lambda^{-1}zt \sum_{1\leq i\leq p-2} \sum_{0\leq j\leq q} \partial_z F_{i,j}(z,1;\lambda) \\ & \times\partial_t F_{p-i-1,q-j} - zt\partial_t F_{p-1,q} \\ & +F(F_{i,j} : (i,j) < (p,q)) \end{aligned} \tag{9.5.17}$$

where the last term is defined by (9.5.16).

Theorem 9.5.1 *Equation* (9.3.18), *and* (9.4.18) *as well, is well-defined in the ring* $\mathcal{L}\{\Re; x,y,z,t\}$ *mentioned in* §1.5.

Proof From Lemmas 9.5.1–2 for $q = 1, 2$, $F_{p,q}$ are known, $p \geq 1$. Then, based on Lemma 9.5.3 we may find $F_{p,q}$, $p \geq 1$, for $q = 3, 4, \cdots$ by (9.5.17) respectively. The procedure allows us to obtain all the coefficients $F_{p,q} = F_{p,q}(z,t;\lambda)$ in (9.5.3) and hence to extract f in the form (9.5.3) as a solution of (9.3.18), which is of course in the ring $\mathcal{L}\{\Re; x,y,z,t\}$ where $\Re$ is seen to be the commutative ring of all polynomials of λ with integers as coefficients.

Further, from the uniqueness of the procedure in the ring $\mathcal{L}\{\Re; x, y, z, t\}$, (9.3.18) is well-defined.

In a way similar to that above, equation (9.4.18) is well defined in $\mathcal{L}\{\Re; y, z, t\}$ as well. □

§9.6 Notes

9.6.1 The general solution of (9.1.13) can also be determined in a way similar to the corresponding enumeration of trees described in §2.1.

9.6.2 Based on the equation (9.2.22), it is no doubt to find the solution of Eq.(9.2.21). However, it seems not simple enough by considering the coloration of outerplanar maps. More about this topic can also be seen in [Liu47] and [Liu50].

9.6.3 The equation (9.3.18) appears in [Liu40] and [Liu46] in which some related problems are also discussed. Of course, the general solution of (9.3.18) is still unknown up to now. Moreover, an equation for general planar map with vertex partition has not yet been found.

9.6.4 The chrosum problems were posed and deeply studied by Tutte for planar triangulations in the seventies [Tut21–24], [Tut27], [Tut32], *etc.* The equation (9.4.18) is in fact equivalent to that in [Tut21]. Many related topics especially to the coloring of planar maps can also be found in [Tut16–18], [Tut26], [Tut29], *etc.*

9.6.5 The procedure described in §9.5 is in principle valid for showing almost all equations appearing in this book especially for those with the linear functional relating to vertex or face partition of maps.

9.6.6 No result has been found for the chrosum problem of maps on a general surface whose genus is not zero.

9.6.7 Chrosum equation for cubic planar maps, for vertex coloring of course, looks very complicated [Liu42, Liu44–45]. Is it possible to simplify so that a more simpler equation can be derived?

9.6.8 The nonrooted case on chrosums has not yet been observed for any kind of maps planar or nonplanar.

Chapter 10

Polysum Equations

§10.1 Polysums for bitrees

A *bitree* is a planar map whose edges are either ithmuses or loops. It is easily seen that the planar dual of a bitree is also a bitree. Let $\mathcal{B}$ be the set of all rooted bitrees, then $\mathcal{B}$ is planar selfdual. For convenience, the vertex map ϑ is defined to be in $\mathcal{B}$.

The polysum function of $\mathcal{B}$ considered here is given by

$$f_{\mathcal{B}}(\Phi, x, \underline{y}) = \sum_{B \in \mathcal{B}} \Phi(B) x^{m(B)} \underline{y}^{\underline{n}(B)} \tag{10.1.1}$$

where $m(B)$ is the valency of the root-vertex,

$$\underline{n}(B) = (n_1(B), n_2(B), \cdots)$$

with $n_i(B)$ being the number of nonrooted vertices of valency $i, i \geq 1$ and $\Phi(B)$ is the polynomial of X, Y and z defined by (1.2.1) with the convention: $\forall e \in E(B)$, the edge set of B,

$$w(e) = 1 \, (\text{mod } 2) \tag{10.1.2}$$

for brevity.

For decomposing $\mathcal{B}$, we should first partition $\mathcal{B}$ into the form as

$$\mathcal{B} = \mathcal{B}_0 + \mathcal{B}_{\text{I}} + \mathcal{B}_{\text{II}} \tag{10.1.3}$$

where $\mathcal{B}_0$ consists of only the vertex map ϑ,

$$\mathcal{B}_{\text{I}} = \{B | \forall B \in \mathcal{B}, B \neq \vartheta \text{ and } e_r(B) \text{ is a loop}\}$$

and hence

$$\mathcal{B}_{\mathrm{II}} = \{B|\forall B \in \mathcal{B}, B \neq \vartheta \text{ and } e_r(B) \text{ is an isthmus}\}.$$

Lemma 10.1.1 *Let $\mathcal{B}_{\langle \mathrm{I}\rangle} = \{B - a|\forall B \in \mathcal{B}_{\mathrm{I}}\}$. Then, we have*

$$\mathcal{B}_{\langle \mathrm{I}\rangle} = \mathcal{B} \odot \mathcal{B} \tag{10.1.4}$$

where $\odot$ is the 1v-production in §3.2.

Proof For any $B \in \mathcal{B}_{\langle \mathrm{I}\rangle}$, let $\tilde{B}$ be the map in $\mathcal{B}_{\mathrm{I}}$ such that $\tilde{B} - a = B$. Because $a = e_r(B)$ is a loop, $B = B_1 \dot{+} B_2$ such that one of B_1 and B_2 is in the inner domain with the other in the outer domain of a. From the heredity of bitrees, both B_1 and B_2 are in $\mathcal{B}$ and hence $B \in \mathcal{B} \odot \mathcal{B}$.

Conversely, for any $B = B_1 \dot{+} B_2 \in \mathcal{B} \odot \mathcal{B}$, we may uniquely construct a map $\tilde{B} \in \mathcal{B}_{\langle \mathrm{I}\rangle}$ by adding an edge which is a loop $\tilde{a}$ such that one of B_1 and B_2 is in the inner domain with the other in the outer domain of $\tilde{a}$. It is easily seen that $B = \tilde{D} - \tilde{a}$. Thus $B \subset \mathcal{B}_{\langle \mathrm{I}\rangle}$. □

Let $\mathcal{B}_{(\mathrm{II})}(m) = \{B \bullet a|\forall B \in \mathcal{B}_{\mathrm{II}}(m)\}$ where $a = e_r(B)$ and $\mathcal{B}_{\mathrm{II}}(m) = \{B|\forall B \in \mathcal{B}_{\mathrm{II}}, m(B) = m\}$. Then we have

$$\mathcal{B}_{(\mathrm{II})} = \sum_{m \geq 0} \mathcal{B}_{(\mathrm{II})}(m) \tag{10.1.5}$$

and

$$\mathcal{B}_{\mathrm{II}} = \sum_{m \geq 0} \mathcal{B}_{\mathrm{II}}(m). \tag{10.1.6}$$

Lemma 10.1.2 *For $m \geq 0$, we have*

$$\mathcal{B}_{(\mathrm{II})}(m) = \mathcal{B}(m-1) \odot \sum_{i \geq 0} \mathcal{B}(i) \tag{10.1.7}$$

where $\mathcal{B}(i) = \{B|\forall B \in \mathcal{B}, m(B) = i\}, i \geq 0$.

Proof For $B \in \mathcal{B}_{(\mathrm{II})}(m)$, let B' be the map in $\mathcal{B}_{\mathrm{II}}(m)$ such that $B = B' \bullet a'$ where $a' = e_r(B')$. Write the root-vertex $v_{r'}$ and the nonrooted vertex $v_{\beta r'}$ of B' in the form as

$$v_{r'} = (r', \mathcal{P}r', R) \text{ and } v_{\beta r'} = (\alpha\beta r', \mathcal{P}\alpha\beta r', S).$$

Then the root-vertex v_r of B is in the form as

$$v_r = (\mathcal{P}r', R, \mathcal{P}\alpha\beta r', S)$$

in which $r = \mathcal{P}r'$ is the root of B. Denote by B_1 and B_2 the submaps of B which are expanded from

$$v_{r_1} = (\mathcal{P}r', R) \text{ and } v_{r_2} = (\mathcal{P}\alpha\beta r', S)$$

as the root-vertices, respectively. By the heredity of bitrees, B_1 and B_2 are in $\mathcal{B}$. It is easily seen that $B = B_1 \dot{+} B_2$. Since $m(B_1) = m(B') - 1$ and there is an integer $i \geq 0$ such that $m(B_2) = i$, we soon see that B is a member of the set on the right hand side of (10.1.7).

Conversely, for B in the set on the right hand side, let the root-vertex of B be

$$(r, R, s, S)$$

such that $B = B_1 \dot{+} B_2$ has (r, R) and (s, S) as the root-vertices of $B_1, m(B_1) = m - 1$ and $B_2, m(B_2) = i \geq 0$, respectively. Then we can uniquely construct a map $B' \in \mathcal{B}_{\mathrm{II}}$ such that $B = B' \bullet a', a' = e_r(B')$, by splitting the root-vertex of B into $v_{r'}$ and $v_{\beta r'}$ with an edge Kr', the root-edge of B', where

$$v_{r'} = (r', r, R) \text{ and } v_{\beta r'} = (\alpha\beta r', s, S).$$

Since $B_1 \in \mathcal{B}(m-1)$, we have $m(B') = m(B_1) + 1 = m$ and hence $B' \in \mathcal{B}_{\mathrm{II}}(m)$. This implies $B \in \mathcal{B}_{(\mathrm{II})}(m)$. $\square$

In what follows, we evaluate the respective contributions of $\mathcal{B}_0, \mathcal{B}_{\mathrm{I}}$ and $\mathcal{B}_{\mathrm{II}}$ to the polysum function $f = f_{\mathcal{B}}(\Phi, x, \underline{y})$ defined by (10.1.1).

Since $m(\vartheta) = 0, \underline{n}(\vartheta) = \underline{0}$ and $\Phi(\vartheta) = 1$, from (1.2.3) we have

$$f_{\mathcal{B}_0}(\Phi, x, \underline{y}) = 1. \tag{10.1.8}$$

Based on (1.2.1) and Lemma 10.1.1, we find that

$$\begin{aligned} f_{\mathcal{B}_{\mathrm{I}}}(\Phi, x, \underline{y}) &= \sum_{B \in \mathcal{B}_I} \Phi(B) x^{m(B)} \underline{y}^{\underline{n}(B)} \\ &= x^2 (Xz + Y) \Big(\sum_{B \in \mathcal{B}} \Phi(B) x^{m(B)} \underline{y}^{\underline{n}(B)} \Big)^2 \end{aligned}$$

$$= x^2(Xz + Y)f^2 \tag{10.1.9}$$

in which the formula $\Phi(B) = \Phi(B_1)\Phi(B_2)$ is applied for $B = B_1 \dot{+} B_2$ by considering (10.1.2).

Based on (1.2.1) and Lemma 10.1.2, we find that

$$\begin{aligned} f_{\mathcal{B}_{\rm II}}(\Phi, x, \underline{y}) &= \sum_{B\in\mathcal{B}_{\rm II}} \Phi(B) x^{m(B)} \underline{y}^{\underline{n}(B)} \\ &= x(X + Yz)\left(\sum_{B\in\mathcal{B}} \Phi(B) x^{m(M)} \underline{y}^{\underline{n}(B)}\right) \\ &\quad \times \left(\sum_{i\geq 0} \int_y y \sum_{B\in\mathcal{B}} \Phi(B) y^{m(B)} \underline{y}^{\underline{n}(B)}\right) \\ &= x(X + Yz) f \int_y y f_y \end{aligned} \tag{10.1.10}$$

where $f_y = f_{\mathcal{B}}(\Phi, y, \underline{y})$ and $\int_y$ is the functional defined in §1.6.

Theorem 10.1.1 *The polysum function $f = f_{\mathcal{B}}(\Phi, x, \underline{y})$ defined by* (10.1.1) *satisfies the following functional equation:*

$$\int_y y f_y = \frac{1 - f^{-1}}{x(X + Yz)} - \frac{x(Xz + Y)}{X + Yz} f \tag{10.1.11}$$

where $f_y = f|_{x=y}$ and $\int_y$ is the functional mentioned in §1.6.

Proof Since

$$f = f_{\mathcal{B}_0}(\Phi, x, \underline{y}) + f_{\mathcal{B}_{\rm I}}(\Phi, x, \underline{y}) + f_{\mathcal{B}_{\rm II}}(\Phi, x, \underline{y})$$

by (10.1.3), from (10.1.8–10) we soon find that

$$f = 1 + x^2(Xz + Y)f^2 + x(X + Yz) f \int_y y f_y.$$

By dividing $x(X+Yz)f$ on the two sides and by rearranging, (10.1.11) is obtained. □

If the parameters X, Y and z are taken to be $X = Y = 1$ and $z = 0$, then f in this case becomes the enufunction of $\mathcal{B}$ with vertex partition and hence the equation

$$\int_y y f_y = \frac{1 - f^{-1}}{x} - xf \tag{10.1.12}$$

can be employed for enumerating rooted bitrees with vertex partition.

For simplification, we investigate the case $y_i = y^i, i \geq 1$, by substituting x and y for $\frac{x}{y}$ and y^2 in $f = f_{\mathcal{B}}(\Phi, x, \underline{y})$ defined by (10.1.1). That is, to consider

$$f_{\mathcal{B}}(\Phi, x, y) = \sum_{B \in \mathcal{B}} \Phi(B) x^{m(B)} y^{n(B)} \tag{10.1.13}$$

where $m(B)$ and $n(B)$ are the valency of root-vertex and the size of B respectively.

Theorem 10.1.2 *The polysum function $f = f_{\mathcal{B}}(\Phi, x, y)$ defined by* (10.1.13) *satisfies the equation as*

$$x^2 y f^2 + \frac{xy(X + Yz)f^* - 1}{Xz + Y} f + \frac{1}{Xz + Y} = 0 \tag{10.1.14}$$

where $f^* = f|_{x=1} = f_{\mathcal{B}}(\Phi, 1, y)$.

Proof Because the corresponding values in (10.1.8–10) are respectively 1, $x^2 y(Xz + Y)f^2$ and $xy(X + Yz)ff^*$, from (10.1.3) we soon find that

$$f = 1 + x^2 y(Xz + Y)f^2 + xy(X + Yz)ff^*.$$

By dividing the two sides by $(Xz+Y)$ and rearranging the terms, the equation (10.1.14) is then soon found. □

Further, we discuss the case of $X = Y = 1$ and $z = 0$ in (10.1.14). That is to solve the following equation:

$$x^2 y f^2 + (xyf^* - 1)f + 1 = 0. \tag{10.1.15}$$

Fortunately, (10.1.15) allows us to set $x = 1$ such that f becomes f^* and hence to obtain the following equation with f^* as the only one unknown function:

$$2yf^{*2} - f^* + 1 = 0. \tag{10.1.16}$$

Therefore we can easily find the solution which is the enufunction of rooted bitrees with the size as the parameter for the equation (10.1.16). That is,

$$f^* = \frac{1 - \sqrt{1 - 8y}}{4y}$$

$$= \frac{1}{4y}\sum_{i\geq 1}\frac{2^{i+1}(2i-2)!}{(i-1)!i!}y^i$$

$$= \sum_{i\geq 1}\frac{2^{i-1}(2i-2)!}{(i-1)!i!}y^{i-1}$$

$$= \sum_{n\geq 0}\frac{2^n(2n)!}{n!(n+1)!}y^n. \tag{10.1.17}$$

From the selfduality of bitrees, according to (10.1.17), we may find an explicit expression of

$$f_{\mathcal{B}}(x,y) = \sum_{B\in\mathcal{B}} x^{i(B)}y^{j(B)} \tag{10.1.18}$$

where $i(B)$ and $j(B)$ are the respective numbers of nonrooted vertices and nonrooted faces of B. By substituting $(x+y)/2$ for y in (10.1.16), we can soon obtain

$$f_{\mathcal{B}}(x,y) = \sum_{i\geq 0}\sum_{j\geq 0}\frac{(2i+2j)!}{i!j!(i+j+1)!}x^iy^j \tag{10.1.19}$$

because $i(B)+j(B)$ is just the size of a map $B \in \mathcal{B}$with $i(B)$ nonrooted vertices and $j(B)$ nonrooted faces from the Euler formula in Theorem 1.1.2 for $p = 0$.

§10.2 Outerplanar polysums

Let $\mathcal{O}$ be the set of all rooted nonseparable outerplanar maps in which the vertex map is excluded and both of the link map L_1 and the loop map L_0 are included. For $O \in \mathcal{O}$, let $m(O), s(O)$ and $n(O)$ be the valency of root-vertex, the valency of root-face and the size of O, respectively.

The polysum function of $\mathcal{O}$ is defined to be as

$$f_{\mathcal{O}}(\chi,x,y,z) = \sum_{O\in\mathcal{O}} \chi(O)x^{m(O)}y^{s(O}z^{n(O)} \tag{10.2.1}$$

where $\chi(O)$ is the dichromate as known in (1.2.12). The function given in (10.2.1) is also called the *dichrosum function.*

In order to determine $f_{\mathcal{O}}(\chi, x, y, z)$, we partition $\mathcal{O}$ as

$$\mathcal{O} = \mathcal{O}_0 + \mathcal{O}_1 + \mathcal{O}_2 \tag{10.2.2}$$

where $\mathcal{O}_0$ consists of only one map which is the loop map L_0, $\mathcal{O}_1$, of only the link map L_1 and $\mathcal{O}_2$, of all others in $\mathcal{O}$.

For $\mathcal{O}_0$, since the loop map L_0 is with $m(L_0) = 2$, $s(L_0) = 1$, $n(L_0) = 1$ and $\chi(L_0) = \nu$ by which y is replaced for avoiding confusion in (1.2.12), we have the contribution of $\mathcal{O}_0$ to $f_{\mathcal{O}}(\chi, x, y, z)$ as

$$f_{\mathcal{O}_0}(\chi, x, y, z) = \nu x^2 yz. \tag{10.2.3}$$

Similarly, for $\mathcal{O}_1$ we have

$$f_{\mathcal{O}_1}(\chi, x, y, z) = \mu x y^2 z \tag{10.2.4}$$

where μ is the substitute of x in (1.2.12) as a parameter of the polynomial χ.

The main part is to evaluate $f_{\mathcal{O}_2}(\chi, x, y, z)$ for which we have to decompose $\mathcal{O}_2$ in a proper way.

Lemma 10.2.1 *Let $\mathcal{O}_{\langle 2\rangle} = \{O - a | \forall O \in \mathcal{O}_2\}$ with $a = e_r(O)$. Then we have*

$$\mathcal{O}_{\langle 2\rangle} = \sum_{k\geq 1} (\mathcal{O} - L_0)^{\hat{\times} k} \tag{10.2.5}$$

where $\hat{\times}$ is the chain 1v-production defined as in (6.1.23) *and $L_0 = \mathcal{O}_0$ for convenience.*

Proof Similar to the proof of Lemma 6.1.4. However, the attention that the loop is not in $\mathcal{O}_{\text{sf}}$ there has to be paid to. □

Lemma 10.2.2 *Let $\mathcal{O}_{(2)} = \{O \bullet a | \forall O \in \mathcal{O}_2\}$ with $a = e_r(O)$. Then, we have*

$$\mathcal{O}_{(2)} = \sum_{k\geq 0} L_0^{\times k} \times\cdot\, (\mathcal{O} - L_1) \tag{10.2.6}$$

where $\times\cdot$ is the inner 1v-production defined as in (6.1.7) *and $L_1 = \mathcal{O}_1$ for convenience.*

Proof In fact, it is the dual form of Lemma 10.2.1. □

Let us write

$$f_A = \sum_{O \in \mathcal{O}_2} \chi(O - a) x^{m(O)} y^{s(O)} z^{n(O)} \tag{10.2.7}$$

and

$$f_B = \sum_{O \in \mathcal{O}_2} \chi(O \bullet a) x^{m(O)} y^{s(O)} z^{n(O)}. \tag{10.2.8}$$

Then we have

$$f_{\mathcal{O}_2}(\chi, x, y, z) = f_A + f_B \tag{10.2.9}$$

by considering (1.2.12) in that for any $O \in \mathcal{O}_2, a = e_r(O)$ is neither loop nor isthmus.

From Lemma 10.2.1 and (1.2.12),

$$\begin{aligned} f_A &= xz \sum_{O \in \mathcal{O} - L_0} \chi(O) x^{m(O)} y^{s(O)} z^{n(O)} \\ &\quad \times \sum_{k \geq 1} \left(\sum_{O \in \mathcal{O} - L_0} \chi(O) y^{s(O)-1} z^{n(O)} \right)^{k-1} \\ &= xz(f - \nu x^2 yz) \frac{1}{1 - \frac{1}{y}(f^+ - \nu yz)} \\ &= \frac{xyz(f - \nu x^2 yz)}{y + \nu yz - f^+} \end{aligned} \tag{10.2.10}$$

where $f = f_{\mathcal{O}}(\chi, x, y, z)$, $f^+ = f_{\mathcal{O}}(\chi, 1, y, z)$.

From Lemma 10.2.2 and (1.2.12),

$$\begin{aligned} f_B &= xyz \left(\sum_{k \geq 0} (\nu xz)^k \right) \\ &\quad \times \left(\sum_{O \in \mathcal{O} - L_1} \chi(O) \Big(\sum_{i=1}^{m(O)-1} x^i \Big) y^{s(O)} z^{n(O)} \right) \\ &= \frac{xyz}{1 - \nu xz} \sum_{O \in \mathcal{O} - L_1} \chi(O) \frac{x - x^{m(O)}}{1 - x} y^{s(O)} z^{n(O)} \\ &= \frac{xyz}{1 - \nu xz} \partial_x f \end{aligned} \tag{10.2.11}$$

where

$$\partial_x f = \frac{x f_{\mathcal{O}}(\chi, 1, y, z) - f_{\mathcal{O}}(\chi, x, y, z)}{1 - x} \tag{10.2.12}$$

which is the $\langle 1, x \rangle$-difference as defined by (1.6.8).

Theorem 10.2.1 *The polysum function $f = f_{\mathcal{O}}(\chi, x, y, z)$ of $\mathcal{O}$ defined by* (10.2.1) *satisfies the equation as*

$$\left(1 - \frac{xyz}{y + \nu yz - f^{+}}\right) f - \frac{xyz\partial_x f}{1 - \nu xz}$$

$$= xyz \left(\nu x + \mu y - \frac{\nu x^2 yz}{y + \nu yz - f^{+}}\right) \tag{10.2.13}$$

where $f^{+} = f_{\mathcal{O}}(\chi, 1, y, z)$.

Proof From (10.2.2), we see that

$$f = f_{\mathcal{O}_0}(\chi, x, y, z) + f_{\mathcal{O}_1}(\chi, x, y, z) + f_{\mathcal{O}_2}(\chi, x, y, z).$$

By (10.2.3–4) and (10.2.7–11),

$$\begin{aligned} f = {} & \nu x^2 yz + \mu x y^2 z \\ & + \frac{xyz(f - \nu x^2 yz)}{y - \nu yz + f^{+}} \\ & + \frac{xyz}{1 - \nu xz} \partial_x f \end{aligned}$$

from which (10.2.13) can be obtained after rearrangement. □

Particularly, let

$$h = h_{\mathcal{O}}(\chi, x, 1, z). \tag{10.2.14}$$

Then, from Theorem 10.2.1, we have that h satisfies the following equation:

$$\left(1 - \frac{xz}{1 + \nu z - h^{+}}\right) h - \frac{xz\partial_x h}{1 - \nu z}$$

$$= xz \left(\nu x + \mu - \frac{\nu x^2 z}{1 + \nu z - h^{+}}\right) \tag{10.2.15}$$

where $h^{+} = f_{\mathcal{O}}(\chi, 1, 1, z)$.

By employing (10.2.12), we may write (10.2.15) in the following form:

$$\left(1-\frac{xz}{1+\nu z-h^+}+\frac{xz}{(1-\nu z)(1-x)}\right)f$$
$$=xz\left(\mu+\nu x-\frac{\nu x^2 z}{1+\nu z-h^+}\right.$$
$$\left.+\frac{xh^+}{(1-\nu z)(1-x)}\right). \tag{10.2.16}$$

Then, in view of $x=\xi(z)=\xi$ as a parameter, the following equation system,

$$1-\frac{\xi z}{1+\nu z-h^+}+\frac{\xi z}{(1-\nu z)(1-\xi)}=0;$$
$$\mu+\nu\xi-\frac{\nu\xi^2 z}{1+\nu z-h^+}+\frac{\xi h^+}{(1-\nu z)(1-\xi)}=0, \tag{10.2.17}$$

can be used to find the expressions of h^+ and z as functions of ξ, and then by Lagrangian inversion to extract h^+ as a power series of z. The equation (10.2.17) is sometimes called the *characteristic equation* of (10.2.15).

§10.3 General polysums

Let $\mathcal{M}$ be the set of all rooted general planar maps in which the vertex map is of course included. For $M\in\mathcal{M}$, write $m(M), s(M)$ and $n(M)$ for the valency of root-vertex, the valency of root-face and the size of M, respectively, as in §10.2. The polysum function considered is

$$f_{\mathcal{M}}(\chi,x,y,z)=\sum_{M\in\mathcal{M}}\chi(M)x^{m(M)}y^{s(M)}z^{n(M)}, \tag{10.3.1}$$

in which $\chi(M)$ with the parameters μ and ν instead of respective x and y in (1.2.12) is the dichromate.

According to the principle of (1.2.12), $\mathcal{M}$ has to be partitioned into the form as

$$\mathcal{M}=\sum_{i=0}^{3}\mathcal{M}_i \tag{10.3.2}$$

where $\mathcal{M}_0=\vartheta$, the set consists of the vertex map ϑ itself,

$$\mathcal{M}_1=\{M|\forall M\in\mathcal{M}, e_r(M)\text{ is a loop}\},$$

$$\mathcal{M}_2 = \{M | \forall M \in \mathcal{M}, e_r(M) \text{ is an isthmus}\}$$

and

$$\mathcal{M}_3 = \{M | \forall M \in \mathcal{M}, e_r(M) \text{ is neither loop nor isthmus}\}.$$

Since ϑ has no edge with $\chi(\vartheta) = 1$ from (1.2.12), the contribution of ϑ to $f_{\mathcal{M}}(\chi, x, y, z)$ is

$$f_0 = f_{\mathcal{M}_0}(\chi, x, y, z) = 1. \tag{10.3.3}$$

For a map $M = (\mathcal{X}_{,}, \mathcal{P}) \in \mathcal{M}_1$, because the root-edge $a = e_r(M)$ is a loop, it is allowed to write for the root-vertex in the form as

$$v_r = (r, \mathcal{P}r, R, \alpha\beta r, s, S). \tag{10.3.4}$$

Then, we have

$$M - a = M_1 \hat{+} M_2 \tag{10.3.5}$$

where $\hat{+}$ is the inner $1v$-addition, M_1 and M_2 are the submaps expanded by (s, S) and $(\mathcal{P}r, R)$ as the respective root-vertices on M. Of course, the root of M_1 is $r_1 = s$ and that of M_2 is $r_2 = \mathcal{P}r$.

Lemma 10.3.1 *Let* $\mathcal{M}_{\langle 1 \rangle} = \{M - a | \forall M \in \mathcal{M}_1\}$ *with* $a = e_r(M)$. *Then, we have*

$$\mathcal{M}_{\langle 1 \rangle} = \mathcal{M} \hat{\times} \mathcal{M} \tag{10.3.6}$$

where $\hat{\times}$ *is the inner* $1v$*-production.*

Proof For $M = M_1 \hat{+} M_2 \in \mathcal{M} \hat{\times} \mathcal{M}$, from that both of M_1 and M_2 in (10.3.5) are allowed to be the vertex map ϑ, we may always find a map $M' \in \mathcal{M}_1$ by adding a loop a' so that $M = M' - a'$. This implies that $\mathcal{M} \hat{\times} \mathcal{M}$ is a subset of $\mathcal{M}_{\langle I \rangle}$.

Conversely, from what is mentioned above, we also see that $\mathcal{M}_{\langle I \rangle}$ is a subset of $\mathcal{M} \hat{\times} \mathcal{M}$. □

By considering that M_2 has no contribution to the valency of root-face of M in (10.3.5), from Lemma 10.3.1 and (1.2.12) we have

$$\begin{aligned} f_1 &= f_{\mathcal{M}_1}(\chi, x, y, z) \\ &= \sum_{M \in \mathcal{M}_1} \chi(M) x^{m(M)} y^{s(M)} z^{n(M)} \end{aligned}$$

$$= \nu x^2 yz \sum_{M \in \mathcal{M}} \chi(M) x^{m(M)} y^{s(M)} z^{n(M)}$$

$$\times \sum_{M \in \mathcal{M}} \chi(M) x^{m(M)} z^{n(M)}$$

$$= \nu x^2 yz f f^* \tag{10.3.7}$$

where $f = f_{\mathcal{M}}(\chi, x, y, z)$ and $f^* = f_{\mathcal{M}}(\chi, x, 1, z)$.

For $M = (\mathcal{X}, \mathcal{P}) \in \mathcal{M}_2$, because the root-edge $a = e_r(M)$ is an isthmus, we are allowed to write

$$v_r = (r, \mathcal{P}r, R) \text{ and } v_{\beta r} = (\alpha\beta r, \mathcal{P}\alpha\beta r, S) \tag{10.3.8}$$

for the root-vertex and the other end of the root-edge of M respectively. Then we have

$$M \bullet a = M_1 \dot{+} M_2 \tag{10.3.9}$$

where $\dot{+}$ is the $1v$-addition, M_1 and M_2 are the submaps of M expanded by

$$(\mathcal{P}r, R) \text{ and } (\mathcal{P}\alpha\beta r, S) \tag{10.3.10}$$

as the respective root-vertices of them with $r_1 = \mathcal{P}r$ and $r_2 = \mathcal{P}\alpha\beta r$ chosen to be the corresponding roots. It is easily checked that both M_1 and M_2 are rooted general planar maps with ϑ as the degenerate case.

Lemma 10.3.2 *Let $\mathcal{M}_{(2)} = \{M \bullet a | \forall M \in \mathcal{M}_2\}$ with $a = e_r(M)$. Then we have*

$$\mathcal{M}_{(2)} = \mathcal{M} \dot{\times} \mathcal{M} \tag{10.3.11}$$

where $\dot{\times}$ is the $1v$-production.

Proof It is, in fact, the dual case of Lemma 10.3.1. □

From Lemma 10.3.2 and (1.2.12), the contribution of $\mathcal{M}_2$ to the chrosum function $f_{\mathcal{M}}(\chi, x, y, z)$ is

$$f_2 = f_{\mathcal{M}_2}(\chi, x, y, z)$$

$$= \mu x y^2 z \sum_{M \in \mathcal{M}} \chi(M) x^{m(M)} y^{s(M)} z^{n(M)}$$

$$\times \sum_{M \in \mathcal{M}} \chi(M) y^{s(M)} z^{n(M)}$$

$$= \mu x y^2 z f f^+ \tag{10.3.12}$$

where $f^+ = f_{\mathcal{M}}(\chi, 1, y, z)$.

For $M = (\mathcal{X}, \mathcal{P}) \in \mathcal{M}_3$, write

$$v_r = (r, \mathcal{P}r, R) \text{ and } v_{\beta r} = (\alpha\beta r, \mathcal{P}\alpha\beta r, S) \tag{10.3.13}$$

for the root-vertex and the other end of the root-edge $a = e_r(M)$ of M respectively. Then $M - a$ and $M \bullet a$ are the maps expanded from M by

$$(\mathcal{P}r, R) \text{ and } (\mathcal{P}\alpha\beta r, S) \tag{10.3.14}$$

instead of v_r and $v_{\beta r}$, and by

$$(\mathcal{P}\alpha\beta r, S, \mathcal{P}r, R) \tag{10.3.15}$$

instead of v_r, respectively. Of course, the root of $M - a$ is $\mathcal{P}r$ and that of $M \bullet a$ is $\mathcal{P}\alpha\beta r$.

For a map $M = (\mathcal{X}, \mathcal{P}) \in \mathcal{M}$, write

$$(r, R), (\mathcal{P}\alpha\beta r, R_1), \cdots, ((\mathcal{P}\alpha\beta)^i, R_i), \cdots \tag{10.3.16}$$

for the vertices on the boundary of root-face of M. Let $\triangle_i M$ as shown in §7.3 be the *i-th increment* of M, *i.e.*, the map which is expanded by

$$(r_i, r, R) \text{ and } (\alpha\beta r_i, (\mathcal{P}\alpha\beta)^i r) \tag{10.3.17}$$

where Kr_i is the new edge added on M and r_i is the root of $\triangle_i M, i \geq 0$.

Let us now write

$$\mathcal{M}_\triangle = \sum_{M \in \mathcal{M}} \left\{ \triangle_i M \mid 0 \leq i \leq s(M) \right\} \tag{10.3.18}$$

where $s(M)$ is the valency of root-face of M.

Lemma 10.3.3 *For $\mathcal{M}_\triangle$, we have*

$$\mathcal{M}_\triangle = \mathcal{M}_3 + \mathcal{M}_1. \tag{10.3.19}$$

Proof For $M \in \mathcal{M}_\triangle$, if $M \notin \mathcal{M}_1$ then the root-edge has to be neither loop nor ithmus by considering that any increment of M is never allowed to have the root-edge which is a loop or an ithmus. This implies $M \in \mathcal{M}_3$. Thus $\mathcal{M}_\Delta$ is a subset of $\mathcal{M}_1 + \mathcal{M}_3$.

Conversely, it is easily checked that for any $M \in \mathcal{M}_3$ or $\mathcal{M}_1$, we can always construct a map $\tilde{M} \in \mathcal{M}$ such that there is an $i \geq 0, M = \Delta_i\tilde{M}$. Thus $\mathcal{M}_1 + \mathcal{M}_3$ is also a subset of $\mathcal{M}_\Delta$. □

Dually, for $M = (\mathcal{X}, \mathcal{P}) \in \mathcal{M}$, the *i-th decrement* $\nabla_i M$ is expanded by

$$(r_i, \mathcal{P}^{m(M)-i+2}r, \cdots, \mathcal{P}^{m(M)}r)$$

and

$$(\alpha\beta r_i, r, \mathcal{P}r, \cdots, \mathcal{P}^{m(M)-i+1}r)$$

as the root-vertex and the other end of the root-edge $Kr_i, 0 \leq i \leq m(M)$, where $\nabla_0 M$ has the root-vertex and the other end of Kr_1 as respective (r_0) and $(\alpha\beta r_0, r, \cdots, \mathcal{P}^{m(M)}r)$.

Similarly, let us write

$$\mathcal{M}_\nabla = \sum_{M \in \mathcal{M}} \left\{\nabla_i M \mid 0 \leq i \leq m(M)\right\} \tag{10.3.20}$$

where $m(M)$ is the valency of root-vertex of M.

Lemma 10.3.4 *For $\mathcal{M}_\nabla$, we have*

$$\mathcal{M}_\nabla = \mathcal{M}_3 + \mathcal{M}_2. \tag{10.3.21}$$

Proof The dual case of Lemma 10.3.3. □

Let $\mathcal{M}_{\langle i\rangle} = \{M - a | \forall M \in \mathcal{M}_i\}$ and $\mathcal{M}_{(i)} = \{M \bullet a | \forall M \in \mathcal{M}_i\}$ for $a = e_r(M), i = 1, 2, 3$.

Lemma 10.3.5 *We have*

$$\mathcal{M}_{\langle 3\rangle} \bigcup \mathcal{M}_{\langle 1\rangle} = \mathcal{M} \tag{10.3.22}$$

and

$$\mathcal{M}_{(3)} \bigcup \mathcal{M}_{(2)} = \mathcal{M}. \tag{10.2.23}$$

Proof From duality, it is only necessary to prove one of (10.3.22) and (10.3.23), the former.

It is easily seen that the set on the left hand side is a subset of that on the right hand side of (10.3.22).

Conversely, by the procedure as shown in (10.3.15–17), the set on the right is also a subset of that on the left. □

From (10.3.20),

$$\begin{aligned}
f_\triangle &= \sum_{M \in \mathcal{M}_\triangle} \chi^(M - a) x^{m(M)} y^{s(M)} z^{n(M)} \\
&= xz \sum_{M \in \mathcal{M}} \chi(M) x^{m(M)} \sum_{i=1}^{s(M)+1} y^i z^{n(M)} \\
&= xz \sum_{M \in \mathcal{M}} \chi(M) x^{m(M)} \frac{y(1 - y^{s(M)+1})}{1-y} z^{n(M)} \\
&= \frac{xyz}{1-y}(f^* - yf)
\end{aligned} \tag{10.3.24}$$

and hence from Lemma 10.3.3, (10.3.22) and (10.3.7),

$$\begin{aligned}
f_A &= \sum_{M \in \mathcal{M}_3} \chi(M - a) x^{m(M)} y^{s(M)} z^{n(M)} \\
&= f_\triangle - \frac{1}{\nu x} f_1 \\
&= \frac{xyz}{1-y}(f^* - yf) - xyzff^*.
\end{aligned} \tag{10.3.25}$$

Dually, we may find that

$$\begin{aligned}
f_B &= \sum_{M \in \mathcal{M}_3} \chi(M \bullet a) x^{m(M)} y^{s(M)} z^{n(M)} \\
&= \frac{xyz}{1-y}(f^+ - xf) - xyzff^+.
\end{aligned} \tag{10.3.26}$$

By (1.2.12), we have

$$\begin{aligned}
f_3 &= \sum_{M \in \mathcal{M}_3} \chi(M) x^{m(M)} y^{s(M)} z^{n(M)} \\
&= f_A + f_B
\end{aligned}$$

from (10.3.25–26),

$$= xyz\left(\frac{f^* - yf}{1-y} + \frac{f^+ - xf}{1-x}\right)$$

$$-xyzf(f^* + f^+). \tag{10.3.27}$$

Theorem 10.3.1 *The polysum function* $f = f_{\mathcal{M}}(\chi, x, y, z)$ *defined by* (10.3.1) *satisfies the following equation:*

$$\left(1 + xyz\left(\frac{y}{1-y} + \frac{x}{1-x}\right)\right.$$

$$\left.-xyz(\nu x - 1)f^* - xyz(\mu y - 1)f^+\right)f$$

$$= 1 + xyz\left(\frac{f^*}{1-y} + \frac{f^+}{1-x}\right) \tag{10.3.28}$$

where $f^* = f_{\mathcal{M}}(\chi, x, 1, z)$ *and* $f^+ = f_{\mathcal{M}}(\chi, 1, y, z)$.

Proof From (10.3.2), $f = f_0 + f_1 + f_2 + f_3$. By (10.3.3), (10.3.7), (10.3.12) and (10.3.27), (10.3.28) is then obtained after grouping and rearranging the terms. □

In principle, the equation (10.3.27) allows us to determine the polysum function $f_{\mathcal{M}}(\chi, x, y, z)$ term by term in a power series form of x, y, z.

§10.4 Nonseparable polysums

Let $\mathcal{N}$ be the set of all rooted nonseparable planar maps with both the loop map and the link map included but the vertex map excluded. The polysum function is adopted here to be

$$f_{\mathcal{N}}(\chi, x, y, z) = \sum_{N \in \mathcal{N}} \chi(N) x^{m(N)} y^{s(N)} z^{n(N)} \tag{10.4.1}$$

where $m(N)$, $s(N)$ and $n(N)$ are the valency of root-vertex, the valency of root-face and the size of $\mathcal{N}$ as in the last section, respectively.

Now the set $\mathcal{N}$ is partitioned as

$$\mathcal{N} = \mathcal{N}_0 + \mathcal{N}_1 + \mathcal{N}_2 \tag{10.4.2}$$

where $\mathcal{N}_0$ consists of the single loop map, $\mathcal{N}_1$ is of the single link map and $\mathcal{N}_2$, of all others in $\mathcal{N}$.

Because the loop map L_0 has that $m(L_0) = 2, s(L_0) = 1$ and $n(L_0) = 1$, from (1.2.12)

$$f_0 = f_{\mathcal{N}_0}(\chi, x, y, z) = \nu x^2 yz. \tag{10.4.3}$$

By the duality, we soon find that

$$f_1 = f_{\mathcal{N}_1}(\chi, x, y, z) = \mu x y^2 z. \tag{10.4.4}$$

The main part for $f = f_{\mathcal{N}}(\chi, x, y, z)$ is to determine $f_2 = f_{\mathcal{N}_2}(\chi, x, y, z)$. Because the root-edge a of any map $N \in \mathcal{N}_2$ is neither loop nor isthmus, from (1.2.12) we have to consider the functions:

$$f_A = \sum_{N \in \mathcal{N}_2} \chi(N - a) x^{m(N)} y^{s(N)} z^{n(N)} \tag{10.4.5}$$

and

$$f_B = \sum_{N \in \mathcal{N}_2} \chi(N \cdot a) x^{m(N)} y^{s(N)} z^{n(N)} \tag{10.4.6}$$

so that

$$f_2 = f_A + f_B \tag{10.4.7}$$

is evaluated.

Lemma 10.4.1 *Let $\mathcal{N}_{\langle 2 \rangle} = \{N - a | \forall N \in \mathcal{N}_2\}$ with $a = e_r(N)$. Then we have*

$$\mathcal{N}_{\langle 2 \rangle}(m) = (\mathcal{N} - L_0)_{m-1} \hat{\times} \sum_{k \geq 0} (\mathcal{N} - L_0)^{\hat{\times} k} \tag{10.4.8}$$

where $\hat{\times}$ is the chain 1v-production,

$$\mathcal{N}_{\langle 2 \rangle}(m) = \{N - a | \forall N \in \mathcal{N}_2, m(N) = m\} \tag{10.4.9}$$

and

$$(\mathcal{N} - L_0)_{m-1} = \{N | \forall N \in \mathcal{N} - L_0, m(N) = m - 1\} \tag{10.4.10}$$

for $m \geq 2$.

Proof Similar to the proof of Lemma 9.3.1 with paying attention to that $\mathcal{N}$ contains the loop map L_0. □

Lemma 10.4.2 *Let* $\mathcal{N}_{(2)} = \{N \bullet a | \forall N \in \mathcal{N}_2\}$ *with* $a = e_r(N)$. *Then, we have*

$$\mathcal{N}_{(2)}(s) = (\mathcal{N} - L_1)_{s-1} \times\cdot \sum_{k\geq 0} (\mathcal{N} - L_1)^{\times k} \tag{10.4.11}$$

where $\times\cdot$ *is the inner* 1v*-production,*

$$\mathcal{N}_2(s) = \{N \bullet a | \forall N \in \mathcal{N}_2, s(N) = s\} \tag{10.4.12}$$

and

$$(\mathcal{N} - L_1)_{s-1} = \{N | \forall N \in \mathcal{N} - L_1, s(N) = s - 1\} \tag{10.4.13}$$

for $s \geq 2$.

Proof In fact, the dual case of Lemma 10.4.1. □

By Lemma 10.4.1,

$$\begin{aligned}
f_A &= xyz \sum_{N\in\mathcal{N}-L_0} \chi(N) x^{m(N)} \sum_{i=1}^{s(N)-1} y^i z^{n(N)} \\
&\quad \times \sum_{k\geq 0} \left(\sum_{N\in\mathcal{N}-L_0} \chi(N) \sum_{i=1}^{s(N)-1} y^i z^{n(N)} \right)^k \\
&= xyz \sum_{N\in\mathcal{N}-L_0} \frac{\chi(N) x^{m(N)} (y - y^{s(N)}) z^{n(N)}}{1-y} \\
&\quad \times \sum_{k\geq 0} \left(\sum_{N\in\mathcal{N}-L_0} \frac{\chi(N)(y - y^{s(N)}) z^{n(N)}}{1-y} \right)^k \\
&= xyz \frac{\partial_y (f - \nu x^2 yz)}{1 - \partial_y (f^+ - \nu yz)} \\
&= xyz \frac{\partial_y f}{1 - \partial_y f^+}
\end{aligned} \tag{10.4.14}$$

where $\partial_y f$ is the $\langle 1, y\rangle$-difference for y as defined in §1.6 and $f^+ = f_{\mathcal{N}}(\chi, 1, y, z)$.

Dually, from Lemma 10.4.2,

$$f_B = xyz \sum_{N \in \mathcal{N} - L_1} \chi(N) \sum_{j=1}^{m(N)-1} x^j y^{s(N)} z^{n(N)}$$

$$\times \sum_{k \geq 0} \left(\sum_{N \in \mathcal{N} - L_1} \chi(N) \sum_{j=1}^{m(N)-1} x^j z^{n(N)} \right)^k$$

$$= xyz \frac{\partial_x (f - \mu x y^2 z)}{1 - \partial_x (f^* - \mu x z)}$$

$$= xyz \frac{\partial_x f}{1 - \partial_x f^*} \tag{10.4.15}$$

in which $\partial_x f$ is the $\langle 1, x\rangle$-difference for x and $f^* = f_{\mathcal{N}}(\chi, x, 1, z)$.

Theorem 10.4.1 *The polysum function* $f = f_{\mathcal{N}}(\chi, x, y, z)$ *defined in* (10.4.1) *satisfies the equation as*

$$f - xyz \left(\frac{\partial_y f}{1 - \partial_y f^+} + \frac{\partial_x f}{1 - \partial_x f^*} \right)$$

$$= \nu x^2 yz + \mu x y^2 z. \tag{10.4.16}$$

Proof From (10.4.2), $f = f_0 + f_1 + f_2 = f_0 + f_1 + f_A + f_B$. Then, by (10.4.3–4) and (10.4.14–15) the theorem can soon be found. □

§10.5 Notes

10.5.1 Based on (10.1.11), it looks possible to extract an expression of the polysum function of bitrees with vertex partition and/or face partition.

10.5.2 For constant weight, one might find some classes of maps for which a functional equation can be extracted for determining the corresponding polysum function as that for bitrees.

10.5.3 An equation satisfied by the dichrosum function of bitree can also be found without much difficulty.

10.5.4 Based on the duality between chromatic polynomials and flow polynomials as shown in §1.2, corresponding equations can be found for the *flow sum function*, a polysum function in which the polynomial is replaced by the flow polynomial on a class corresponding to those used for chrosums.

10.5.5 Could we find a simpler way for solving the equation (10.2.16) instead of what has be shown in (10.2.17). More about dichrosum for outerplanar maps can be seen in [Liu54, Liu49].

10.5.6 In [Tut19], one can see an equation which is for the dichrosum function of general planar maps with four parameters, except for those in the polynomial. However, the decomposition in §10.3 is the same as that by Tutte.

10.5.7 Although the equation (10.4.16) looks much simpler, no explicit expression has been obtained. The original one which depends on more parameters can be seen in [Liu48].

10.5.8 What happens for all the problems mentioned in this chapter with nonrooted planar maps or maps on surfaces.

Chapter 11

Chromatic Solutions

§11.1 General solutions

The set $\mathcal{M}$ of maps considered here is the same as that in §9.2. And $\mathcal{M}$ is still partitioned into two classes $\mathcal{M}_0$ and $\mathcal{M}_{\rm I}$. However, the chrosum function is given by

$$f_{\mathcal{M}}(P, x.y.z) = \sum_{M \in \mathcal{M}} P(M, \lambda) x^{m(M)} y^{s(M)} z^{n(M)} \tag{11.1.1}$$

where $m(M), s(M)$ and $n(M)$ are the valency of root-vertex, the valency of root-face, and the size of M, respectively, instead of those dependent on the face partition in §9.2, in order to find explicit expressions for more general cases of $f = f_{\mathcal{M}}(P, x, y, z)$, $f^+ = f_{\mathcal{M}}(P, 1, y, z)$, $f^* = f_{\mathcal{M}}(P, x, 1, z)$, and

$$f^{+*} = f_{\mathcal{M}}(P, 1, 1, z) = f^{*+}.$$

The last is just h in §9.2, by substituting z for y.

Because $\mathcal{M}_0$ consists of the link map L only, from $m(L) = 1$, $s(L) = 2$, $n(L) = 1$ and $P(L) = \lambda(\lambda - 1)$ we have

$$f_{\mathcal{M}_0}(P, x, y, z) = \lambda(\lambda - 1) x y^2 z. \tag{11.1.2}$$

By Lemma 9.2.1,

$$\begin{aligned} f_A &= \sum_{M \in \mathcal{M}_{\rm I}} P(M - a) x^{m(M)} y^{s(M)} z^{n(M)} \\ &= \frac{xyz}{\lambda} \left(\sum_{M \in \mathcal{M}} P(M) x^{m(M)} y^{s(M)-1} z^{n(M)} \right) \end{aligned}$$

$$\times \sum_{k\geq 1} \frac{1}{\lambda^{k-1}} \left(\sum_{M\in\mathcal{M}} P(M) y^{s(M)-1} z^{n(M)} \right)^k$$

$$= \frac{xz\frac{f}{\lambda y} f^+}{1 - \frac{1}{\lambda y} f^+}$$

$$= \frac{xzf}{\lambda y - f^+}. \qquad (11.1.3)$$

In order to evaluate

$$f_B = \sum_{M\in\mathcal{M}_{\mathrm{I}}} P(M \bullet a) x^{m(M)} y^{s(M)} \underline{y}^{\underline{n}(M)}, \qquad (11.1.4)$$

the set $\mathcal{M}_{(\mathrm{I})} = \{M \bullet a | \forall M \in \mathcal{M}_{\mathrm{I}}\}$ should be partitioned into $\mathcal{M}_{\mathrm{S}}$ and $\mathcal{M}_{\mathrm{N}}$ as well. Let

$$\mathcal{M}_{\mathrm{I}}^{\mathrm{N}} = \{M | \forall M \in \mathcal{M}_{\mathrm{I}}, M \bullet a \in \mathcal{M}_{\mathrm{N}}\}$$

and

$$\mathcal{M}_{\mathrm{I}}^{S} = \{M | \forall M \in \mathcal{M}_{\mathrm{I}}, M \bullet a \in \mathcal{M}_{\mathrm{S}}\},$$

then

$$\mathcal{M}_{\mathrm{I}} = \mathcal{M}_{\mathrm{I}}^{\mathrm{N}} + \mathcal{M}_{\mathrm{I}}^{S}. \qquad (11.1.5)$$

By Lemma 9.2.2,

$$f_{\mathrm{N}} = \sum_{M\in\mathcal{M}_{\mathrm{I}}^{\mathrm{N}}} P(M \bullet a) x^{m(M)} y^{s(M)} z^{n(M)}$$

$$= xyz^2 \sum_{M\in\mathcal{M}_{\mathrm{N}}} P(M) x^{m(M)} y^{s(M)} z^{n(M)}$$

$$= \frac{xyz^2}{\lambda(\lambda-1)} \sum_{M\in\mathcal{M}} P(M) x^{m(M)} y^{s(M)-1} z^{n(M)}$$

$$\times \sum_{M\in\mathcal{M}} P(M) y^{s(M)-1} z^{n(M)-1}$$

$$= \frac{xz}{\lambda(\lambda-1)y} f f^+. \qquad (11.1.6)$$

However, for $\mathcal{M}_{\mathrm{S}}$ we adopt another kind of decomposition instead of Lemma 9.2.3.

Lemma 11.1.1 *For $\mathcal{M}_\mathrm{S}$, we have*

$$\mathcal{M}_\mathrm{S} = \mathcal{M} - L \tag{11.1.7}$$

and

$$\mathcal{M}_\mathrm{I}^\mathrm{S} = \sum_{M \in \mathcal{M}-L} \Big\{\nabla_i M \mid 1 \leq i \leq m(M)\Big\} \tag{11.1.8}$$

where $\nabla_i M$ is the i-th decrement of M defined in §7.1.

Proof It is easily seen that $\mathcal{M}_\mathrm{S}$ does not contain L and that (11.1.7) holds.

For the second formula, it can be similarly proved to the procedure in the proof of Lemma 7.1.6 with the attention of the nonseparability, outerplanarity and simplicity. □

From Lemma 11.1.1,

$$\begin{aligned} f_\mathrm{S} &= \sum_{M \in \mathcal{M}_\mathrm{I}^\mathrm{S}} P(M) x^{m(M)} y^{s(M)} z^{n(M)} \\ &= xyz \sum_{M \in \mathcal{M}-L} P(M) \sum_{i=1}^{m(M)-1} x^i y^{s(M)} z^{n(M)} \\ &= xyz \sum_{M \in \mathcal{M}-L} P(M) \frac{x - x^{m(M)}}{1-x} y^{s(M)} z^{n(M)} \\ &= \frac{xyz}{1-x}\Big(xf^+ - f\Big). \end{aligned} \tag{11.1.9}$$

Theorem 11.1.1 *The chrosum function $f = f_\mathcal{M}(P, x, y, z)$ defined by* (11.1.1) *satisfies the following equation:*

$$\begin{aligned} &\left(1 - xz\left(\frac{f^+}{\lambda y - f^+} + \frac{y}{1-x} - \frac{f^+}{\lambda(\lambda-1)y}\right)\right) f \\ &\quad = xyz\left(\lambda(\lambda-1)y - \frac{x}{1-x} f^+\right) \end{aligned} \tag{11.1.10}$$

where $f^+ = f_\mathcal{M}(P, 1, y, z)$.

Proof Because

$$f = f_0 + f_A - (f_\mathrm{S} + f_\mathrm{N}),$$

from (11.1.2),(11.1.3), (11.1.6) and (11.1.9) the theorem can be derived within rearrangement. □

Imagine that $x = \xi$ is a function of y and z. Then, we may find z and f^+ as functions of y and ξ by the characteristic equation system of (11.1.10):

$$1 - z\xi\left(\frac{f^+}{\lambda y - f^+} + \frac{f^+}{\lambda(\lambda-1)y} - \frac{zy}{1-\xi}\right) = 0;$$

$$\lambda(\lambda-1)y - \frac{\xi}{1-\xi}f^+ = 0. \tag{11.1.11}$$

From the second of (11.1.11), f^+ can be expressed into a function of ξ and y. Then according to the first of (11.1.11) z is expressed as a function of y and ξ as well. In consequence, we can find that

$$f^+ = \frac{\lambda(\lambda-1)y(1-\xi)}{\xi};$$

$$z = \frac{1-\xi}{y\xi - \dfrac{\lambda-1-\xi}{\lambda-1-\lambda\xi}(1-\xi)^2}. \tag{11.1.12}$$

By virtue of (11.1.12), we are allowed to employ Corollary 1.5.2 at $\xi = 1$ for obtaining the explicit expression of

$$f^+ = \lambda(\lambda-1)y\left(zy + \sum_{n\geq 2}\sum_{s=\lceil\frac{n+1}{2}\rceil}^{n} D_{n,s}(\lambda)y^s z^n\right) \tag{11.1.13}$$

where

$$D_{n,s}(\lambda) = \sum_{j=0}^{2s-n-1}\frac{C(n,s,j)(s-s)!}{(2s-n-j-1)!(s-j-2)!} \times A_{n-s}^{n-s}(2s-n-j-1;\lambda); \tag{11.1.14}$$

$$C(m,s,j) = \frac{(-1)^j(n-1)!}{s!(n-s)!j!}; \tag{11.1.15}$$

$$A_k^k(r;\lambda) = \sum_{j=0}^{r}\frac{\lambda^{r-j}(\lambda-2)^{k-j}r!k!(k+r-j-1)!}{j!(r-j)!(k-j)!(k-1)!}, \tag{11.1.16}$$

particularly

$$A_k^r(r;2) = \begin{cases} \frac{2^{r-k} r!(r-1)!}{(r-k)!(k-1)!}, \ 0 \le k \le r; \\ \\ 0, \text{ otherwise}. \end{cases} \tag{11.1.17}$$

Based on (11.1.12), we may further find an explicit expression of the general function $f = f_{\mathcal{M}}(P, x, y, z)$ through

$$f = \frac{\lambda(\lambda-1)xy^2z\frac{\xi-x}{\xi}}{1 - x\left(1 + zy - z\frac{1-\xi}{\xi}\alpha(\xi,\lambda)(1-x)\right)} \tag{11.1.18}$$

where

$$\alpha(\xi, \lambda) = \frac{\lambda - 1 - \xi}{\lambda - 1 - \lambda\xi} \tag{11.1.19}$$

by applying Corollary 1.5.2 at $\xi = 1$.

When $y = 1$, (11.1.13) becomes

$$f^{+*} = \lambda(\lambda - 1)z + \sum_{n \ge 2} H_n(\lambda) z^n \tag{11.1.20}$$

where

$$H_n(\lambda) = \lambda(\lambda - 1) \sum_{s=\lceil \frac{n+1}{2} \rceil}^{n} D_{n,s}(\lambda) \tag{11.1.21}$$

in which $D_{n,s}(\lambda)$ is given by (11.1.14) for $n \ge 2$.

Although $H_n(\lambda)$ involves alternative terms, some closer formulae can be extracted for particular values of λ.

When $\lambda = 2$, from the uniqueness of 2-coloring bipartite outerplanar maps in $\mathcal{M}$, $\frac{1}{2}f^{+*}$ is in fact the enufunction of bipartite maps in $\mathcal{M}$ with the size as the parameter. By suitable treatment, we have found that

$$H_n(2) = 2 \sum_{\substack{i=1 \\ n-1=i (\mathrm{mod}\ 2)}}^{\lfloor \frac{n-1}{3} \rfloor} \frac{1}{n}\binom{n}{i}\binom{i + \frac{n-1-3i}{2} - 1}{i-1} \tag{11.1.22}$$

for $n \ge 2$.

When $\lambda = 3$, we have

$$H_n(3) = 6 \sum_{s=\lceil \frac{n+1}{2} \rceil}^{n} \sum_{j=0}^{2s-n-1} B(n,s,j)E(n,s,j) \tag{11.1.23}$$

where

$$B(n,s,j) = \frac{(-1)^j(n-1)!(s-2)!}{s!(n-s-1)!(s-j-2)!j!}; \tag{11.1.24}$$

$$E(n,s,j) = \sum_{i=1}^{2s-n-j-1} \frac{3^{2s-n-j-i-1}}{i!(2s-n-j-i-1)!} \times \frac{(s-j-i-2)!}{(n-s-i)!} \tag{11.1.25}$$

for $n \geq 2$.

When $\lambda = \infty$, we may see that f in this case becomes the enufunction of $\mathcal{M}$ with the same three parameters. Since the polynomial

$$Q(M,\mu) = \mu^{\nu(M)}P(M,\mu^{-1}), \tag{11.1.26}$$

where $\nu(M)$ is the order of M, has the properties:

$$Q(M;\mu) = Q(M-a;\mu) - \mu Q(M \bullet a;\mu) \tag{11.1.27}$$

for $a = e_r(M)$, and

$$Q(M;\mu) = Q(M_1;\mu)Q(M_2;\mu) \tag{11.1.28}$$

for $M = M_1 \cup M_2$, provided $M_1 \cap M_2 = v$, a vertex, from $\lambda = \infty$ we have $\mu = 0$. In fact,

$$f_0 = \sum_{M \in \mathcal{M}} x^{m(M)} y^{s(M)} z^{n(M)} \tag{11.1.29}$$

for $f_0 = f_{\mathcal{M}}(Q,x,y,z)|_{\mu=0}$ since $Q(M,0) = 1$.

Theorem 11.1.2 *The function f_0 defined by (11.1.29) satisfies the following equation:*

$$\left(1 - \frac{xzf_0^+}{y - f_0^+}\right) f_0 = xy^2z \tag{11.1.30}$$

where $f_0^+ = f_0|_{x=1}$.

Proof By the same procedure as in the proof of Theorem 11.1.1, we can soon find that

$$f_0 = f_{\mathcal{M}_0}|_{\mu=0} + f_A|_{\mu=0}$$
$$= xy^2z + \frac{xzf_0^+}{y - f_0^+} f_0.$$

This implies (11.1.30). □

First, by setting $x = 1$ on the equation (11.1.30), we have

$$f_0^+ = z\left(\frac{f_0^+}{y - f_0^+} + y^2\right). \tag{11.1.31}$$

By employing Corollary 1.5.1,

$$f_0^+ = \sum_{n\geq 1} \sum_{s=\lceil \frac{n+3}{2} \rceil}^{n+1} \frac{(n-1)!}{(s-1)!(2s-n-3)!} \times \frac{(s-3)! y^s z^n}{(n-s-1)!(n-s)!}. \tag{11.1.32}$$

From (11.1.30), we have

$$f_0 = xy^2z \sum_{n\geq 0} \left(\frac{xzf_0^+}{y - f_0^+}\right)^n. \tag{11.1.33}$$

According to (11.1.31), it is also allowed to apply Corollary 1.5.1 for finding

$$f_0 = xy^2z + \sum_{n\geq 3} \sum_{m=2}^{\lfloor \frac{n+1}{2} \rfloor} \sum_{s=\lceil \frac{n+3}{2} \rceil}^{n-m+2} F_{n,m,s} x^m y^s z^n \tag{11.1.34}$$

where

$$F_{n,m,s} = \frac{(m-1)}{(n-m-s+2)!(2s-n-3)!} \times \frac{(n-m-1)!}{(n-s+1)!} \tag{11.1.35}$$

for $n \geq 3$.

Further, from (11.1.32) we have

$$f_0^{+*} = \sum_{n\geq 1} z^n \sum_{s=\lceil\frac{n+3}{2}\rceil}^{n+1} \frac{(n-1)!}{(s-1)!(2s-n-3)!} \times \frac{(s-3)!}{(n-s-1)!(n-s)!}. \quad (11.1.36)$$

§11.2 Cubic triangles

Consider the solutions of equation (9.3.18) for nonseparable planar maps when λ in the chromial P is taken to be particular values. The case $\lambda = 1$ has no significance for only the vertex map is 1-colorable. However, the differentiation of P with respect to λ at $\lambda = 1$ is indeed meaningful in graph theory.

Let the edges of a map M be enumerated as $e_1, e_2, \cdots, e_{n(M)}$, $n(M)$ is the size of M. For a spanning tree T of M, an edge $e_j \in T$ is said to be *in-active* if no cotree edge on the fundamental cocircuit generated by e_j has its suffix greater than j. And, an edge e_j not on T is said to be *out-active* if no tree edge on the fundamental circuit has its suffix greater than j. Write $i(T)$ and $o(T)$ respectively as the numbers of in-active and out-active edges on T. Although $i(T)$ and $o(T)$ depend on the way of enumerating edges, the number of spanning trees with given numbers of in-active and out-active edges is independent of the way. Because it is known that

$$P(M) = -(-1)^{p(M)+1}\lambda \sum_{T\in\mathcal{T}(M)} (1-\lambda)^{i(M)} \quad (11.2.1)$$

where $p(M)+1$ is the order of M and $\mathcal{T}(M)$ is the set of all spanning trees on M, we soon find that

$$\left.\frac{\mathrm{d}P(M)}{\mathrm{d}\lambda}\right|_{\lambda=1} = (-1)^{p(M)+1} J_1(M) \quad (11.2.2)$$

where $J_1(M)$ is the number of spanning trees on M which has exactly 1 in-active edge without out-active edge. Such a spanning tree is called a $(1,0)$-*tree*.

Let us write

$$f' = \frac{\partial}{\partial\lambda} f_{\mathcal{M}}(P, x, y, z, t)\Big|_{\lambda=1}. \tag{11.2.3}$$

Then from the linearity of the operators ∂_z and ∂_t, we have

$$\frac{\partial\partial_z f}{\partial\lambda}\Big|_{\lambda=1} = \partial_z f'; \ \frac{\partial\partial_t f}{\partial\lambda}\Big|_{\lambda=1} = \partial_t f'. \tag{11.2.4}$$

Since $f|_{\lambda=1} = \partial_z f|_{\lambda=1} = \partial_t f|_{\lambda=1} = 0$, by differentiating the two sides of (9.3.18) with respect to λ at $\lambda = 1$, we obtain

$$f' = xz^2t + zt\Big(y\partial_z f' - x\partial_t f'\Big). \tag{11.2.5}$$

If f' is expressed as a power series of x and y, that is

$$f' = \sum_{p\geq 1}\sum_{q\geq 1} \alpha_{p,q} x^p y^q \tag{11.2.6}$$

where $\alpha_{p,q} = \alpha_{p,q}(z,t)$ is a function of z and t, then from (11.2.5) we have

$$\alpha_{p,q} = zt\Big(\partial_z \alpha_{p,q-1} - \partial_t \alpha_{p-1,q}\Big) \tag{11.2.7}$$

for $p \geq 2, q \geq 1$ while it is easily checked that $\alpha_{1,0} = z^2t$.

Lemma 11.2.1 *We have*

$$\alpha_{1,q} = z^2 t^{q+1} \tag{11.2.8}$$

for $q \geq 0$.

Proof Because $\alpha_{0,q} = 0$ for any $q \geq 0$, (11.2.8) becomes

$$\alpha_{1,q} = zt\partial_z \alpha_{1,q-1}. \tag{11.2.9}$$

By considering $\alpha_{1,0} = z^2t$, (11.2.8) is true for $q = 0$. Assume for $q-1, \alpha_{1,q-1} = z^2t^q$, $q \geq 1$. Then according to (11.2.9)

$$\begin{aligned}\alpha_{1,q} &= zt\left(\frac{zt^q - z^2t^q}{1-z}\right) \\ &= zt \cdot zt^q \\ &= z^2t^{q+1}.\end{aligned}$$

By induction, (11.2.8) is always true for any $q \geq 0$. □

Since $\alpha_{1,0} = z^2t$, from (11.2.7) for $p \geq 2, q = 0$, we may see that

$$\alpha_{p,0} = \begin{cases} z^2t, & \text{if } p = 1; \\ \\ = 0, & \text{otherwise.} \end{cases} \tag{11.2.10}$$

Lemma 11.2.2 *We have*

$$\alpha_{p,1} = (-1)^{p+1} z^{p+1} t^2 \tag{11.2.11}$$

for $p \geq 1$.

Proof First, from Lemma 11.2.1 for $q = 1, \alpha_{11} = z^2t^2$. Assume that (11.2.11) is true for the first suffix less than $p \geq 2$. According to (11.2.10) and (11.2.7), we see that

$$\begin{aligned} \alpha_{p,1} &= -zt\partial_t \alpha_{p-1,1} \\ &= -zt\partial_t (-1)^p z^p t^2 \\ &= -zt \frac{t(-1)^p z^p - (-1)^p z^p t^2}{1-t} \\ &= -zt \cdot (-1)^p t z^p \\ &= (-1)^{p+1} z^{p+1} t^2. \end{aligned}$$

By induction on p, the Lemma is obtained. □

In general, for any $q \geq 2$, $\alpha_{p,q}$ involves the terms of $t^j, 2 \leq j \leq q+1$, and terms of $z^i, 2 \leq i \leq p+1$ because at most $p+1$ (the order of maps) vertices are on the root-face boundary and at most $q+1$ (the dual order of maps) faces are incident with the root-vertex. That implies

$$\alpha_{p,q} = \sum_{2 \leq i \leq p+1} \sum_{2 \leq j \leq q+1} A_{i,j}^{(p,q)} z^i t^j \tag{11.2.12}$$

where $A_{i,j}^{(p,q)}$ is a number only dependent on p, q, i, j themselves.

Theorem 11.2.1 *For $p \geq 2$ and $q \geq 2$, we have the following recursion as*

$$A_{i,j}^{(p,q)} = \sum_{k=i}^{p+1} A_{k,j-1}^{(p,q-1)} - \sum_{k=j}^{q+1} A_{i-1,k}^{(p-1,q)} \tag{11.2.13}$$

with (11.2.8) *and* (11.2.11) *as the initial conditions.*

Proof According to (11.2.12),

$$\begin{aligned}
\partial_z \alpha_{p,q-1} &= \sum_{i=2}^{p+1} \sum_{j=2}^{q} A_{i,j}^{(p,q-1)} \frac{z - z^i}{1-z} t^j \\
&= \sum_{i=2}^{p+1} \sum_{j=2}^{q} A_{i,j}^{(p,q-1)} \sum_{k=1}^{i-1} z^k t^j \\
&= \sum_{k=1}^{p} \sum_{j=2}^{q} \left(\sum_{i=k+1}^{p+1} A_{i,j}^{(p,q-1)} \right) z^k t^j \\
&= \sum_{i=1}^{p} \sum_{j=2}^{q} \left(\sum_{k=i+1}^{p+1} A_{k,j}^{(p,q-1)} \right) z^i t^j
\end{aligned}$$

and

$$\begin{aligned}
\partial_t \alpha_{p-1,q} &= \sum_{i=2}^{p} \sum_{j=2}^{q+1} A_{i,j}^{(p-1,q)} z^i \frac{t - t^j}{1-t} \\
&= \sum_{i=2}^{p} \sum_{j=2}^{q+1} A_{i,j}^{(p-1,q)} z^i \sum_{k=1}^{j-1} t^k \\
&= \sum_{i=2}^{p} \sum_{k=1}^{q+1} \left(\sum_{j=k+1}^{q+1} A_{i,j}^{(p-1,q)} \right) z^i t^k \\
&= \sum_{i=2}^{p} \sum_{j=1}^{q+1} \left(\sum_{k=j+1}^{q+1} A_{i,k}^{(p-1,q)} \right) z^i t^j.
\end{aligned}$$

Write

$$\begin{aligned}
S_{i,j}^{(p-1,q)} &= \sum_{k=j+1}^{q+1} A_{i,k}^{(p-1,q)}; \\
S_{i,j}^{(p,q-1)} &= \sum_{k=i+1}^{p+1} A_{k,j}^{(p,q-1)}.
\end{aligned} \tag{11.2.14}$$

Then, by (11.2.7) we have

$$\alpha_{p,q} = zt\left(\sum_{i=1}^{p}\sum_{j=2}^{q} S_{i,j}^{(p,q-1)} z^i t^j - \sum_{i=2}^{p}\sum_{j=1}^{q} S_{i,j}^{(p-1,q)} z^i t^j\right)$$

$$= \sum_{i=2}^{p+1}\sum_{j=3}^{q+1} S_{i-1,j-1}^{(p,q-1)} z^i t^j - \sum_{i=3}^{p+1}\sum_{j=2}^{q+1} S_{i-1,j-1}^{(p-1,q)} z^i t^j$$

$$= \sum_{i=2}^{p+1}\sum_{j=2}^{q+1} \left(S_{i-1,j-1}^{(p-1,q)} - S_{i-1,j-1}^{(p-1,q)}\right) z^i t^j \tag{11.2.15}$$

where

$$S_{i-1,1}^{(p,q-1)} = S_{1,j-1}^{(p-1,q)} = 0 \tag{11.2.16}$$

from (11.2.12).

In virtue of (11.2.14–15), the theorem is proved. □

If it is taken that $x = y$ in (9.3.18), then from $p(M) + q(M) = n(M)$, thc size of $M \in \mathcal{M}$, the chrosum function f becomes

$$g = f_{\mathcal{M}}(P, x, x, z, t)$$
$$= \sum_{M\in\mathcal{M}} P(M) x^{n(M)} z^{m(M)} t^{s(M)} \tag{11.2.17}$$

where $n(M), m(M)$ and $s(M)$ are the size, the valency of root-facc and the valency of root-vertex of M respectively. The function g satisfies the equation as

$$\left(1 - \frac{1}{\lambda}\partial_z g^+\right)\left(1 - \frac{1}{\lambda}\partial_t g^*\right)(g - \lambda(\lambda-1)xz^2t)$$
$$= xzt\left(\left(1 - \frac{1}{\lambda}\partial_t g^*\right)\partial_z g - \left(1 - \frac{1}{\lambda}\partial_z g^+\right)\partial_t g\right) \tag{11.2.18}$$

where $g^* = g|_{z=1}$ and $g^+ = g|_{t=1}$.

Write $g' = \frac{\partial}{\partial\lambda} g|_{\lambda=1}$, then g' satisfies the equation as

$$g' = xz^2t + xzt\left(\partial_z g' - \partial_t g'\right). \tag{11.2.19}$$

If g' is expressed in a power series of x as

$$g' = \sum_{n\geq 1} \alpha_n(z,t) x^n, \tag{11.2.20}$$

then $\alpha_n = \alpha_n(z,t)$ is a polynomial as

$$\alpha_n = \sum_{2\le i\le n}\ \sum_{2\le j\le n-i+2} A_{i,j}^{(n)} z^i t^j \tag{11.2.21}$$

for $n \ge 2$ where

$$A_{2,2}^{(n)} = \begin{cases} 1, & \text{if } n = 2; \\ 0, & \text{otherwise.} \end{cases} \tag{11.2.22}$$

From (11.2.19–20), we have

$$\alpha_n = zt(\partial_z \alpha_{n-1} - \partial_t \alpha_{n-1}) \tag{11.2.23}$$

with the initial condition $\alpha_1 = z^2 t$.

Lemma 11.2.3 *We have*

$$\alpha_n(z,t) = \begin{cases} -\alpha_n(t,z), & n = 1 \,(\text{mod } 2); \\ \alpha_n(t,z), & n = 0 \,(\text{mod } 2) \end{cases} \tag{11.2.24}$$

for $n \ge 2$.

Proof First, we see that from (11.2.23)

$$\begin{aligned} \alpha_2(z,t) &= zt(\partial_z \alpha_1 - \partial_t \alpha_1) \\ &= zt(zt - 0) \\ &= z^2 t^2 \end{aligned}$$

that implies $\alpha_2(z,t) = \alpha_2(t,z)$. Further, for any $n \ge 3$, from (11.2.23) we have

$$\begin{aligned} \alpha_n(z,t) &= zt\big(\partial_z \alpha_{n-1}(z,t) - \partial_t \alpha_{n-1}(z,t)\big) \\ &= zt\big(-\partial_z \alpha_{n-1}(t,z) + \partial_t \alpha_{n-1}(t,z)\big) \\ &= zt\big(\partial_t \alpha_{n-1}(t,z) - \partial_z \alpha_{n-1}(t,z)\big) \\ &= +\alpha_n(t,z) \end{aligned}$$

when $n = 0 \,(\text{mod } 2)$ by induction, and

$$\alpha_n(z,t) = zt\big(\partial_z \alpha_{n-1}(t,z) - \partial_t \alpha_{n-1}(t,z)\big)$$

$$= -zt\Big(\partial_t\alpha_{n-1}(t,z) - \partial_z\alpha_{n-1}(t,z)\Big)$$
$$= -\alpha_n(t,z)$$

when $n = 1 \pmod 2$ by induction as well. □

From (11.2.2), we may see that if the minus is instead of plus in (11.2.19) then $\alpha_n(z,t)$ in (11.2.20) is replaced by

$$\beta_n = \sum_{2\le i\le n}\ \sum_{2\le j\le n-i+2} B_{i,j}^{(n)} z^i t^j \tag{11.2.25}$$

for $n \ge 2$ such that

$$\beta_n = zt\Big(\partial_z\beta_{n-1} + \partial_t\beta_{n-1}\Big) \tag{11.2.26}$$

with the initial condition $\beta_1 = z^2 t$.

Now, $B_{i,j}^{(n)}$ is the total sum of (1,0)-trees on all the maps in $\mathcal{M}$ such that the size is n, the valency of the root-face is i and the valency of root-vertex is j.

Lemma 11.2.4 *We have*

$$\beta_n(z,t) = \beta_n(t,z) \tag{11.2.27}$$

for $n \ge 2$.

Proof Similar to the proof of Lemma 11.2.3. □

Let $\mathcal{A}_n$ be the set of all $n \times n$ down-triangular matrices. A transformation $\nabla_{+1} : \mathcal{A}_n \longrightarrow \mathcal{A}_{n+1}$ is called an *expander* if for $A = \{a_{i,j}\} \in \mathcal{A}_n$, *i.e.*, $a_{i,j} = 0, j > i,\ 1 \le i,j \le n$, we have

$$\nabla_{+1}A = B \tag{11.2.28}$$

where $B = (b_{i,j})$ such that

$$b_{i,j} = \sum_{l=j}^{i} a_{i,l} + \sum_{l=j-1}^{i-1} a_{l,j-1} \tag{11.2.29}$$

for $1 \le i,j \le n+1$. Further, we define

$$\nabla_{+1}^{m}A = \nabla_{+1}\Big(\nabla_{+1}^{m-1}A\Big) \tag{11.2.30}$$

for $m \geq 2$ where $\nabla_{+1}^{1} = \nabla_{+1}$.

Write $X_{(m)} = (x^m, x^{m-1}, \cdots, x)$. Then $X_{(m)}^{\mathrm{R}} = (x, x^2, \cdots, x^m)$ is called the *reverse* of $X_{(m)}$.

Theorem 11.2.2 *For $n \geq 2$, we have*

$$\beta_n = ztZ_{(n-1)}\, \nabla_{+1}^{n-3} \begin{pmatrix} 1 & 0 \\ 0 & 1 \end{pmatrix} T_{(n-1)}^{\mathrm{RT}} \tag{11.2.31}$$

where T *represents the transpose,* $\nabla_{+1}^{-1} \begin{pmatrix} 1 & 0 \\ 0 & 1 \end{pmatrix} = 1$ *and*

$$\nabla_{+1}^{0} \begin{pmatrix} 1 & 0 \\ 0 & 1 \end{pmatrix} = \begin{pmatrix} 1 & 0 \\ 0 & 1 \end{pmatrix}$$

itself.

Proof When $n = 2$, we have

$$\beta_2 = ztz\, \nabla_{+1}^{-1} \begin{pmatrix} 1 & 0 \\ 0 & 1 \end{pmatrix} t = z^2 t^2.$$

From (11.2.26), the theorem is true for $n = 2$.

In general, assume

$$\nabla_{+1}^{n-3} \begin{pmatrix} 1 & 0 \\ 0 & 1 \end{pmatrix} = \left(b_{i,j}^{(n-1)}\right)$$

for $1 \leq i, j \leq n-1$ and $b_{i,j}^{(n-1)} = 0$, if $i < j$. Then,

$$\beta_n = ztZ_{(n-1)}(b_{i,j}^{(n-1)})T_{(n-1)}^{\mathrm{RT}}.$$

Since

$$\partial_z \beta_n = t^2 \sum_{i=1}^{n-1} b_{i,1}^{(n-1)} \sum_{l=1}^{n-i} z^l + t^3 \sum_{i=2}^{n-1} b_{i,2}^{(n-1)} \sum_{l=1}^{n-i} z^i$$
$$+ \cdots + t^n b_{n-1,n-1}^{(n-1)} z^n;$$

$$\partial_t \beta_n = t \sum_{i=1}^{n-1} b_{i,1}^{(n-1)} \sum_{l=1}^{n-i} z^l + (t + t^2) \sum_{i=2}^{n-1} b_{i,2}^{(n-1)} \sum_{l=1}^{n-i} z^i$$

$$+\cdots+(t+t^2+\cdots+t^{n-1})b_{n-1,n-1}^{(n-1)}z^2,$$

we have

$$\partial_z\beta_n+\partial_t\beta_n = t\Big(\sum_{i=1}^{n-1} b_{i,1}^{(n-1)}z^{n-i+l}+\sum_{i=2}^{n-1} b_{i,2}^{(n-1)}z^{n-i+1}$$

$$+\cdots+b_{n-1,n-1}^{(n-1)}z^2\Big)$$

$$+t^2\Big(\sum_{i=1}^{n-1} b_{i,1}^{(n-1)}\sum_{l=1}^{n-i} z^l+\sum_{i=2}^{n-1} b_{i,2}^{(n-1)}z^{n-i+1}$$

$$+\cdots+b_{n-1,n-1}^{(n-1)}z^2\Big)+\cdots$$

$$+t^{n-1}\Big(\sum_{i=n-2}^{n-1} b_{i,2}^{(n-1)}\sum_{l=1}^{n-i} z^i+b_{n-1,n-1}^{(n-1)}z^2\Big)$$

$$+t^n b_{n-1,n-1}^{(n-1)}z$$

$$= t\Big(b_{1,1}^{(n-1)}z^n+\big(b_{2,1}^{(n-1)}+b_{2,2}^{(n-1)}\big)z^{n-1}$$

$$+\cdots+\sum_{j=1}^{n-1} b_{n-1,j}^{(n-1)}z^2\Big)$$

$$+t^2\Big(\big(b_{1,1}^{(n-1)}+b_{2,2}^{(n-1)}\big)z^{n-1}+\big(b_{1,1}^{(n-1)}+b_{2,1}^{(n-1)}$$

$$+b_{3,2}^{(n-1)}+b_{3,3}^{(n-1)}\big)z^{n-2}+\cdots+\sum_{i=1}^{n-1} b_{i,1}^{(n-1)}z\Big)$$

$$+\cdots$$

$$= Z_{(n)}\big(b_{i,j}^{(n)}\big)T_{(n)}^{\mathrm{RT}}$$

where

$$b_{i,j}^{(n)} = \sum_{l=j-1}^{i-1} b_{l,j-1}^{(n-1)}+\sum_{l=j}^{i} b_{i,l}^{(n-1)}$$

for $1\le i,j\le n$, and $b_{i,j}^{(n-1)}=0$, when $j>i$. From (11.2.26) and (11.2.28–30), we soon see that

$$\beta_{n+1} = ztZ_{(n)}\,\nabla_{+1}^{n-2}\begin{pmatrix}1&0\\0&1\end{pmatrix}T_{(n)}^{\mathrm{RT}}.$$

Therefore, by induction the theorem is true. □

Because

$$(1), \begin{pmatrix} 1 & 0 \\ 0 & 1 \end{pmatrix}, \nabla_{+1}\begin{pmatrix} 1 & 0 \\ 0 & 1 \end{pmatrix}, \nabla_{+1}^2\begin{pmatrix} 1 & 0 \\ 0 & 1 \end{pmatrix}, \cdots$$

have the property that their diagonals form the combinatorial triangle which consists of the binomial coefficients, they are treated as a *cubic triangle.*

Similarly, from (11.2.23) we may find another cubic triangle as

$$\begin{pmatrix} 1 & 0 \\ 0 & -1 \end{pmatrix}, \nabla_{+1}\begin{pmatrix} 1 & 0 \\ 0 & -1 \end{pmatrix}, \nabla_{+1}^2\begin{pmatrix} 1 & 0 \\ 0 & -1 \end{pmatrix}, \cdots.$$

§11.3 Invariants

Two objectives are assumed to be here. One is the attempt that the general equation (9.4.18) for the chromatic sum of triangulations is replaced by one involving fewer variables and the other is to establish the special cases with the Beraha numbers as the values of λ.

Because for any triangulation T, we always have that $m(T) \leq q(T)+2$ and $s(T) \leq q(M)+1$, all the coefficients of y in h as a power series are polynomials of z, t and λ. By Tutte's terminology, h is *y-restricted.* Since $\delta_t(tf) = f + \delta_t f$, by doing δ_t on the equation (9.4.18) twice, we find that

$$\begin{aligned}\left(y\Big(1+\frac{h^+}{\lambda}\Big) - yz^2\right)\delta_t h \\ + \left(y\Big(1+\frac{h^+}{\lambda}\Big) - z(1+2yz)\right)\delta_t^2 h \\ -yz^2\delta_t^3 h = z^2\delta_t^2(th^\theta).\end{aligned} \tag{11.3.1}$$

For convenience, let us write

$$A = y\Big(1+\frac{h^+}{\lambda}\Big) - yz^2;$$

$$B = y\Big(1+\frac{h^+}{\lambda}\Big) - z(1+2yz);$$

$$C = -yz^2 \tag{11.3.2}$$

where all of them are independent of t.

By doing δ_t for $n-1$ times, multiplying (11.3.1) by u^{n+2}, and then summing over $n \geq 1$, from (11.3.2) we soon obtain that

$$\begin{aligned}\left(A + Bu^{-1} + Cu^{-2}\right) \sum_{n\geq 1} u^n \delta_t^n h - B\delta_t h \\ -C(u^{-1}\delta_t h + \delta_t^2 h) = -C \sum_{n\geq 1} u^n \delta_t^{n+1}(th^\theta).\end{aligned} \tag{11.3.3}$$

Let ψ and ϕ be the two roots of the quadratic equation

$$Au^2 + Bu + C = 0. \tag{11.3.4}$$

From Theorem 1.4.1, assume ψ and ϕ as power series, in fact polynomials of y and z with the constant terms -1 and 0, respectively. That is no loss of generality within an interchangement between ψ and ϕ. By considering that

$$\begin{aligned}-B\delta_t h - C(u^{-1}\delta_t h + \delta_t^2 h) \\ = -(Au + B + Cu^{-1})\delta_t h, \\ +Au\delta_t h - C\delta_t^2 h,\end{aligned} \tag{11.3.5}$$

from (11.3.3) we find that

$$(\psi - \phi)A\delta_t h = -C \sum_{n\geq 1} (\psi^n - \phi^n)\delta_t^{n+1}(th^\theta). \tag{11.3.6}$$

Since ψ and ϕ are the roots of (11.3.4), we have $\psi\phi A = C$. Substituting it into (11.3.6), and then dividing the two sides by A, we may obtain

$$(\psi - \phi)\delta_t h = -\psi\phi \sum_{n\geq 1} (\psi^n - \phi^n)\delta_t^{n+1}(th^\theta). \tag{11.3.7}$$

Since h^θ is a power series of t, y with λ, we may write

$$h^\theta = \sum_{r\geq 0} f_r (t-1)^r \tag{11.3.8}$$

where $f_r, r \geq 0$, is a power series of y with λ.

According to (11.3.8) which can be rewritten as

$$(\psi - \phi)(\delta_t^h + th^\theta) = -\phi \sum_{n\geq 0} \psi^n \delta_t^n (th^\theta)$$

$$+\psi \sum_{n\geq 0} \phi^n \delta_t^n (th^\theta), \tag{11.3.9}$$

from

$$\sum_{n\geq 0} \psi^n \delta_t^n (th^\theta) = \sum_{r\geq 0} f_r \sum_{n=0}^{r} \psi^n (t-1)^{r-n}$$

we have

$$(1 - t + \psi) \sum_{n\geq 0} \psi^n \delta_t^n (th^\theta)$$

$$= \sum_{n\geq 0} f_r \Big(\psi^{r-1} - (t-1)^{r+1}\Big)$$

$$= \psi(1+\psi)h^\theta(1+\psi, y, \lambda)$$

$$-(t-1)th^\theta(t, y, \lambda). \tag{11.3.10}$$

Moreover, (11.3.9) remains valid when ψ is replaced by ϕ.

Combining (11.3.9–10), we may find that

$$(\psi - \phi)(1 - t + \psi)(1 - t + \phi)\delta_t h$$

$$= -\psi\phi\Big((1 - t + \phi)(1 + \psi)h^\theta(1 + \psi, y, \lambda)$$

$$-(1 - t + \psi)(1 + \phi)h^\theta(1 + \phi, y, \lambda)$$

$$+(\psi - \phi)th^\theta(t, y, \lambda)\Big). \tag{11.3.11}$$

By considering that $h = h^+ + (t-1)\delta_t h$, from (9.4.18) with the replacement of $z = ys$ we find that

$$\left(s^2y^2t^2 + (y-1)\Big(s - \frac{th^+}{\lambda} - t\Big)\right)\delta_t h$$

$$= sy^2t\lambda(\lambda - 1) - y^2s^2th^\theta(t, y, \lambda)$$

$$-\Big(s - \frac{th^+}{\lambda} - t\Big)h^+. \tag{11.3.12}$$

According to the relationship between roots and the coefficients of the quadratic equation (11.3.4), from (11.3.3) we can derive that

$$y^2 s(1+\psi)(1+\phi) = -\psi\phi;$$

$$y^2\Big(1+\frac{h^+}{\lambda}\Big)(1+\psi)^2(1+\phi)^2 = (\psi\phi-1)\psi\phi. \tag{11.3.13}$$

Then (11.3.12) can be rewritten as

$$\begin{aligned}
&y^2(1+\psi)^2(1+\phi)^2(1+\psi-t)(1+\phi-t)\delta_t h\\
&= -y^2\psi\phi(1+\psi)^2(1+\phi)^2 t h^\theta(t,y,\lambda)\\
&\quad +t\Big(-\psi^2\phi^2(1+\psi)(1+\phi)\lambda(\lambda-1)\\
&\qquad +\lambda(\psi\phi-1)^2\psi\phi - \lambda y^2(1+\psi)^2\\
&\qquad \times(1+\phi)^2(\psi\phi-1)\Big)\\
&\quad +\Big(\lambda\psi\phi(1+\psi)(1+\phi)(\psi\phi-1)\\
&\qquad -\lambda y^2(1+\psi)^3(1+\phi)^3\Big).
\end{aligned} \tag{11.3.14}$$

From (11.3.11) and (11.3.14), by eliminating $\delta_t h$ while $th^\theta(t,y,\lambda)$ is eliminated as well, we are left with the following identity:

$$\begin{aligned}
&-\psi\phi y^2(1+\psi)^2(1+\phi)^2\Big((1-t+\phi)(1+\psi)h^\theta(1+\psi,y,\lambda)\\
&\qquad -(1-t+\psi)(1+\phi)h^\theta(1+\phi,y,\lambda)\Big)\\
&= (\psi-\phi)\Big(t\Big(-\psi^2\phi^2(1+\psi)(1+\phi)\lambda(\lambda-1)\\
&\quad +\lambda(\psi\phi-1)^2\psi\phi - \lambda y^2(1+\psi)(1+\phi)^2(\psi\phi-1)\Big)\\
&\qquad +\Big(\lambda\phi\psi(1+\psi)(1+\phi)(\psi\phi-1)\\
&\qquad -\lambda y^2(1+\psi)^3(1+\phi)^3\Big)\Big).
\end{aligned} \tag{11.3.15}$$

By equating terms independent of t and the coefficients of t in (11.3. 15), we may obtain two relations. From them, by eliminating the terms involving $h^\theta(1+\psi, y, \lambda)$ and $h^\theta(1+\phi, y, \lambda)$, we may respectively find the following two equation:

$$\begin{aligned}
&y^2\psi(1+\psi)(1+\phi)^2(\psi-\phi)h^\theta(1+\psi, y, \lambda)\\
&\quad = \lambda(\psi-\phi)\Big(-\psi^2\phi(1+\phi)(\lambda-1)\\
&\qquad +\psi\phi(\psi\phi-1) - y^2(1+\psi)^2(1+\phi)^2\Big) \qquad (11.3.16)
\end{aligned}$$

and

$$\begin{aligned}
&y^2\phi(1+\phi)(1+\psi)^2(\phi-\psi)h^\theta(1+\phi, y, \lambda)\\
&\quad = \lambda(\phi-\psi)\Big(-\phi^2\psi(1+\psi)(\lambda-1)\\
&\qquad +\phi\psi(\psi\phi-1) - y^2(1+\phi)^2(1+\psi)^2\Big). \qquad (11.3.17)
\end{aligned}$$

Theorem 11.3.1 *The function $h^\theta(t, y, \lambda)$ can be determined by the following pair of relations:*

$$\begin{aligned}
&(1-t)t_1t_2^2y^2h^\theta(t_1, y, \lambda) = \lambda t_1^2t_2^2y^2\\
&\qquad +\lambda(1-t_1)(1-t_2)(t_1-\mu t_2+\mu t_1t_2);\\
&(1-t_2)t_1^2t_2y^2h^\theta(t_2, y, \lambda) = \lambda t_1^2t_2^2y^2\\
&\qquad +\lambda(1-t_1)(1-t_2)(t_2-\mu t_1+\mu t_1t_2) \qquad (11.3.18)
\end{aligned}$$

where $t_1 = 1+\psi, t_2 = 1+\phi$ and $\mu = \lambda - 1$.

Proof The relation (11.3.18) holds because of (11.3.16–17) by setting $\psi = t_1 - 1$ and $\phi = t_2 - 1$ and $\mu = \lambda - 1$.

The procedure of determining $h^\theta(t, y, \lambda)$ is as follows. First, t_2 is considered as a rational function of t_1 and λ. Because the constant term of t_2 is 1 while that of ϕ is 0 and $t_2 = 1+\phi$, we can equate the coefficients of y^2 in the second of (11.3.18) and then find the coefficient of y^2 in t_2 as $t_1^2/(1-t_1)$. Then, by equating the coefficients of y^2 in the first of (11.3.18), we find the constant term of $h^\theta(t, y, \lambda)$. Having this, we may determine the coefficient of y^4 in t_2 by equating coefficients of y^4 in the second of (11.3.18), and so on, in alternation between the pair of relations (11.3.18). □

Let $\mathcal{W}(z,t,y,\lambda)$ be the ring of power series of z,t,y and λ and $\mathcal{W}_1(z,t,y,\lambda)$, the subring of y-restricted ones in $\mathcal{W}(z,t,y,\lambda)$. The fields of all quotients of members in $\mathcal{W}(z,t,y,\lambda)$ and $\mathcal{W}_1(z,t,y,\lambda)$ is respectively denoted by $\mathcal{F}(z,t,y,\lambda)$ and $\mathcal{F}_1(z,t,y,\lambda)$. And likewise, $\mathcal{W}(t,y,\lambda)$, $\mathcal{W}(y,\lambda)$, $\mathcal{F}(t,y,\lambda)$, $\mathcal{F}_1(t,y,\lambda)$, $\mathcal{F}(y,\lambda)$, $\mathcal{F}_1(y,\lambda)$ and so on. Consider an element

$$f(t,y,\lambda) = \frac{p_1(t,y,\lambda)}{p_2(t,y,\lambda)}$$

of $\mathcal{F}_1(t,y,\lambda)$, where $p_1(t,y,\lambda)$ and $p_2(t,y\lambda)$ are elements of $\mathcal{W}_1(t,y,\lambda)$. If t in $p_1(t,y,\lambda)$ and $p_2(t,y,\lambda)$ is replaced by $u \in \mathcal{W}(t,y,\lambda)$, then $f(u,y,\lambda) \in \mathcal{F}(t,y,\lambda)$. In particular, u may be t_1 or t_2. If $f(t_1,y,\lambda) = f(t_2,y,\lambda)$, then $f(t,y,\lambda)$ is said to be an *invariant* of $\mathcal{F}_1(t,y,\lambda)$. Of course, any member of $\mathcal{W}_1(y,\lambda)$ is an invariant of $\mathcal{F}_1(t,y,\lambda)$ and it is called a *trivial* one.

Lemma 11.3.1 *The number*

$$J = \frac{y^2 h^\theta(t,y,\lambda)}{\lambda} - \frac{ty^2}{1-t} + \frac{1-t}{t^2} \tag{11.3.19}$$

of $\mathcal{F}_1(t,y,\lambda)$ is an invariant.

Proof Let J_1 and J_2 members of $\mathcal{F}_1(t,y,\lambda)$ be obtained by substituting t_1 and t_2 for t in J respectively. Let i and j be the integers 1 and 2 in some order. Then, by Theorem 11.3.1, from the symmetry between t_i and t_j the Lemma follows. □

Now, let us define a sequence $v_1, v_2, v_3, \cdots$ recursively by

$$v_{n+2} = \mu v_{n+1} - v_n + 1 - \mu,$$

$$v_1 = t_1^{-1} \text{ and } v_2 = t_2^{-1} \tag{11.3.20}$$

for $n \geq 1$.

Lemma 11.3.2 *Write*

$$\Lambda(u,v) = u^2 - \mu uv + v^2 + (\mu-1)(u+v) - \mu \tag{11.3.21}$$

as a polynomial of two variables u and v with μ as some constant value. Then,

$$\Lambda(v_n, v_{n+1}) = J \tag{11.3.22}$$

for all $n \geq 1$.

Proof By the same procedure of the proof of Lemma 11.3.1, it can be shown that

$$\Lambda(v_1, v_2) = \Lambda(t_1^{-1}, t_2^{-1}) = J_1 = J_2 = J.$$

Then for any $n \geq 2$ from (11.3.21) we can verify that $\Lambda(v_{n+1}, v_{n+2}) = \Lambda(v_n, v_{n+1})$ and hence the lemma. □

Further, we define some polynomials Q_n and A_n in μ for n being any integer positive, negative or zero. The definitions are recursively given by

$$Q_n = \mu Q_{n-1} - Q_{n-2},\ Q_0 = 0,\ Q_1 = 1; \tag{11.3.23}$$

$$A_{n+1} = A_n + (\mu - 1)Q_n,\ A_0 = 1. \tag{11.3.24}$$

In particular, we have $Q_2 = \mu, Q_3 = \mu^2 - 1, Q_4 = \mu^3 - 2\mu,\ A_1 = 1, A_2 = \mu, A_3 = \mu^2$ and $A_4 = \mu^3 - \mu + 1$.

It can be shown from (11.3.23–24) that $A_n - \mu A_{n-1} + A_{n-2}$ is independent of n. Then, A_n can be determined by

$$A_n = \mu A_{n-1} - A_{n-2} + 1,\ A_0 = A_1 = 1. \tag{11.2.25}$$

Theorem 11.3.2 *For $n \geq 1$, we have*

$$v_n = Q_{n-1}v_2 - Q_{n-2}v_1 + 1 - A_{n-1}. \tag{11.3.26}$$

Proof Write $Y_n = v_n - Q_{n-1}v_2 + Q_{n-2}v_1 - 1 + A_{n-1}$. It is easily checked that $Y_1 = Y_2 = 0$. However, for $n \geq 2$, we have $Y_n - \mu Y_{n-1} + Y_{n-2} = 0$ from (11.3..25), (11.3.26) and (11.3.20). Hence $Y_n = 0$ for each n. □

Let us take

$$\mu = 2\cos\alpha.$$

It is well known that

$$Q_n = \frac{\sin(n\alpha)}{\sin\alpha}. \qquad (11.3.27)$$

Then it can be verified that

$$A_n = \frac{\mu-1}{\mu-2}\left(\frac{\cos\frac{(2n-1)\alpha}{2}}{\cos\frac{\alpha}{2}} - 1\right) + 1 \qquad (11.3.28)$$

by using basic trigonometrical identities.

Lemma 11.3.3 *Let i and j be the integers 1 and 2 in some order. Then, for any positive integer m,*

$$\begin{aligned} &Q_m^2\left(\frac{y^2h^\theta(t_i,y,\lambda)}{\lambda} - \frac{t_iy^2}{1-t_i}\right) \\ &-(Q_{m-1}v_i + A_m - Q_m) \\ &\times(-Q_{m+1}v_i + A_{m+1} + A_m) \\ &-(Q_mv_j - A_m - Q_{m-1}v_i) \\ &\times(Q_mv_j + A_{m+1} - Q_{m+1}v_i) = 0. \end{aligned} \qquad (11.3.29)$$

Proof By expanding (11.3.29) into a form with powers in y_i and y_j, each coefficient of $y_i^ly_j^s$ is shown based on (11.3.23) or (11.3.24) to be zero and hence the lemma. □

Now write

$$\begin{aligned} H(m) = Q_m^2&\left(\frac{y^2h^\theta(t,y,\lambda)}{\lambda} - \frac{ty^2}{1-t}\right) \\ &-(Q_{m-1}v + A_m - Q_m) \\ &\times(-Q_{m+1}v + A_{m+1} - Q_m) \end{aligned} \qquad (11.3.30)$$

where $v = t^{-1}$ and $m \geq 0$. Of course, $H(m) \in \mathcal{F}_1(t,y,\lambda)$. Let $H_i(m) = H(m)|_{t=t_i}, i = 1,2$. Naturally,

$$H_i(m) \in \mathcal{F}(t,y,\lambda),\ i = 1,2.$$

From (11.3.29), we have

$$H_i(m) = -(Q_m v_j + A_{m+1} - Q_{m+1} v_i)$$
$$\times (Q_{m-1} v_i + A_m - Q_m v_j). \qquad (11.3.31)$$

Moreover, μ is assigned to be $2\cos\alpha, \alpha = \frac{2\pi}{n}$ for $n \geq 3$. From (11.3.18) we see ψ and ϕ, therefore t_1 and t_2, remain well-defined as power series in s, t and y when λ is assigned to be a constant. They are still in $\mathcal{W}(t, y, z, \lambda)$, even so no term involves the variable λ. Likewise, the substitution transforms members of $\mathcal{W}_1(t, y, \lambda)$ and $\mathcal{F}_1(t, y, \lambda)$ into themselves, respectively.

For $n \geq 2$, let us denote

$$I(n) = \begin{cases} (1-v) \prod_{m=1}^{M-1} H(m), & n = 2M-1; \\ (1-v)(A_M + v) \prod_{m=1}^{M-1} H(m), & n = 2M. \end{cases} \qquad (11.3.32)$$

Lemma 11.3.4 *For $n \geq 2, I(n)$ defined by (11.3.32) is an invariant of $\mathcal{F}_1(t, y, \lambda)$.*

Proof When $n = 2M$, by (11.3.27) $Q_{M+m} = -Q_m$ for each integer m. Thus,

$$Q_M = -Q_0 = 0 \text{ and } Q_{M-1} = -Q_{-1} = 1.$$

Noticing that if $r = M - 1$, then

$$Q_r v_j + A_{r+1} - Q_{r+1} v_i = A_M + v_j. \qquad (11.3.33)$$

Whether t is replaced by t_1 or t_2 in $I(2M)$, by (11.3.31) the result is

$$(-1)^{M-1}(1-v_1)(1-v_2)(A_M + v_1)(A_M + v_2)$$
$$\times \prod_{m=1}^{M-2} (Q_m v_1 + A_{m+1} - Q_{m+1} v_2)$$
$$\times (Q_m v_2 + A_{m+1} - Q_{m-1} v_1)$$

where an empty product is to be interpreted as unity. Thus, $I(2M)$ is an invariant of $\mathcal{F}_1(t, y, \lambda)$. Similarly, for the case $n = 2M + 1$. $\square$

In fact, it can be shown that each of $I(2M-1)$ and $I(2M)$, $M \geq 2$, is a polynomial in J of degree $\leq M-2$ with the coefficients in this polynomial being power series in y alone.

Theorem 11.3.3 *For $n \geq 3$, we have*

$$I(n) = (-1)^{\lfloor \frac{n-1}{2} \rfloor} \prod_{j=1}^{n} (1 - v_j) \tag{11.3.34}$$

when $t = t_1$ or t_2 and $\mu = 2\cos\frac{2\pi}{n}$.

Proof By (11.2.27–28), the sequence $v_1, v_2, v_3, \cdots$ is periodic with period n. Moreover, from these formulae, it can be deduced that

$$Q_{n-m} = -Q_m \text{ and } A_{n-m} = A_{m+1}$$

for any integer m. We have

$$Q_m v_1 + A_{m+1} - Q_{m+1} v_2 = 1 - v_{m+2}$$

for $m \geq 0$ from Lemma 11.3.3. Here, $0 \leq m \leq n$. By applying these results to the formulae of $I(2M)$ and $I(2M-1)$ symmetrically in terms of t_1 and $t_2, t = t_1$ or t_2, from (11.3.33) the theorem is found. □

§11.4 Four color solutions

Let $h^{\triangleright}(y, \lambda)$ be the chrosum function of rooted planar triangulations with the face number and the color number as the parameters, *i.e.*,

$$h^{\triangleright}(y, \lambda) = \sum_{T \in \mathcal{T}_{\mathrm{Tri}}^{\triangle}} P(T, \lambda) y^{m(T)} \tag{11.4.1}$$

where $m(T)$ is the number of faces including the root-face, $P(T, \lambda)$ is the chromial of T, and $\mathcal{T}_{\mathrm{Tri}}^{\triangle}$ is the set of all rooted nonseparable planar triangulations. From the triangularity, for any $T \in \mathcal{T}_{\mathrm{Tri}}^{\triangle}$, $m(T) = 0 (\mathrm{mod}\ 2)$. For convenience, we are allowed to consider $h^{\triangle}(z, \lambda) = h^{\triangleright}(\sqrt{z}, \lambda)$, *i.e.*,

$$h^{\triangle}(z, \lambda) = \sum_{T \in \mathcal{T}_{\mathrm{Tri}}} P(T, \lambda) z^{n(T)} \tag{11.4.2}$$

where $n(T)$ is half of the face number of T.

By observing the relationship between $\mathcal{T}^{\triangle}_{\mathrm{Tri}}$ and $\mathcal{T}^{\theta}_{\mathrm{Tri}}$, the set of all maps in $\mathcal{T}_{\mathrm{Tri}}$ (which is given in §9.4) with the root-face being a 2-gon, since $h^{\theta}(t,y,\lambda)$ is the chrosum function of $\mathcal{T}^{\theta}_{\mathrm{Tri}}$, we have $h^{\triangleright}(y,\lambda) = h^{\theta}(1,y,\lambda)$.

The purpose of this section is to determine $h^{\triangleright}(y,\lambda)$, or $h^{\triangle}(z,\lambda)$ instead, based on what has been discussed on $h(t,y,\lambda)$ in the last section.

First, from a differential equation satisfied by $h^{\triangle}(z,\lambda)$ for all λ which are set to be the *Beraha numbers*

$$B_n = 2 + 2\cos\left(\frac{2\pi}{n}\right),$$

$n \geq 2$, because $h^{\triangle}(z,\lambda)$ is a power series of z with all the coefficients being polynomials of λ, the equation is inferred to be held for all values of λ except 4 which causes to be infinite for some coefficients of the equation.

Then the limiting case of $\lambda = 4$ is investigated.

Now let us write

$$\beta = 3z + \nu\frac{\mathrm{d}}{\mathrm{d}z}\left(\frac{z^2 h}{\lambda} + \nu z\right) \tag{11.4.3}$$

where $h = h^{\triangle}(z,\lambda)$, $\nu = (4-\lambda)^{-1}$.

We consider the following partial differential equation which is suggested by the observation on $h^{\triangle}(l,y,\lambda)$ at the Beraha numbers for λ as

$$(-\lambda + \beta u)\frac{\partial W}{\partial z} = 2u^2\frac{\partial W}{\partial u} \tag{11.4.4}$$

with the initial condition

$$W_0 = n\cos^{-1}\left(\frac{\nu\sqrt{u}}{2}\right). \tag{11.4.5}$$

The inverse cosine is to be defined as a power series in $\sqrt{u}$ with the initial term $\frac{\pi}{2}$. Of course, $W_0 = W|_{z=0}$.

It can be seen that the equation is well-defined and that there exists a polynomial Y of degree 4 in u with the form

$$Y = u(A + Bu + Cu^2 + Du^3) \tag{11.4.6}$$

such that

$$u^4 = -Y\left(\frac{\partial W}{\partial z}\right)^2 \tag{11.4.7}$$

by the study of $h^{\triangleright}(t, y, \lambda)$ at all $\lambda = B_n, n \geq 2$, the Beraha numbers [Tut33, Tut36].

According to (11.4.4) and (11.4.7), we are allowed to eliminate W. Rewrite (11.4.7) in the form

$$\frac{\partial W}{\partial z} = u^2(-Y)^{\frac{1}{2}}. \tag{11.4.8}$$

Then, by (11.4.4),

$$\frac{\partial W}{\partial u} = \tfrac{1}{2}(-\lambda + \beta u)(-Y)^{\frac{1}{2}}. \tag{11.4.9}$$

Because

$$\frac{\partial}{\partial z}\left(\frac{\partial W}{\partial u}\right) = \frac{\partial}{\partial u}\left(\frac{\partial W}{\partial z}\right),$$

and then from (11.4.8–9)

$$\frac{\partial}{\partial u}\left(u^2(-Y)^{\frac{1}{2}}\right) = \frac{1}{2}\frac{\partial}{\partial z}\left((-\lambda + \beta u)(-Y)^{\frac{1}{2}}\right), \tag{11.4.10}$$

we may derive

$$\left(8 - 2\frac{\mathrm{d}\beta}{\mathrm{d}z}\right)uY - 2u^2\frac{\partial Y}{\partial u} + (-\lambda + \beta u)\frac{\partial Y}{\partial z} = 0. \tag{11.4.11}$$

In view of (11.4.6), we have

$$\begin{aligned}&\left(8 - 2\frac{\mathrm{d}\beta}{\mathrm{d}z}\right)(Au^2 + Bu^3 + Cu^4 + Du^5)\\ &\quad -2(Au^2 + 2Bu^3 + 3Cu^4 + 4Du^5)\\ &\quad +(-\lambda + \beta u)\left(u\frac{\mathrm{d}A}{\mathrm{d}z} + u^2\frac{\mathrm{d}B}{\mathrm{d}z}\right.\\ &\quad \left. +u^3\frac{\mathrm{d}C}{\mathrm{d}z} + u^4\frac{\mathrm{d}D}{\mathrm{d}z}\right) = 0.\end{aligned}$$

By equating coefficients of the same powers of u on the two sides, we obtain the following differential equation system:

$$-\lambda\frac{\mathrm{d}A}{\mathrm{d}z}=0, \tag{11.4.12}$$

$$\Big(6-2\frac{\mathrm{d}\beta}{\mathrm{d}z}\Big)A-\lambda\frac{\mathrm{d}B}{\mathrm{d}z}+\beta\frac{\mathrm{d}A}{\mathrm{d}z}=0, \tag{11.4.13}$$

$$\Big(4-2\frac{\mathrm{d}\beta}{\mathrm{d}z}\Big)B-\lambda\frac{\mathrm{d}C}{\mathrm{d}z}+\beta\frac{\mathrm{d}B}{\mathrm{d}z}=0, \tag{11.4.14}$$

$$\Big(2-2\frac{\mathrm{d}\beta}{\mathrm{d}z}\Big)C-\lambda\frac{\mathrm{d}D}{\mathrm{d}z}+\beta\frac{\mathrm{d}C}{\mathrm{d}z}=0, \tag{11.4.15}$$

$$-2\frac{\mathrm{d}\beta}{\mathrm{d}z}D+\beta\frac{\mathrm{d}D}{\mathrm{d}z}=0. \tag{11.4.16}$$

The constant terms of A, B, C and D as power series of z have to be determined as the initial conditions. From (11.4.4) and (11.4.8),

$$u^4(-\lambda+\beta_0u)^2=-Y_0\cdot 4u^4\left(\frac{\partial W_0}{\partial u}\right)^2.$$

From (11.4.4), we see that $\beta_0=\nu^2$. According to (11.4.5), we have

$$(\lambda-\nu^2u)^2=-\frac{n^2\nu^2Y_0}{u(4-\nu^2u)},$$

i.e.,

$$n^2\nu^2Y_0=-u(4-\nu^2u)(\lambda-\nu^2u)^2. \tag{11.4.17}$$

By virtue of (11.4.6) and (11.4.17), by equating coefficients of the same powers of u, we obtain that

$$A_0=\frac{-\lambda^2}{n^2\nu^2},\quad B_0=\frac{8\lambda+\lambda^2}{n^2},$$

$$C_0=\frac{-(4+2\lambda)\nu^2}{n^2},\quad D_0=\frac{\nu^4}{n^2}. \tag{11.4.18}$$

In what follows, we solve the equations (11.4.12–16) with the initial conditions shown in (11.4.18).

First, from (11.4.12) we find that

$$A=A_0=-\frac{\lambda^2}{n^2\nu^2}. \tag{11.4.19}$$

Second, integrating (11.4.13) we find that

$$6Az - 2A\beta - \lambda B + c_1 = 0$$

where c_1 is a constant. Taking coefficients of z^0 in this and applying (11.4.18), we have that $n^2 c_1 = \lambda^3$. Then we deduce that

$$n^2\nu^2 B = \lambda(\lambda\nu^2 - 24z + 8\beta), \tag{11.4.20}$$

$$n\nu^2\frac{\mathrm{d}B}{\mathrm{d}z} = \lambda\left(-24 + 8\frac{\mathrm{d}B}{\mathrm{d}z}\right). \tag{11.4.21}$$

Third, substituting (11.4.19–21) in (11.4.14) we find

$$n^2\nu^2\frac{\mathrm{d}C}{\mathrm{d}z} = 4\lambda\nu^2 - 96z + 8\beta + 48z\frac{\mathrm{d}\beta}{\mathrm{d}z}$$

$$-2\lambda\nu^2\frac{\mathrm{d}\beta}{\mathrm{d}z} - 8\beta\frac{\mathrm{d}\beta}{\mathrm{d}z}. \tag{11.4.22}$$

By integrating (11.4.22), we have

$$n^2\nu^2 C = 4\lambda\nu^2 z - 48z^2 + 8\gamma + 48z\beta$$

$$-48\gamma - 2\lambda\nu^2\beta - 4\beta^2 + c_2$$

where c_2 is a constant and

$$\gamma = \int_0^z \beta(x)\mathrm{d}x. \tag{11.4.23}$$

By taking coefficients of z^0, from (11.4.18)

$$-(4 - 2\lambda)\nu^4 = -2\lambda\nu^4 - 4\nu^4 + c_2.$$

Hence $c_2 = 0$. That leads to

$$n^2\nu^2 C = 4\lambda\nu^2 z - 48z^2 - 40\gamma$$

$$+48z\beta - 2\lambda\nu^2\beta - 4\beta^2. \tag{11.4.24}$$

Fourth, solving (11.4.16) we obtain

$$D = c_3\beta^2$$

for some constant c_3. By (11.4.18), $n^2c_3 = 1$. Thus

$$n^2D = \beta^2, \tag{11.4.25}$$

$$n^2\frac{\mathrm{d}D}{\mathrm{d}z} = 2\beta\frac{\mathrm{d}\beta}{\mathrm{d}z}. \tag{11.4.26}$$

Finally, multiplying (1.4.15) by $n^2\nu^2$ and then substituting (11.4. 22), (11.4.24) and (11.4.26) in it, we find that

$$\begin{aligned}
2\left(1-\frac{\mathrm{d}\beta}{\mathrm{d}z}\right)&(4\lambda\nu^2z-48z^2-40\gamma-2\lambda\nu^2\beta+48z\beta-4\beta^2)\\
&-2\lambda\nu^2\beta\frac{\mathrm{d}\beta}{\mathrm{d}z}+\beta\Big(4\lambda\nu^2-96z+8\beta+48z\frac{\mathrm{d}\beta}{\mathrm{d}z}\\
&-2\lambda\nu\frac{\mathrm{d}\beta}{\mathrm{d}z}-8\beta\frac{\mathrm{d}\beta}{\mathrm{d}z}\Big)=0.
\end{aligned}$$

This simplifies to

$$\left(1-\frac{\mathrm{d}\beta}{\mathrm{d}z}\right)(\lambda\nu^2z-12z^2-10\gamma+6z\beta)=6z\beta,$$

and hence

$$\left(1-\frac{\mathrm{d}^2\gamma}{\mathrm{d}z^2}\right)\left(-\lambda\nu^2z+12z^2+10\gamma-6z\frac{\mathrm{d}\gamma}{\mathrm{d}z}\right)+6z\frac{\mathrm{d}\gamma}{\mathrm{d}z}\frac{\mathrm{d}^2\gamma}{\mathrm{d}z^2}=0,$$

or

$$\left(1-\frac{\mathrm{d}^2\gamma}{\mathrm{d}z^2}\right)\left(-\lambda\nu^2z+12z^2+10\gamma\right)+6z\frac{\mathrm{d}\gamma}{\mathrm{d}z}=0. \tag{11.4.27}$$

By integrating (11.4.4), we have

$$\gamma=\frac{\nu z^2h}{\lambda}+\nu^2z+\frac{3z^2}{2}, \tag{11.4.28}$$

and hence

$$\gamma_0=\gamma|_{z=0}=0 \text{ and } \gamma_0'=\frac{\mathrm{d}\gamma}{\mathrm{d}z}\Big|_{z=0}=\nu^2. \tag{11.4.29}$$

Theorem 11.4.1 *The differential equation* (11.4.27) *with the initial conditions shown in* (11.4.29) *is well-defined when* γ *is a power series of* z *with the coefficients being polynomials in* ν (*or* λ *as well*) *whose coefficients are rational numbers.*

Proof By the similar procedure shown in §11.2. □

Since $\nu = \frac{1}{4-\lambda}$, λ is not allowed to be 4 in (11.4.28–29). We have to discuss the particular case. Let us write

$$H = z^2 h^{\triangle}. \tag{11.4.30}$$

Then, from (11.4.28)

$$\gamma = \frac{\nu H}{\lambda} + \nu^2 z + \frac{3z^2}{2}, \tag{11.4.31}$$

$$\frac{\mathrm{d}\gamma}{\mathrm{d}z} = \frac{\nu}{\lambda}\frac{\mathrm{d}H}{\mathrm{d}z} + \nu^2 + 3z^2, \tag{11.4.32}$$

$$\frac{\mathrm{d}^2\gamma}{\mathrm{d}z^2} = \frac{\nu}{\lambda}\frac{\mathrm{d}^2H}{\mathrm{d}z^2} + 3. \tag{11.4.33}$$

Substituting (11.4.31–33) in (11.4.27), we may find that

$$\begin{aligned}\lambda^{-1}\frac{\mathrm{d}^2H}{\mathrm{d}z^2}&\left(z + 9\nu^{-1}z^2 + 10\lambda^{-1}H - 6\lambda^{-1}z\frac{\mathrm{d}H}{\mathrm{d}z}\right)\\ &= -2\nu^{-1}z + 6z - 20\lambda^{-1}\nu^{-1}H\\ &\quad +18\lambda^{-1}\nu^{-1}\frac{\mathrm{d}H}{\mathrm{d}z}.\end{aligned}$$

Since $\lambda \longrightarrow 4, \nu^{-1} \longrightarrow 0$, the equation becomes

$$\frac{\mathrm{d}^2H}{\mathrm{d}z^2}\left(2z + 5H - 3z\frac{\mathrm{d}H}{\mathrm{d}z}\right) = 48z \tag{11.4.34}$$

for $\lambda = 4$. By observing (11.4.30),

$$H|_{z=0} = 0, \quad \left.\frac{\mathrm{d}H}{\mathrm{d}z}\right|_{z=0} = 0. \tag{11.4.35}$$

Theorem 11.4.2 *The differential equation* (11.4.34) *with the initial conditions shown in* (11.4.35) *is well-defined when H is a power series of z^2 with all the coefficients being integers.*

Proof Similarly to the proof of Theorem 11.4.1. □

This allows us to determine the chrosum function $h^{\triangle}$ of planar triangulations with only the face number as the parameter for four colors.

§11.5 Notes

11.5.1 More detailed discussions on the general solution of chromatic equation (11.1.10) for rooted non-separable outerplanar maps in $\lambda = 2, \lambda = 3$ and $\lambda = \infty$ can be seen in [Liu47] and [Liu50].

11.5.2 In [Liu40] and [Liu46], one can also see the cubic triangles. However, the definition of the expander by (11.2.29) is a correction of the confusion with suffixes in those papers. More results can also be seen in [Liu46].

11.5.3 Results appearing in §11.3 can be seen in [Tut27].

11.5.4 The whole theory described in §11.4 can also be seen in [Tut33, Tut36–37], especially [Tut41].

11.5.5 Although the equation (11.4.34) looks much simpler, no one has found a close form of H and hence $h^{\triangle}$ up to now.

Chapter 12

Stochastic Behaviors

§12.1 Asymptotics for outerplanar maps

Let $P_n^{(c)}$ and $Q_n^{(c)}$ be the numbers of general planar and outerplanar maps of n edges with the property c respectively. Here, c='g','nl','s' and 'ns' mean that the maps are respectively the most general, loopless, simple and nonseparable maps.

Theorem 12.1.1 *If the size n is sufficiently large, then almost all the most general planar maps are not outerplanar.*

Proof From Theorem 8.1.2, we know that

$$P_n^{(\mathrm{g})} = \frac{2 \cdot 3^n (2n)!}{n!(n+2)!} \tag{12.1.1}$$

by (1.7.6),

$$\begin{aligned} &\sim \frac{2 \cdot 3^n \sqrt{2\pi} e^{-2n} (2n)^{2n-\frac{1}{2}}}{\sqrt{2\pi} e^{-n} n^{n-\frac{1}{2}} \sqrt{2\pi} e^{-n(n+2)} (n+2)^{n+\frac{3}{2}}} \\ &\sim \frac{2^{2n+\frac{1}{2}} \cdot 3^n}{\sqrt{2\pi}} \cdot e^2 n^{-\frac{5}{2}} \\ &= O(n^{-\frac{5}{2}} 12^n). \end{aligned}$$

Similarly, based on §2.4 or [DoY1], we have

$$\begin{aligned} Q_n^{(\mathrm{g})} &= \frac{2^n (2n)!}{n!(n+1)!} \\ &= O(n^{-\frac{3}{2}} 8^n). \end{aligned} \tag{12.1.2}$$

Hence,

$$\frac{Q_n^{(g)}}{P_n^{(g)}} = O\Big(n\Big(\frac{2}{3}\Big)^n\Big) = O(n(0.667)^n). \tag{12.1.3}$$

From $\lim_{n\to\infty} n\left(\frac{3}{2}\right)^n = 0$, the theorem is obtained. □

Since it is known that

$$Q_n^{(\mathrm{nl})} = \frac{1}{n(n+1)} \sum_{j=0}^{n-1} A_j \tag{12.1.4}$$

where

$$A_j = \frac{(2n-j)!}{j!(n-j)!(n-j-1)!}$$

and it can be checked that A_j in the range from 0 to $n-1$ is unimodular with the maximum at

$$j \sim (\alpha n + \beta), \ \alpha = \frac{2-\sqrt{2}}{2},$$

$$\beta = -\Big(\frac{1}{2} + \frac{\sqrt{2}}{4}\Big),$$

by setting $k = (\alpha n + \beta) - j$ and employing

$$\log(1+\epsilon) = \epsilon - \frac{\epsilon^2}{2} + O(\epsilon^3)$$

we may, under the condition of $k = o(n)$, obtain

$$\left(\frac{n}{2\pi n(n-j)}\right)^{\frac{1}{2}} \sim (2\pi)^{-\frac{1}{2}} n^{-\frac{1}{2}} (\alpha(1-\alpha))^{-\frac{1}{2}},$$

$$\Big(\frac{n}{j}\Big)^j \sim \alpha^{-(\alpha n - \beta - k)} \exp\Big\{(k-\beta)\Big(1 - \frac{k-\beta}{\alpha n}\Big)\Big\},$$

$$\Big(\frac{n}{n-j}\Big)^{n-j} \sim (1-\alpha)^{-((1-\alpha)n + k - \beta)}$$

$$\times \exp\Big\{-(k-\beta)\left(1 + \frac{k-\beta}{(1-\alpha)n}\right)\Big\},$$

$$\left(\frac{2\pi - j}{2\pi n(n-j)}\right)^{\frac{1}{2}} \sim (\pi n)^{\frac{1}{2}} \left(\frac{2-\alpha}{2(1-\alpha)}\right)^{\frac{1}{2}},$$

$$\left(\frac{2n-j}{n}\right)^n \sim (2-\alpha)^n \exp\Big\{\frac{k-\beta}{2-\alpha}\Big\},$$

$$\left(\frac{2n-j}{n-j}\right)^{n-j} \sim \left(\frac{2-\alpha}{1-\alpha}\right)^{(1-\alpha)n+k-\beta}$$

$$\times \exp\Big\{-\frac{k-\beta}{2-\alpha}\left(1+\frac{k-\beta}{(1-\alpha)n}\right)\Big\},$$

$$\frac{n-j}{n(n+1)} \sim (1-\alpha)n^{-1}. \tag{12.1.5}$$

By (1.7.6) and (12.1.5), it can be shown that

$$A_j \sim \frac{1+\sqrt{2}}{4\pi} n^{-2}(1+\sqrt{2})^{2n} \exp\Big\{-4\sqrt{2}\frac{k^2}{n}\Big\}. \tag{12.1.6}$$

Further, if $|k| > n^s, s > \frac{1}{2}$, then

$$A_j = o\Big(n^{-1}\sum_{i=1}^{n-1} A_i\Big).$$

Thus

$$\sum_{|k|>n^s} A_j = \sum_{|k|>n^s} o\Big(n^{-1}\sum_{i=1}^{n-1} A_i\Big) < o\Big(\sum_{i=1}^{n-1} A_i\Big).$$

This allows us to estimate

$$\sum_{j=0}^{n-1} A_j \sim \sum_{|k|\leq n^s} A_j \sim \frac{1+\sqrt{2}}{4}\pi^{-1}n^{-2}$$

$$\times (1+\sqrt{2})^{2n} \sum_{|k|\leq n^s} \exp\Big\{-4\sqrt{2}\frac{k^2}{n}\Big\}. \tag{12.1.7}$$

The last sum in (12.1.7) is an approximation with step size $\frac{1}{\sqrt{n}}$ to

$$\sum_{|k|\leq n^s} \exp\Big\{-4\sqrt{2}\frac{k^2}{n}\Big\} \sim 2\sqrt{n}\int_0^{n^{s-\frac{1}{2}}} e^{-4\sqrt{2}x^2}\,\mathrm{d}x$$

$$\sim 2\sqrt{n}\int_0^\infty e^{-4\sqrt{2}x^2}\,\mathrm{d}x$$
$$= 2^{-\frac{5}{4}}\pi^{\frac{1}{2}}n^{\frac{1}{2}}.$$

Therefore

$$Q_n^{(\mathrm{nl})} \sim \frac{1+\sqrt{2}}{82^{\frac{1}{4}}}\pi^{-\frac{1}{2}}n^{-\frac{3}{2}}(1+\sqrt{2})^{2n}$$
$$= O(n^{-\frac{3}{2}}(1+\sqrt{2})^{2n}). \tag{12.1.8}$$

Theorem 12.1.2 *If the size is sufficiently large, then almost all loopless planar maps are not outerplanar.*

Proof From Theorem 7.1.2, we have

$$P_n^{(\mathrm{nl})} = \frac{6(4n+1)!}{(3n+3)!n!}$$
$$= O\Big(n^{-\frac{5}{2}}\Big(\frac{256}{27}\Big)^n\Big).$$

By (12.1.8),

$$\frac{Q_n^{(\mathrm{nl})}}{P_n^{(\mathrm{nl})}} = O\Big(n\Big(\frac{27(1+\sqrt{2})^2}{256}\Big)^n\Big)$$
$$= O(n(0.615)^n).$$

This implies the theorem. □

On simple planar maps, from (7.3.28) we have

$$P_n^{(\mathrm{s})} = \sum_{i=0}^{n-2} B_i$$

where

$$B_i = \frac{4(2n+1)!(2n-i-4)!}{i!(n-i-2)!(2n-i+1)!}.$$

It is easily checked that B_i on the interval $[0, n-2]$ is unimodular and that the maximum occurs at

$$i = l,\ \text{if } n = 2l+1;$$
$$= l-1,\ \text{if } n = 2l.$$

By the similar procedure as mentioned above, we have

$$B_i \sim \frac{64\sqrt{2}}{243} e^2 \pi^{-1} l^{-3} 64^l \exp\Big\{-\frac{2k^2}{l}\Big\}.$$

Employing the integration to estimate the summation under the condition that $|k| > l^s, s > \frac{1}{2}$, we find that

$$P_n^{(s)} \sim \frac{64e^2}{243\sqrt{\pi}} l^{-\frac{5}{2}} 64^l$$
$$= O(l^{-\frac{5}{2}} 64^l). \qquad (12.1.9)$$

Theorem 12.1.3 *If the size is sufficiently large, then almost all simple planar maps are not outerplanar.*

Proof First it is seen that

$$Q_n^{(s)} = \sum_{j=0}^{\lfloor \frac{n-1}{2} \rfloor} B'_j$$

where

$$B'_j = \frac{1}{n(n+1)} \cdot \frac{(2n-2j)!}{j!(n-j)!(n-1-2j)!}$$

and that B'_j on the interval $[0, \lfloor \frac{n-1}{2} \rfloor]$ is unimodular as well. The maximum locates approximately at

$$j = \xi n + \eta,\ \xi = \frac{2-\sqrt{2}}{4},\ \eta = -\frac{7}{8}.$$

In view of the procedure above, we may have

$$Q_n^{(s)} \sim \frac{(1+\sqrt{2})^{\frac{1}{2}}}{2^{\frac{3}{4}} \pi^{\frac{1}{2}}} n^{-\frac{3}{2}} 2^n (1+\sqrt{2})^n$$
$$= O(n^{-\frac{3}{2}} 2^n (1+\sqrt{2})^n).$$

Then from (12.1.9), we obtain

$$\frac{Q_n^{(s)}}{P_n^{(s)}} \sim \frac{243(1+\sqrt{2})^{\frac{1}{2}}}{64} n \left(\frac{1+\sqrt{2}}{4}\right)^n$$

$$= O(n(0.604)^n).$$

This leads to the theorem. □

Theorem 12.1.4 *If the size is sufficiently large, then almost all nonseparable planar maps are not outerplanar*

Proof From what has been discussed in Chapter 6, we may find that

$$P_n^{(\mathrm{ns})} = \frac{2(2n-3)!}{(2n-1)!n!}$$

by (1.7.6),

$$\sim \frac{\sqrt{3}}{27}\pi^{-\frac{1}{2}}n^{-\frac{5}{2}}\left(\frac{27}{4}\right)^n$$
$$= O\left(n^{-\frac{5}{2}}\left(\frac{27}{4}\right)^n\right)$$

and

$$Q_n^{(\mathrm{ns})} = \frac{(2n-2)!}{(n-1)!n!}$$

by (1.7.6),

$$\sim \frac{1}{4}\pi^{-\frac{1}{2}}n^{-\frac{3}{2}}4^n$$
$$= O(n^{-\frac{3}{2}}4^n). \qquad (12.1.10)$$

Thus, we obtain

$$\frac{Q_n^{(\mathrm{ns})}}{P_n^{(\mathrm{ns})}} \sim \frac{27}{4\sqrt{3}}n\left(\frac{16}{27}\right)^n$$
$$= O(n(0.593)^n).$$

This is the theorem. □

Trees are considered as planar maps in §2.1. Of course, they are a special kind of outerplanar maps. The number of rooted plane trees of size n is

$$T_n = \frac{(2n)!}{n!(n+1)!}, \qquad (12.1.11)$$

the *Catalan number*. The following theorem shows how much they occupy among outerplanar maps.

Theorem 12.1.5 *If the size is sufficiently large, then almost all outerplanar maps are not trees.*

Proof According to (12.1.11), by (1.7.6) we have

$$T_n \sim \frac{e}{\sqrt{2\pi}} 4^n n^{-\frac{3}{2}}. \tag{12.1.12}$$

By considering (12.1.2),

$$\frac{T_n}{Q_n^{(g)}} = O(2^{-n}).$$

This is the theorem. □

§12.2 The average on tree-rooted maps

For a structure which usually is a map, to determine the frequency of occurrences of the structure in a given map from a particular category of maps is encountered very often in combinatorics, or even in some other branches of mathematics. This causes to investigate the average number of the structure per map all over the category.

A rooted map is said to be *tree-rooted* if a spanning tree is distinguished as a rooted tree in the map. A tree-rooted map in which each face (loop and multiedge are allowed) is an $(r+2)$-gon, with the only possible exception of the outerface (*i.e.*,the root-face) which is a t-gon, and in which the number of vertices is v, is said to be in type $\langle r, t, v\rangle$. Write $\mathcal{R}(r,t,v)$ be the set of all tree-rooted planar maps in type $\langle r,t,v\rangle$ for $r, t-1, v \geq 1$.

For a map $M = (\mathcal{X}_{\alpha,\beta}, \mathcal{P}) \in \mathcal{R}(r,t,v)$, let T be a spanning tree which is treated as a rooted map as well. The root $r(T) = \mathcal{P}^k r(M)$ such that

$$k = \min\{i|\ K\mathcal{P}^i r(M) \in E(T)\}.$$

Because for each edge $e_i = K(\mathcal{P}\alpha\beta)^{i-1} r(M), i = 1, 2, \cdots, l$, on the outerface boundary, there is a unique cycle with the tree T, which is denoted by P_i. If e_i is on the tree, then e_i itself is treated as the degenerate cycle of 2-gon. The common part $P(r, n_i)$ of M with P_i and its inner domain form a rooted outerplanar map with all inner faces $(r+2)$-gons and the outerface (n_i+1)-gon. Such $P(r,n_i)$ is called an $(r+2)$*-partitioned* (n_i+1)*-gon* . Let $\mathcal{P}(r,n)$ be the set of all $(r+2)$-partitioned $(n+1)$-gons. For any tree T of order v, the sequence $(P_1, P_2, \cdots, P_t; T)$, $P_i = P(r,n_i) \in \mathcal{P}(r,n_i)$, such that

$$\sum_{i=1}^{t} n_i = 2v - 2, \tag{12.2.1}$$

is called a (r,t,v)*-tree perm.*

Lemma 12.2.1 *There is a* 1*-to-*1 *correspondence between the maps in type* $\langle r,t,v\rangle$ *and the* (r,t,v)*-tree perms.*

Proof From what is just mentioned, for any $M \in \mathcal{R}(r,t,v)$, there is a unique $(r.t,v)$-tree perm which corresponds to M by showing that all the P_i satisfy (12.2.1) because the total contribution of all $n_i, i = 1, \cdots, t$, is just 2 times the size $v-1$ of T.

Conversely, for an (r,t,v)-tree perm $(P_1, \cdots, P_t; T)$ of (12.2.1), we are allowed to locate

$$P(r,n_1), P(r,n_2), \cdots, P(r,n_t)$$

with T such that $n_1, n_2, \cdots, n_t$ edges are on the tree T according to the order defined by the root-face boundary of the map T, $P(r,n_i) \in \mathcal{P}(r,n_i), i = 1, 2, \cdots, t$. That is to find a tree-rooted map M in type $\langle r,t,v\rangle$ and hence $M \in \mathcal{R}\langle r,t,v\rangle$. From the well definedness the lemma is obtained. □

Let $\mathcal{R}(r) = \sum_{n\geq 1} \mathcal{R}(r,n)$ be the set of all rooted outerplanar maps with all inner faces $(r+1)$-gons. Of course, loops and multiedges are allowed. Because

$$\mathcal{R}(r) = \mathcal{R}_0(r) + \mathcal{R}_1(r) \tag{12.2.2}$$

where $\mathcal{R}_o(r)$ consists of only the link map without inner face and $\mathcal{R}_1(r)$ is the chain $1v$-production of $\mathcal{R}(r)$ in $r+1$ time, *i.e.*,

$$\mathcal{R}_1(r) = \mathcal{R}(r)^{\hat{\times}(r+1)} \tag{12.2.3}$$

as shown in §6.1, we have

$$f(x) = x + f^{r+1}(x) \tag{12.2.4}$$

where

$$f(x) = \sum_{M \in \mathcal{R}(r)} x^{n(M)} \tag{12.2.5}$$

is the enufunction of $\mathcal{R}(r)$ with $n(M)+1$ being the valency of the outerface of M.

Lemma 12.2.2 *The number of (r,t,v)-tree perms with a specified rooted tree is*

$$\frac{t\big(k(r+1)+t-1\big)!}{k!(kr+t)!} \tag{12.2.6}$$

where $2v = kr+t+2$.

Proof From (12.2.4–5), we see that the number of elements in $\mathcal{R}(r,n)$ is the coefficient of x^n in $f(x)$ as a power series of x. By Lemma 12.2.1, the number of (r,t,v)-tree perms should be the coefficient of x^{2v-2} in the expansion of $f^t(x)$. However, (12.2.4) allows us to employ Corollary 1.5.2 at $f = x$ for finding

$$f^t(x) = x^t + \sum_{k \geq 1} \frac{t}{k!} \frac{\mathrm{d}^{k-1}}{\mathrm{d}x^{k-1}} \Big\{ x^{k(r+1)+t-1} \Big\}$$

$$= \sum_{k \geq 0} \frac{t\big(k(r+1)+t-1\big)!}{k!(kr+t)!} x^{kr+t}. \tag{12.2.7}$$

This is the Lemma. □

Because it is known that the number of rooted plane trees of order v is

$$\frac{(2v-2)!}{v!(v-1)!}, \tag{12.2.8}$$

from Lemma 12.2.2 we have the number of tree-rooted maps in type $\langle r,t,v\rangle$ is

$$\big|\mathcal{R}(r,t,v)\big| = \frac{(2v-2)!t(k(r+1)+r-1)!}{v!(v-1)!k!(k+r+t)!}$$

$$= \frac{t(2v+k-3)!}{(v-1)!v!k!} \tag{12.2.9}$$

where $2v = kr+t+2$.

By Euler formula, the parameter k in (12.2.9) is in fact the number of nonrooted faces in a map of type $\langle r,t,v\rangle$. In the dual case, we have that the number of tree-rooted maps with v vertices, the nonrooted vertices of valency u, and the root-vertex of valency t is

$$\frac{t(2x+v-4)!}{x!(x-1)!(v-1)!} \tag{12.2.10}$$

where $x=\frac{1}{2}((v-1)(u-2)+t+2)$.

From (5.2.9), the number of rooted planar maps of order v, each nonrooted vertex of valency $2s$ and the root-vertex of valency $2t$, is

$$\frac{((v-1)s+t-1)!(2t)!}{(v-1)!((v-1)s+t-v+2)!t!(t-1)!} \times\left(\frac{(2s-1)!}{s!(s-1)!}\right)^{v-1}. \tag{12.2.11}$$

Then the statistical average of trees per map in this kind of maps should be

$$\frac{\big(v(2s-1)+2(t-s)\big)!t!(t-1)!}{\big((v-1)(s-1)+t\big)!\big((v-1)s+t-1\big)!(2t-1)!} \times\left(\frac{s!(s-1)!}{(2s-1)!}\right)^{v-1}. \tag{12.2.12}$$

Specifically for $t=s$, that is for regular Eulerian maps, this average becomes

$$\frac{\big(v(2s-1)\big)!}{\big(v(s-1)+1\big)!(vs-1)!}\left(\frac{s!(s-1)!}{(2s-1)!}\right)^{v}. \tag{12.2.13}$$

Since for $s=t$, (12.2.11) becomes

$$\frac{(2s)!(vs-1)!}{(v-1)!\big(v(s-1)+2\big)!}\left(\frac{(2s-1)!}{s!(s-1)!}\right)^{v} \tag{12.2.14}$$

which, for fixed $s\neq 1$, as v large enough, is approximately

$$\sqrt{\frac{2s}{\pi}}\frac{\big(v(s-1)\big)^{-\frac{5}{2}}}{(s-1)^{v(s-1)}}\left(\frac{(2s-1)!}{s!(s-1)!}\right)^{v}, \tag{12.2.15}$$

by considering (1.7.6) the average number of trees for a rooted 2s-valent planar map of order v is asymptotically approximated by

$$\frac{v^{-\frac{1}{2}}(2s-1)^{v(2s-1)+\frac{1}{2}}}{\sqrt{2\pi}s^{sv-\frac{1}{2}}(s-1)^{v(s-1)+\frac{3}{2}}}\left(\frac{s!(s-1)!}{(2s-1)!}\right)^{v}. \tag{12.2.16}$$

§12.3 Hamiltonian circuits per map

A rooted planar map is said to be a *Hamiltonian rooted map* if it is Hamiltonian and has a Hamiltonian circuit distinguished. Here, we only consider the case of all inner faces triangles. If a Hamiltonian rooted map has the Hamiltonian circuit as the root-face boundary, then it is called a *sliced polygon*. Let $\mathcal{H}_{\text{out}}$ be the set of all the sliced polygons. Then we find the number $N_{\mathcal{H}_{\text{out}}}(m)$ of sliced polygons of order m for $m \geq 3$. For convenience, the link map L is assumed to be the degenerate sliced polygon. Since it can be seen that

$$\mathcal{H}_{\text{out}} = \mathcal{H}_0 + \mathcal{H}_1, \tag{12.3.1}$$

where $\mathcal{H}_0$ consists of only the link map, and that

$$\begin{aligned}\mathcal{H}_{\langle 1\rangle} &= \{H - a|\forall H \in \mathcal{H}_1\}\\ &= \mathcal{H}_{\text{out}}^{\hat{\times}2}\end{aligned} \tag{12.3.2}$$

in which $\hat{\times}$ is the chain 1v-production described in §6.1, we may soon find that the enufunction

$$f(x) = \sum_{H\in\mathcal{H}_{\text{out}}} x^{m(H)} \tag{12.3.3}$$

where $m(H)$ is the order of H satisfies the following equation:

$$f(x) = x^2 + x^{-1}f(x)^2. \tag{12.3.4}$$

Because the available solution of (12.3.4) is

$$f(x) = \frac{x(1-\sqrt{1-4x})}{2}, \tag{12.4.5}$$

the number $N_{\mathcal{H}_{\text{out}}}(m)$ is the coefficient of x^m in $f(x)$ as a powers series of x. That is

$$N_{\mathcal{H}_{\text{out}}}(m) = \frac{1}{m-1}\binom{2m-4}{m-2}. \tag{12.3.6}$$

Let $\mathcal{H}(n,m)$ be the set of Hamiltonian rooted maps of order $n+m$ with the outerface of valency $m \geq 3$. For $M = (\mathcal{X}_{\alpha,\beta}, \mathcal{P}) \in \mathcal{H}(n,m)$, the vertices of M which are not on the outerface boundary are called *inner vertices*, while others which are on the outerface boundary *outer vertices.* Write $n(M)$ and $m(M)$ are, respectively, the numbers of inner and outer vertices of M. Then $n(M) = n$ and $m(M) = m$. Let H be the Hamiltonian circuit rooted with the root $r_H = \mathcal{P}^s r(M)$,

$$s = \min\{j|\ K\mathcal{P}^j r(M) \in E(H)\}.$$

Assume the m outer vertices of M is $v_1, v_2, \cdots, v_m$ in the order determined by the root-face boundary with $v_1 = v_r$ as the root-vertex. For each edge $K\mathcal{P}^{j+1}r(M) = (v_j, v_{j+1}), j = 1, \cdots, m$, $v_{m+1} = v_1$, there is a dipath $P_H(v_j, v_{j+1})$ on the Hamiltonian circuit H which is rooted. Then (v_j, v_{j+1}) with $P_H(v_j, v_{j+1})$ forms a sliced polygon P_j. Meanwhile, the Hamiltonian circuit H itself forms a sliced polygon as well.

In general, for m sliced polygons $P_1, P_2, \cdots, P_m$, we may form a sequence $(P_1, P_2, \cdots, P_m; H)$ which is called an *H-perm* in which H is a sliced polygon with

$$\sum_{i=1}^{m} n(P_i) - m \tag{12.3.7}$$

vertices where $n(P_i)$ is the order of $P_i, i = 1, 2, \cdots, m$. Further, $(P_1, P_2, \cdots, P_m)$ is called an *m-perm.*

Lemma 12.3.1 *Let $\mathcal{P}(n,m)$ be the set of all the H-perms $(P_1, P_2, \cdots, P_m; H)$ such that*

$$\sum_{i=1}^{m} n(P_i) = 2m + n. \tag{12.3.8}$$

Then we have

$$\left|\mathcal{H}(n,m)\right| = \left|\mathcal{P}(n,m)\right| \tag{12.3.9}$$

for $n \geq 0, m \geq 3$.

Proof From what is just mentioned above, we see that for each $M \in \mathcal{H}(n,m)$ there is a unique H-perm $P \in \mathcal{P}(n,m)$ which corresponds to M.

Conversely, for a H-perm $P = (P_1, \cdots, P_m; H) \in \mathcal{P}(n,m)$ we can uniquely construct a Hamiltonian rooted map $M \in \mathcal{H}(n,m)$ in the following way. Assume the rooted edge of P_i is $(v_i, v_{i+1}), i = 1, 2, \cdots, m$, $v_{m+1} = v_1$. Choose $v_1, v_2, \cdots, v_m$ to be the outerface boundary with v_1 as the root-vertex. Let $P(v_i, v_{i+1})$ be the path not including (v_i, v_{i+1}) on the polygon of P_i, $i = 1, \cdots, m$. Then choose the polygon of the sliced polygon H to be consists of $P(v_1, v_2)$, $P(v_2, v_3)$, $\cdots$, $P(v_m, v_1)$. Because of (12.3.8) it is easily verified that $M \in \mathcal{H}(n,m)$. □

From this lemma, in order to determine the number of Hamiltonian rooted maps with n inner vertices and m outer vertices, it is only necessary to determine the number of H-perms which satisfy (12.3.8).

Because of (12.3.6), the number of m-perms with (12.3.8) is

$$a_{n,m} = \sum_{\substack{\alpha_1+\cdots+\alpha_m=n \\ \alpha_i \geq 0 \\ 1 \leq i \leq m}} \prod_{i=1}^{m} \binom{2\alpha_i}{\alpha_i} \frac{1}{\alpha_i + 1}. \tag{12.3.10}$$

Write

$$A(x) = \sum_{i \geq 1} \binom{2i}{i} \frac{1}{i+1} x^n. \tag{12.3.11}$$

Then from (12.3.4), we see $A(x)$ satisfies the equation as

$$A(x) = 1 + xA^2(x) \tag{12.3.12}$$

by considering

$$A(x) = \frac{1}{x^2} f(x). \tag{12.3.13}$$

Since $a_{n,m}$ is the coefficient of x^n in $A^m(x)$, on the basis of (12.3.12) by employing Corollary 1.5.2 (the Lagrangian inversion) we have

$$a_{n,m} = \frac{m(2n+m-1)!}{n!(n+m)!}. \tag{12.3.14}$$

Lemma 12.3.2 *We have*

$$|\mathcal{P}(n,m)| = \frac{m(2n+2m-4)!(2n+m-1)!}{(n+m-1)!(n+m-2)!(n+m)!} \tag{12.3.15}$$

for $n \geq 0, m \geq 3$.

Proof Because

$$\Big|\mathcal{P}(n,m)\Big| = a_{n,m}\Big|\mathcal{H}_{\text{out}}(m+n)\Big|,$$

where $\mathcal{H}_{\text{out}}(m+n)$ is the set of sliced polygons of order $m+n$, by (12.3.6)

$$\Big|\mathcal{H}_{\text{out}}(m+n)\Big| = \frac{(2n+2m-4)!}{(n+m-1)!(n+m-2)!},$$

and by (12.3.14) the lemma is soon found. □

According to Lemma 12.3.1, the number of Hamiltonian rooted maps in $\mathcal{H}(n,m)$ is

$$p_{n,m} = \Big|\mathcal{P}_{(n,m)}\Big|, \tag{12.3.16}$$

as shown by (12.3.15).

From (3.3.18), we see that the number of rooted planar triangulations of n inner vertices and m outer vertices is

$$\frac{2^{n+1}(2m-3)!(3n+2m-4)!}{(m-2)!(m-2)!n!(2n+2m-2)!}. \tag{12.3.17}$$

Hence the average number of Hamiltonian circuits per map is

$$\frac{m}{2^{n+1}}\frac{((m-2)!)^2(2n+m-1)!(2n+2m-4)!}{(2m-3)!(n+m-2)!(n+m-1)!}$$

$$\times\frac{(2n+2m-2)!}{(n+m)!(3n+2m-3)!}. \tag{12.3.18}$$

Now we keep the number m of outer vertices fixed, and observe the asymptotic behavior of (12.3.18) as n sufficiently large. By (1.6.7), we can find that

$$p_{n,m} \sim \frac{m}{\pi}2^{4n+3m-5}n^{-3}, \tag{12.3.19}$$

and then the average number given by (12.3.18) is asymptotically

$$\frac{8m\big((m-2)!\big)^2}{(2m-3)!\sqrt{3\pi}}\left(\frac{32}{7}\right)^{m-2}\left(\frac{32}{27}\right)^{n} n^{-\frac{1}{2}}. \tag{12.3.20}$$

A Hamiltonian rooted map in $\mathcal{H}(n,m)$ with all the edges on the Hamiltonian circuit not on the outface boundary is said to be an *inner H-map*.

According to (12.3.6), the number of inner H-maps with n inner vertices and m outer vertices should be

$$b_{n,m} = \sum_{\substack{\alpha_1+\cdots+\alpha_m=n\\ \alpha_i>0\\ 1\le i\le m}} \prod_{i=1}^{m}\frac{1}{\alpha_i+1}\binom{2\alpha_i}{\alpha_i}. \tag{12.3.21}$$

By considering (12.3.3) and(12.3.5), we may see that $b_{n,m}$ is the coefficient of x^n in the expansion of

$$B(x) = \left(\frac{1-2x-\sqrt{1-4x}}{2x}\right)^m. \tag{12.3.22}$$

It looks with some complication for determining the power series of $B(x)$ by applying Lagrangian inversion. However, it is seen that $B(x)$ satisfies the ordinary differential equation as

$$x^2(4x-1)\frac{\mathrm{d}^2B}{\mathrm{d}x^2}+x(6x-1)\frac{\mathrm{d}B}{\mathrm{d}x}+m^2B=0. \tag{12.3.23}$$

From this equation, a recursion formula for the coefficients of $B(x)$ as a power series of x is found:

$$b_{n,m} = \frac{2(n-1)(2n-1)}{(n+m)(n-m)}b_{n-1,m} \tag{12.3.24}$$

with $b_{n,m}=1$. Thus

$$b_{n,m} = \frac{m}{n}\binom{2n}{n+m} \tag{12.3.25}$$

for $n \geq m$.

Further, by considering (12.3.6) for the sliced polygon H on the H-perm $(P_1, \cdots, P_m; H)$, we soon find that the number of inner H-maps with n inner vertices and m outer vertices is

$$q_{n,m} = \frac{m(2n+2m-4)!}{n(n+m-1)!(n+m-2)!} \times \frac{(2n)!}{(n+m)!(n-m)!}. \tag{12.3.26}$$

Write

$$r_{n,m} = \frac{q_{n,m}}{p_{n,m}}. \tag{12.3.27}$$

Then from (12.3.15–16) we have

$$r_{n,m} = \frac{(2n)!(n-1)!}{(n-m)!(2n+m-1)!}. \tag{12.3.28}$$

By (1.7.6), the approximate value of $q_{n,m}$ for m being fixed is asymptotically

$$q_{n,m} \sim \frac{m}{\pi} 2^{4n+2m-4} n^{-3} \tag{12.3.29}$$

as n sufficiently large.

Combining (12.3.19) and (12.3.29), we have

$$r_{n,m} \sim \frac{1}{2^{m-1}}. \tag{12.3.30}$$

Theorem 12.3.1 *If the numbers of inner and outer vertices are sufficiently large, then almost all Hamiltonian rooted maps are not an inner H-map.*

Proof A direct result of (12.3.30). □

§12.4 The asymmetry on maps

As mentioned at the beginning of the book, rooted maps do not have symmetry. That is to say, a root chosen on a map destroys the symmetry of the map. However, in view of asymptotic behavior, it looks reasonable that almost all maps are asymmetric when the order of maps is large enough. One might think of that most 3-connected planar maps, in other words, nonrooted c-nets is with symmetry. However, the main purpose in this section is to show that almost all nonrooted c-nets are asymmetric when they are sufficiently large.

For a map $M = (\mathcal{X}_{\alpha,\beta}, \mathcal{P})$ not necessary to be a nonrooted c-net, let α be a nontrivial automorphism of minimal order (which will be a prime) on M. From Theorem 1.1.4, α has no fixed element in $\mathcal{X}$. In view of topology, α has the following three possibilities:

Po1 *Orientation-preserving, i.e.*, that the rotation at each vertex (each face as well) does not change under α;

Po2 *Orientation-reversing, i.e.*, not orientation-preserving (always of order 2), with at least an *invariant, i.e.*, a fixed vertex, edge or face of α on M;

Po3 Orientation-reversing without invariant.

Automorphisms with Po1,Po2 or Po3 are respectively said to be *rotative, plane-reflective* or *pole-reflective.*

A map can also be considered as the set of *cells* which are vertices, edges and faces called 0-*cells*,1-*cells* and 2-*cells* respectively.

If a set S of cells has each cell incident with exactly two cells in S, then S is called a *cell-cycle.* A cell-cycle which is without a proper subset that is also a cell-cycle is called a *cell-circuit.*

The three kinds of nontrivial automorphisms have the following properties:

Fact 1 *The set of invariants for a plane-reflective automorphism α of a nonrooted c-net M is a cell-circuit itself.*

Fact 2 *For a pole-reflective automorphism α of a nonrooted c-net M, there is a cell-circuit without vertex* (*i.e.*, 0-cell)*such that it is fixed under α itself.*

Fact 3 *A rotative automorphism α has precisely two invariants.*

An edge is said to be *flipped* by the automorphism α if it is an invariant for α but its incident vertices are not. All other invariant edges are said to be *transfixed* by α.

For a nonrooted c-net M which has an automorphism α of minimal order, if α is plane-reflective, by Fact 1 the boundary of the set of invariants consists of two simple closed curves which are allowed to have edges (of course, transfixed) in common without crossing. Let $\hat{R}_1$ and $\hat{R}_2$ be the closures of the two simply-connected domains of the plane π cut off by the two curves without inner point in common. Naturally, they do not contain any invariant except for the boundaries.

Define M_1 to be the submap of M which is restricted on $\hat{R}_1$. It is easily seen that all transfixed edges and fixed vertices are included in M_1.

Let C be a simple closed curve contained in the union of the invariants of α and intersecting each of them, in particular, intersecting each flipped edge at a unique point.

Define $M_2(C)$ be the map obtained from M_1 by adding the intersecting point as vertices and adding each interval of C contained in a (fixed) face of M and each of the portions of the flipped edges between the new vertices on C and the vertices of M on the boundary of $\hat{R}_1$.

Of course, $\hat{R}_1$ and $\hat{R}_2$ are interchanged by α.

Claim 1 *The map M_1 is nonseparable and the map $M_2(C)$ is 3-connected.*

For α being pole-reflective or rotative, let C be a simple closed curve on π passing through no vertex of M, intersecting no edge in more than one point, and intersecting at least two faces (in fact at least 3 faces since M is 3-connected) each of which has at most two incident edges of one point in common with C.

Write R_1 and R_2 as the two domains of C which are always assumed to be open. Of course, C is the common boundary of R_1 and R_2.

Define $N_1(C)$ to be the submap of M in R_1 and $N_2(C)$ to be the map obtained from $N_1(C)$ by adding a new vertex v which lies in R_2, together with new edges joining v to each of the vertices of $N_1(C)$ which is incident with an edge of M intersecting C.

Then, $N_3(C)$ is defined to be the map obtained from $N_1(C)$ by adding new vertices at each point of C which intersects an edge of

M and new edges consisting of each portion of C between two new vertices as well as each portion of an edge of M lying between one of the new vertices and one of $N_1(C)$, so that R_2 is a face of $N_3(C)$.

Claim 2 *$N_2(C)$ is nonseparable and $N_3(C)$ is 3-connected.*

In what follows, we estimate the number of the nonrooted c-nets by considering the possibilities of producing M by the pieces M_1, M_2, N_2, N_3 and the likes.

Lemma 12.4.1 *Let $B_{m,l}$ be the number of rooted nonseparable planar maps of size m with the valency l of the root-face. Then*

$$\log B_{m,l} < (A_1c^2 + A_2c + A_3)m + \sigma \tag{12.4.1}$$

where σ is a constant, $A_1 = -0.979, A_2 = -0.098$ and $A_3 = 1.931$ for $0.45 \le c \le 1$ and $l = cm$.

Proof Let

$$F_{j,l,m}^{-1} = \frac{(j-2)!(3m-j-l-1)!}{(m-j)!\big((j-l)!\big)^2(2l-j)!(2m-l)!}. \tag{12.4.2}$$

Then from §6.3 we find

$$B_{m,l} = \frac{l}{(2m-l)!}\sum_{j=l}^{\min\{m,2l\}}\frac{(3l-2j+1)(2j-l)}{F_{j,l,m}}. \tag{12.4.3}$$

If x is chosen to be j/m such that $c < x < 1$ and $F_{j,l,m}^{-1}$ is maximized, then from (1.7.6) it can be shown that

$$F_{j,l,m}^{-1} = O(m^D \exp\big(mG(x,c)\big)) \tag{12.4.4}$$

where D is an absolute constant and

$$\begin{aligned} G(x,c) &= x\log x + (3-x-c)\log(3-x-c) \\ &\quad -(1-x)\log(1-x) - 2(x-c)\log(x-c) \\ &\quad -(2c-x)\log(2c-x) - (2-c)\log(2-c). \end{aligned} \tag{12.4.5}$$

In view of (12.4.3), it follows that

$$B_{m,l} = O(m^E \exp\left(mG(x,c)\right))$$

where E is a constant as the cases $j = l, j = m$, and $j = 2l$. Because it is known that

$$G(x,c) < A_1c^2 + A_2c + A_3$$

in the case considered, the lemma is obtained. □

Lemma 12.4.2 *For α being plane-reflective on a nonrooted c-net M, the number of flipped edges is at most $\frac{m}{3}$ and the number of transfixed edges is at most $\frac{n}{2}$ where m and n are the size and the order of M respectively.*

Proof Since M is 3-connected, any vertex is incident with at least 3 edges. Because M has no multi-edge, any two flipped edges are not allowed to be incident with the same vertex. This implies that the number of flipped edges is not more than half of the order n. By virtue of $3n \leq 2m$, the first statement is true.

Since each invariant vertex is incident with at most two transfixed edges and at each vertex which is incident with two transfixed edges there are at most two noninvariant edges with it for the 3-connectedness, each vertex corresponds at most one transfixed edge, and hence the number of transfixed edges is not more than half of the order of M. This implies the last statement of the lemma. □

Lemma 12.4.3 *Let R_m be the number of c-nets* (*i.e.*, rooted 3-connected planar maps) *of size m. Then, we have*

$$R_m \sim \frac{2m^{-\frac{5}{2}}4^m}{243\sqrt{\pi}} \tag{12.4.6}$$

when m is sufficiently large.

Proof According to (8.2.21), from (1.6.7) by the method used in §12.1, the lemma can be soon obtained. □

First, by Claim 1 M_1 is nonseparable for α plane-reflective on a nonrooted c-net M. Let the root of M_1 be chosen such that $F = \pi - \hat{R}_1$

is the root-face. The valency of F is denoted by l. Put $t = \frac{l}{m}$. Suppose $t \geq 0.3$. Let a and b denote the respective numbers of transfixed and flipped edges by α.

Then, M_1 contains all a transfixed edges, no flipped edge, and half of the remaining $m - a - b$ edges. Thus, M_1 is with the size $\frac{m+a-b}{2}$. Put $r = \frac{a-b}{l}$. Hence $rt = \frac{a-b}{m}$ so $|rt| < \frac{1}{3}$ by Lemma 12.4.2. The number of distinct possibilities for M_1 is at most $B_{p,l}$, where

$$p = \frac{m+a-b}{2} = \frac{m(1+rt)}{2}.$$

If the rooting is ignored except for specifying F, the number of possibilities for M_1 is still at most $B_{p,l}$.

Because M can be reconstructed when the locations of the a transfixed and b flipped edges are known. The number of choices for the transfixed edges is at most $\binom{l}{a}$. Once these are chosen, the vertices incident with them are fixed and there are at least a such vertices, so the number of choices for the b vertices incident with the flipped edges is at most $\binom{l-a}{b}$. The remaining vertices fixed by α can be chosen at most 2^{l-a-b} ways. Thus the number of possibilities for M, given l, a and b, is

$$\begin{aligned} p_\alpha(M; l, a, b) &\leq B_{p,l}\binom{l}{a}\binom{l-a}{b}2^{l-a-b} \\ &= B_{p,l}R(l, a, a-b) \end{aligned} \tag{12.4.7}$$

where

$$\begin{aligned} R(l, a, d) &= \binom{l}{a}\binom{l-a}{a-d}2^{l-2a+d} \\ &= \frac{l!2^{l-2a+d}}{a!(a-d)!(l-2a+d)!}. \end{aligned} \tag{12.4.8}$$

By estimating the maximum of $R(l, a, d)$, it follows that

$$R(l, a, d) = O(l^\sigma R\big(l, \big(\frac{r+1}{2}\big)^2 l, rl\big)) \tag{12.4.9}$$

when $d = a - b = rl$. By (1.7.6),

$$R(l.a, a-b) = O(l^\sigma \exp(l f_r)) \tag{12.4.10}$$

where

$$f_r = 2\log 2 - (1+r)\log(1+r) - (1-r)\log(1-r) \qquad (12.4.11)$$

as long as $r \neq \pm 1$, *i.e.*, $d \neq \pm l$. (treated later!)

Let $c = \frac{l}{p} = \frac{2t}{1+rt}$. Then since $rt \leq \frac{1}{3}$, we have

$$c \geq \frac{6t}{4} = \frac{3}{2}t \geq 0.45.$$

By Lemma 12.4.1 and (12.4.10–11),

$$\log\left(B_{p,l}R(l,a,a-b)\right) < \sigma \log m$$
$$+m\left(\left(\frac{1+rt}{2}\right)(A_1c^2 + A_2c + A_3) + tf_r\right).$$

Since $-1 < r < 1, 0.3 \leq t \leq 1$ and $|rt| \leq \frac{1}{3}$, we can now have that

$$\log\left(B_{p,l}R(l,a,a-b)\right) < 1.375n + \sigma \log m. \qquad (12.4.12)$$

In the case $d = l$ or $-l$, this still holds because $B_{p,l}$ and $R(l,a,a-b)$ change by a factor of at most a polynomial in m of bounded degree when d is changed by ± 1, and hence (12.4.12) holds when $d = l-1$ and $d = -l+1$. From (12.4.12), the logarithm of the sum of $p_\alpha(M;l,a,b)$ over all appropriate l, a and $b(l \geq 0.3m)$ is less than $1.375m + \sigma \log m$. Hence, this is a bound on the logarithm of the number of distinct nonrooted c-nets in the case under consideration.

Now suppose that $t < 0.3$. Let C be a simple closed curve contained in the union of the invariants of α and intersecting each of them. By Claim 1, $M_2(C)$ is 3-connected. Let F' be the face of $M_2(C)$ whose boundary is C, and suppose that α transfixes a edges, flips b edges, and fixes s faces of M. Then the edges of $M_2(C)$ incident with F' are the a transfixed edges together with the s intervals of C contained in faces of M fixed by α. So F' has valency $a+s$. The size of $M_2(C)$ is

$$\frac{m+a+b+2s}{2}.$$

The number of distinct possibilities for $M_2(C)$ with the face F' distinguished is at most the number of c-nets of the same size. From Lemma 12.4.3, it is

$$O(2^{m+a+b+2s}). \qquad (12.4.13)$$

For $M_2(C)$ given with F', we still need to know precisely which vertices incident with F' are invariants of α and which edges incident with F' are transfixed by α before determining M. It is, however, enough to be told merely which edges incident with F' are transfixed by α, for any invariant vertex not incident with such an edge has valency at least 4 in $M_1(C)$, while the vertices incident with F' which are not invariants of α are not vertices of M and have valency 3. Thus, given $M_2(C)$ with a and F' specified, the number of possibilities for M is at most $\binom{a+s}{a}$ and b is actually determined. This implies that the number of possibilities for M is

$$O(2^{m+a+b+2s}\binom{a+s}{a}) \tag{12.4.14}$$

for given a, b, s and l.

Observe that each transfixed edge is incident with the face F in M_1, and each invariant face is incident in M with a noninvariant edge incident with F in M_1. Hence $s + a \le l$. Also, each edge of M flipped by α is incident with invariant faces, each of which is incident with at most one other flipped edge by Fact 1 and so $b \le s$. Thus (12.4.14) is

$$O(2^{m+a+3s}\binom{a+s}{s})$$

which is maximized over a such that $s + a \le l$ by putting $a + s = l$, *i.e.*, it is

$$O(2^{m+l+2s}\binom{l}{s}).$$

Now $2^{2s}\binom{l}{s}$ is maximized when s is approximately $4(l - s)$, *i.e.*, $s = \frac{4}{5}l \pm 1$. So, (12.4.14) is

$$O(l^\sigma 2^{m+\frac{13l}{5}}\binom{l}{l/5}) \ (\sigma \text{ the ubiquitous constant})$$

$$= O(l^\sigma\left(2^{\frac{13}{5}}5^{\frac{1}{5}}\left(\frac{5}{4}\right)^{\frac{4}{5}}\right)^l)$$

$$= O(m^\sigma 2^m 10^{tm}) \ (l = tm)$$

$$= O(m^\sigma 2^m 10^{0.3m}) \ (t < 0.3). \tag{12.4.15}$$

Summing over all appropriate a, b, s, and l only multiplying by m^σ and consequently it does not change the last bound. Thus the logarithm of the number of possibilities for M in case α is plane-reflective and $t < 0.3$, and hence for all t in view of our treatment above of $l \geq 0.3m$, is at most

$$m(\log 2 + 0.3 \log 10) + \sigma \log m < 1.384m. \tag{12.4.16}$$

Then by Fact 2 for α pole-reflective, there is a cell-cycle S, of faces and edges of M, which is fixed on the whole by α. Let C be a simple closed curve described for the condition of Claim 2. Since α has no invariants, it must permute the cells in S in the cyclic fashion, and this permutation must correspond to a half turn because α is of order 2. Since α is orientation-reversing, it follows that the cells of M which are subsets of R_1 are mapped by α onto R_2, and vice versa.

Since every edge of $N_2(C)$ incident with v corresponds to an edge in S and since the edges in S are permuted by α in a cyclic half-turn fashion by α, it follows that $N_2(C)$ determines M if it is known which vertex of $N_2(C)$ is v. Hence the planar dual $N_2(C)^*$ of $N_2(C)$ determines M, if it is known which face F of $N_2(C)^*$ corresponds to v. For similar reasons, $N_2(C)$ determines M if it is known which face, F' say, of $N_2(C)$ is R_2.

Faces F and F' have the same valency, say l. The size of $N_2(C)$, and hence that of $N_2(C)^*$, is

$$\frac{m-l}{2} + l = \frac{m+l}{2},$$

and $N_3(C)$ has l more edges, *i.e.*, $\frac{m+3l}{2}$. Thus, the number of distinct possibilities of M is at most the number of rooted nonseparable maps with $\frac{m+l}{2}$ edges in which the face to the left of the root-edge has valency l, and is also at most the number of c-nets with $\frac{m+3l}{2}$ edges, for the face to the left of the root-edge can be specified to be F or F' is required. Then the logarithm of that number for M is at most

$$\frac{m+l}{2}(A_1c^2 + A_2c + A_3) + \sigma \tag{12.4.17}$$

by Claim 2 if $c = \frac{2l}{m+l} \geq 0.45$, and it is also bounded by

$$\frac{m+3l}{2} \log 4 + \sigma \tag{12.4.18}$$

from (12.4.6).

Let $t = \frac{l}{m}$ and suppose that $t \geq 0.3$. Then $c = \frac{2t}{1+t} > 0.45$ and so (12.4.17) is valid for t in this range. But now (12.4.17) is

$$\frac{m(t+1)}{2}\Big(A_1\Big(\frac{2t}{1+t}\Big)^2 + A_2\Big(\frac{2t}{1+t}\Big) + A_3\Big) + \sigma$$

$$\leq m\Big(A_1 t^2 + A_2 t + \frac{t+1}{2}A_3\Big) + \sigma$$

as $1 + t \leq 2, A_1 < 0$. The minimum of the quadratic form in parentheses is

$$-\frac{(A_1 + \frac{1}{2}A_3)^2}{4A_1} + \frac{A_3}{2} < 1.16.$$

This implies that (12.4.17) is less than $1.16m + \sigma$.

Now suppose $t < 0.3$. Then (12.4.18) is less than $(0.95 \log 4)m + \sigma < 1.317m + \sigma$. Multiplying by the m possibilities for l changes σ only, and so the later is a valid bound for the logarithm of the number of maps M which possess a pole-reflective automorphism.

Finally, by Fact 3 for α rotative, it has just two invariants denoted by I_1 and I_2. Let S be a shortest cell-path in M which begins at I_1 and ends at I_2 and contains no other vertices of M than, perhaps, I_1 and I_2.

Let $S' = \alpha^i(S)$ be the image of S under the i-th power of α where i is less than the order of α. Suppose S and S' both contain the same cell, say W, other than I_1 and I_2. Then, $\alpha^i(W)$ is in S'. Since α^i is not the identity and is rotative as well, it fixes just I_1 and I_2 by Fact 3 and hence $\alpha^i(W) \neq W$. Thus S and S' are internally disjoint.

Let i be chosen such that the number of cells adjacent to I_1 between the one in S and the one in $S' = \alpha^i(S)$ is minimized. Let C^+ be a simple closed curve intersecting each of cells in S and S' but no other cells of M, and intersecting each edge in at most one point. Since S and S' are internally-disjoint, no nonterminal cell in S can even be incident with a nonterminal cell in S'.

Choose R to be the domain of C^+ containing the minimal number of edges of M with the two ends of each. Then by the choice of i and R, each power of α which is not the identity must map each cell of M in R to one outside R. In fact, the part of M in the closure of R, together with the order of α, determines the type of M if the two cells

I_1 and I_2, or the parts of them which are inside the closure of R are known.

If both I_1 and I_2 are faces or edges, $C = C^+$ will then be a cell-cycle without vertex. But, if I_1, say, is a vertex, we form C from C^+ by rooting it around the faces and edges of M within R and incident with I_1 in order to eliminate I_1 from C, and doing the same for I_2 if it is a vertex (I_1 and I_2 can not both be incident with the same face).

Now C is a cell-cycle containing no vertex of M. If the structure of the part of M in the domain R' of C which lies within R is given and known that (i) the order of α, (ii) which of any of I_1 and I_2 is a vertex, (iii) if I_1 or I_2 is a vertex, precisely which set of consecutive edges of M in R' intersecting C is the set of edges incident with it, and (iv) if I_1 and I_2 is a face or an edge, precisely which face or edge intersecting C it is, then the structure of the closure of R can be reconstructed, and hence the type of M. The number of possibilities for the combination of (i), (ii), (iii) and (iv) is at most m^σ.

Further, if for the constructions of $N_2(C)$ and $N_3(C)$, R_1 is chosen to be R', then the type of $N_2(C)$ or $N_3(C)$ determines the structure of the part of M within R' if the face of $N_2(C)$ containing R_2 and the vertex v of $N_3(C)$ lying in R_2 are given. Hence the number of distinct type of M is at most m^σ times the number of distinct possibilities for the dual map $N_2(C)^*$ of $N_2(C)$, and similar bound applies in terms of $N_3(C)$.

In the case $C = C^+$, *i.e.*, neither I_1 nor I_2 is a vertex, $N_2(C)^*$ has $\frac{m}{j} + \frac{l}{2}$ edges where j is the order of α and l is the valency of the vertex v of $N_2(C)$ and of the corresponding face to $N_2(C)^*$. This is at most $\frac{m+l}{2}$. Also the number of edges in $N_3(C)$ is

$$\frac{n}{j} + \frac{3l}{2} \le \frac{m+3l}{2}.$$

It follows as in the discussion of the pole-reflective case that the logarithm of the number of distinct types of M in this case at hand exceeds (12.4.17) by no more than $\sigma \log m$ if $\frac{2l}{m+l} \ge 0.45$ and also exceeds (12.4.18) by no more than $\sigma \log m$ as well otherwise.

On the other hand, if $C \ne C^+$ then the numbers of edges in $N_2(C)^*$ and $N_3(C)$ are strictly less than $\frac{n}{j} + \frac{l}{2}$ and $\frac{n}{j} + \frac{3l}{2}$, respectively, and the conclusion of the preceding paragraphs again applies. The argument in the pole-reflective case gives the bound $\exp(1.318m + \sigma)$ for the number of possibilities of M in case α is rotative.

Theorem 12.4.1 *Almost all 3-connected planar maps are with asymmetry when the size of them are sufficiently large.*

Proof From what has been discussed above, we see that the number of 3-connected planar maps of size m and of a symmetry of minimal or prime order is at most $\exp(1.384m)$ for m sufficiently large. Each such map can be rooted in less than $4m$ distinct ways while each of the map without nontrivial automorphism gives rise to $4m$ distinct rooted maps. Thus

$$4mN_m = R_m + o(\exp(1.384m)) \qquad (12.4.19)$$

where N_m and R_m are respectively the numbers of nonrooted and rooted 3-connected planar maps of size m for $m \geq 6$.

By Lemma 12.4.3, we have

$$4mN_m = R_m(1 + o(0.998)^n). \qquad (12.4.20)$$

This implies the theorem. □

§12.5 Asymptotics via equations

When contour integration techniques are applied to map enumeration, one starts from the functional equation defining the generating, or in other words, enumerating function, or in short, enufunction. However, it very rarely happens that one can solve the functional equation to express the enufunction in terms of elementary functions. One may often proceed as follows. Determine the singularities of the enufunction on its circle of convergence. Usually there is only one, or at least after a simple transformation there is only one. Next, one determines an asymptotic expansion for enufunction in a neighborhood of these singularities. Since the number of maps being enumerated is infinite in the interesting classes, the radius of convergence will be finite. Since the number of rooted maps of size m is

$$\frac{2(2m)!3^3}{m!(m+2)!} = o(12^m), \qquad (12.5.1)$$

the radius of convergence is positive and the analytic methods can be applied. Further, since the functional equations are algebraic, the enufunctions have algebraic singularities. It is fortunate that so often one of the enufunctions has coefficients with a simple expression in terms of a few binomial coefficients. Such a function is a hypergeometric one and the branch of interest has only one singularity in the complex plane.

Suppose for now the enufunction has one variable. Since the coefficients are all non-negative, it follows that if R denotes the radius of convergence of the enufunction then R will be a singularity. Furthermore, the enufunction will be algebraic near R. This implies that around R the enufunction will have an expansion of the form as

$$\sum_{n\geq 0} f_n z^n = h(z) + u^{-k} g(u) \tag{12.5.2}$$

where $h(z)$ is analytic at $z = R$, $g(u)$ is analytic at $u = 0$, $g(0) = a_0 \neq 0$,

$$u = \left(1 - \frac{z}{R}\right)^{\frac{1}{j}},\ j > 0,\ \text{and } a = \frac{k}{j}$$

is not $0, -1, -2, \cdots$. In such a case, Darboux's method applies and gives

$$f_n \sim \frac{a_0 n^{a-1}}{\Gamma(a)} R^{-n} \tag{12.5.3}$$

as $n \to \infty$. Thus, it is only necessary to locate the positive singularity nearest the origin and the behavior of the function near this singularity to obtain the asymptotics.

An example illustrating some of these ideas is provided here. Consider the equation

$$H(x) = h\Big(xH^3(x)\Big) \tag{12.5.4}$$

where $x = z^2$ and $H(x)$ is known as

$$H(x) = \sum_{n\geq 0} \frac{(3n)!2^n}{(n+1)!(2n+1)!} x^n \tag{12.5.5}$$

for deriving the asymptotic of $h(y)$ with

$$y = xH^3(x). \tag{12.5.6}$$

Let R be the radius of convergence of H. Since H has non-negative coefficients,

$$\frac{\mathrm{d}y}{\mathrm{d}x} > 0, \; 0 < x < R,$$

and so x is an analytic function of y for $0 < y < RH^3(R)$. We now show that $y = RH^3(R)$ is a singularity of h, from which $S = RH^3(R)$ follows.

By (12.5.5) and (1.6.7),

$$\frac{(3n)!2^n}{(n+1)!(2n+1)!} \sim \frac{1}{4}\sqrt{\frac{3}{\pi}n^5}\left(\frac{27}{2}\right)^n. \tag{12.5.7}$$

Thus $R = \frac{2}{27}$. Note that $H(R)$ and

$$H'(R) = \frac{\mathrm{d}H}{\mathrm{d}x}\Big|_{x=R}$$

exist and are positive but

$$H''(x) = \frac{\mathrm{d}^2H}{\mathrm{d}x^2} \to \infty$$

as $x \to R^-$. From (12.5.4),

$$h''(xH^3(x)) = \frac{H'(x)H^3(x) - 6A(x)}{B^3(x)} \tag{12.5.8}$$

where

$$h''(y) = \frac{\mathrm{d}^2h(y)}{\mathrm{d}y^2},$$

$$A(x) = \big(H'(x)\big)^2 H(x)\big(H(x) + H'(x)\big),$$

$$B(x) = H^3(x) + 3xH^2(x)H'(x).$$

Thus $h''(y) \to \infty$ as $y \to (RH^3(R))^-$ and so $S = RH^3(R)$.

Looking for other singularities of $h(y)$ on $|y| = S$, by (12.5.6) they are due to singularities of $H(x)$ and the values of y near which $y = xH^3(x)$ can not be uniquely invertible.

On the singularities of $H(x)$, we have

$$\frac{(-\frac{1}{3})_{n+1}(-\frac{2}{3})_{n+1}z^n}{(\frac{1}{2})_{n+1}} = \frac{4z}{9}\frac{(3n)!}{(2n+1)!}\left(\frac{2z}{27}\right)^n \tag{12.5.9}$$

where $(a)_n = a(a+1)\cdots(a+n-1)$, the *rising factorial.* Thus (12.5.5) can be rewritten using a hypergeometric function as shown in §1.7 as

$$H(Rz) = \frac{9}{4z}\Big(F\Big(-\frac{1}{3},\frac{2}{3};\frac{1}{2};z\Big)-1\Big). \tag{12.5.10}$$

The hypergeometric function has only the singularity $z = 1$. This leads to $y = S$.

Now consider inverting $y = xH^3(x)$ on $|y| = S$ with $y \neq S$. Thus, $x \neq R$ and so H is defined. By the implicit function theorem, any singularities must be owed to $\frac{dy}{dx} = 0$. We use $h'(S) \neq \infty$. Since h has non-negative coefficients, no singularity on $|y| = S$ can dominate $y = S$ and so $h'(y) \neq \infty$ on $|y| = S$. By (12.5.4),

$$h'(y) = \frac{dH}{dx}\Big/\frac{dy}{dx}. \tag{12.5.11}$$

At a singularity $\frac{dy}{dx} = 0$ and so $H' = 0$. Since

$$\frac{dy}{dx} = H^3 + 3xH'H^2,$$

we have $H = 0$. If $y = \theta \neq S$ is a singularity with $|\theta| = S$ and ρ is the corresponding value of x, we have shown that $h = (x-\rho)^k g(x)$ where $k \geq 1$, $g(\rho) \neq 0$, and g is analytic near ρ. But then

$$\begin{aligned} h'(y) &= \frac{dH}{dx}\Big/\frac{d(xH^3(x))}{dx} \\ &= \frac{(x-\rho)^{k-1}}{(x-\rho)^{3k-1}} f(x) \end{aligned}$$

where $f(\rho) \neq 0$. This gives $h'(\rho) = \infty$, a contradiction. Hence no such singularity θ exists.

In order to apply Darboux's Theorem, we need to know the value of S and the behavior of $h(y)$ near S. This will come from (12.5.8) and properties of H deduced from (12.5.10). Since

$$\begin{aligned} F\Big(-\frac{1}{3},-\frac{2}{3};\frac{1}{2};z\Big) &= \frac{3}{2}F\Big(-\frac{1}{3},-\frac{2}{3};-\frac{1}{2};1-z\Big) \\ &+\frac{4}{27}\sqrt{3}(1-z)^{\frac{3}{2}}F\Big(\frac{5}{6},\frac{7}{6};\frac{5}{2};1-z\Big), \end{aligned}$$

this has a Taylor series expansion about $z = 1$ in power of $(1-z)^{-\frac{1}{2}}$ beginning

$$\frac{3}{2} - \frac{2}{3}(1-z) + \frac{4\sqrt{3}}{27}(1-z)^{\frac{3}{2}}. \tag{12.5.12}$$

Let $x = R(1-\xi)$ and $y = S(1-\eta)$. Since $y = xH^3(x)$ is invertible near $y = S$, ξ is analytic in η near 0 and by the above, (12.5.4) and (12.5.10), h has a series expansion in powers of $\eta^{\frac{1}{2}}$. From (12.5.8) with $T = H(R) + 3RH'(R)$ and $H = H(R)$,

$$h''(y) \sim H''(x)/(HT)^3 \tag{12.5.13}$$

as $y \to S$. From $y = xH^3(x)$ we have $R\xi \sim \big(S\eta/H^2(R)\big)/T$ and so

$$\xi \sim \frac{\eta H}{T}. \tag{12.5.14}$$

From (12.5.10) and (12.5.12),

$$\begin{aligned} H''(x) &\sim \frac{9}{4}\frac{4\sqrt{3}}{27}\frac{\mathrm{d}^2}{\mathrm{d}x^2}\xi^{\frac{3}{2}} \\ &= \frac{\sqrt{3}}{3R^2}\frac{\mathrm{d}^2\xi^{\frac{3}{2}}}{\mathrm{d}x^2} \\ &= \frac{\sqrt{3}}{4R^2}\xi^{-\frac{1}{2}}. \end{aligned} \tag{12.5.15}$$

Combining (12.5.13–15), we obtain

$$\begin{aligned} h''(y) &\sim \frac{\sqrt{3}\eta^{-\frac{1}{2}}}{4R^2H^{\frac{7}{2}}T^{\frac{5}{2}}} \\ &= \frac{\sqrt{3}S^2}{3R^2H^{\frac{7}{2}}T^{\frac{5}{2}}}\frac{\mathrm{d}^2\eta^{\frac{3}{2}}}{\mathrm{d}y^2}. \end{aligned}$$

Thus,

$$h(y) = a + by + \frac{1}{\sqrt{3}}\left(\frac{H}{T}\right)^{\frac{5}{2}}\eta^{\frac{3}{2}} + \cdots$$

where '$\cdots$' is a power series in $\eta^{\frac{1}{2}}$ beginning with η^2. By Darboux's Theorem

$$h_n \sim \frac{1}{\sqrt{3}}\left(\frac{H}{T}\right)^{\frac{5}{2}}\frac{S^{-n}}{\Gamma(-\frac{3}{2})n^{\frac{5}{2}}}.$$

From (12.5.10) and (12.5.12), $H(R) = \frac{9}{8}$ and $H'(R) = \frac{81}{16}$. Thus

$$S = RH^3(R) = \frac{27}{256}$$

and

$$T = H(R) + 3RH'(R) = \frac{9}{4}.$$

Since $\Gamma(-\frac{3}{2}) = \frac{4}{3}\sqrt{\pi}$,

$$h_n \sim \frac{3}{16n^2\sqrt{6\pi n}}\left(\frac{256}{27}\right)^n. \tag{12.5.16}$$

If an enufunction is with two variable or more, the procedure of estimating the coefficients will be in general much more complicate. We are not allowed to occupy more space here.

Readers who are interested in this topic are referred to read the survey articles of Bender [Ben1] and Bender and Richmond [BeR1].

§12.6 Notes

12.6.1 The asymptotic behavior to which we refer as the subject of random graphs has made popular, of the fraction of those graphs which are subcategory. In spite of what have been mentioned [YaL1] in §12.1, one might ask how many graphs or maps are Hamiltonian, having perfect matching or some other properties in a class of maps or graphs given. On the Hamiltonicity, we have known that almost all cubic planar maps are not Hamiltonian [RRW1]. However, almost all bipartite cubic graphs are indeed Hamiltonian [RoW1].

12.6.2 The problem on average behavior often appears in number theory. However, it has been paid attention to probably since the seventies when Tutte intended to find the average number of four colorings on a planar map [Tut14–19], [Tut21–24], [Tu27–29], [Tut31–32], [Tut36–37]. Because of the difficulty the so called four color average problem has not yet been solved up to now.

12.6.3 The average behavior on tree-rooted maps in §12.2 is based on Mullin's [Mul1–3]. And the average number of Hamiltonian circuits per triangular map appearing in §12.3 can be seen in [Mul5].

12.6.4 Suppose $N_h(I)$ is the number of nonisomorphic maps with a set $I = \{I_1, I_2, \cdots\}$ of invariants and the order h of automorphism group in a class $\mathcal{N}$ of maps. Because there are 4ϵ distinct ways of rooting on a map which is asymmetric, *i.e.*, with the only trivial automorphism, where ϵ is the size of the map. Then, the number of corresponding rooted maps is

$$N_0(I) = 4\epsilon \sum_{h=1}^{4\epsilon} \frac{1}{h} N_h(I) \qquad (12.6.1)$$

and $N_0(I) = 4\epsilon N(I)$,

$$N(I) = \sum_{h=1}^{4\epsilon} N_h(I),$$

iff $h = 1$, *i.e.*, all the maps of size ϵ with an invariant set I in $\mathcal{N}$ are asymmetric.

The result in §12.4 tells us that in probabilistic point of view we are allowed to consider maps (3-connected planar ones) without symmetry. The discussions in §12.4 are based on the articles of Tutte [Tut34] and Bender and Wormald [BeW1]. More recent results can be seen in [RiW1].

12.6.5 The formula (12.5.16) gives the asymptotic estimation of the number of rooted 3-connected triangulations with $2n$ faces. It can also be obtained from finding a close explicit expression of h and then by using (1.7.6) [Tut41].

12.6.6 More about enumerations and asymptotics for maps on general surfaces can be seen in [BeC3–4], [BCR1–2], [BCr2], [BRW1–2], [BGW1], [BeW4], [Gao1–4], [GaR1], [WaL1–3], *etc.*

12.6.7 On nonrooted kinds of map enumerations and asymptotics, refer to [Lis1–2], [LiW1–2], [PoR1], [Wor1], *etc.*

12.6.8 Many other kinds of asymptotic results for map enumeration have been found on the sphere [BeC1–2], [BGM1], [BRW2], [Bro1–2], [Tut3–5], [Wor2] etc., on the projective plane [BCr1], [Gao3], *etc.*

Bibliography

Abramovitz, M. and Stegun, I. A.

[AbS1] Handbook of Mathematical Functions, National Bureau of Standards, 1965.

Andrews, G.E.

[And1] The Theory of Partitions, Encycl. Math. **2**, Addison–Wesley, 1976.

Arques, D.

[Arq1] Relations fonctionelles et denombrement des cartes pointees sur le tore, *J. Comb. Theory* **B43** (1987), 253–274.

[Arq2] Hypercartes pointees sur le tore: decompositions et denombrements, *J. Comb. Theory* **43** (1987), 275–286.

[Aqr3] Relations fonctionnelles et denombrement des hypercartes planaires pointees, Comb. Enum. Montreal, 1985, Lect. Notes in Math., **1234**, Springer, 1980, 5–26.

Baldwin, S.

[Bal1] Toward a theory of forcing on maps of trees. Thirty years after Sharkovskii's theorem: new perspectives (Murcia, 1994), World Sci. Ser. Nonlinear Sci. Ser. B Spec. Theme Issues Proc., 8, World Sci. Publishing, River Edge, NJ, 1995, 45–56.

Baxter, R. J.

[Bax1] Exactly Solved Models in Statistical Mechanics, Acad. Press, 1982.

Bender, E.A.

[Ben1] Asymptotic methods in enumeration, *SIAM Rev.* **16** (1974), 485–515.

[Ben2] Some unsolved problems in map enumeration, *Bull. Inst. Combin. Appl.* **3** (1991), 51–56.

Bender, E.A. and E.R. Canfield

[BeC1] Enumeration of degree restricted maps on the sphere, *DIMACS Ser. in Disc. Math. & Theor. Comp. Sci.*, **9** (1993), 13– 16.

[BeC2] The number of degree-restricted rooted maps on the sphere, *SIAM J. Discrete Math.* **7** (1994), 9–15.

[BeC3] The number of rooted maps on an orientable surface, *J. Combin. Theory* **B53** (1991), 293–299.

[BeC4] The asymptotic number of rooted maps on a surface, *J. Combin. Theory* **A43** (1986), 244–257.

Bender, E.A., E.R. Canfield and L.B. Richmond

[BCR1] The asymptotic number of rooted maps on a surface I, *J. Comb. Theory*, **A43** (1986), 244–257.

[BCR2] The asymptotic number of rooted maps on a surface II, *J. Comb. Theory*, **A63** (1993), 318– 329.

Bender, E. A., R.E. Canfield and R.W. Robinson

[BCr1] The enumeration of maps on the torus and the projective plane, *Canad. Math. Bull.* **31** (1988), 257–271.

[BCr2] The asymptotic number of tree-rooted maps on a surface, *J. Combin. Theory* **A48** (1988), 156–164.

Bender, E.A., Z.C. Gao, W.D. McCuaig and L.B. Richmond

[BGM1] Submaps of maps II: cyclically k- connected planar cubic maps, *J. Combin. Theory* **B55** (1992), 118–124.

Bender, E.A., Z.C. Gao and L.B. Richmond

[BGR1] Submaps of maps. I. General 0-1 laws. *J. Combin. Theory Ser.* **B55** (1992), 104–117.

[BGR2] Almost all rooted maps have large representativity, *J. Graph Theory* **18** (1994), 545–555.

Bender, E.A., Z.C. Gao, L.B. Richmond and N.C. Wormald

[BRW1] Asymptotic properties of rooted 3-connected maps on surfaces, *J. Austral. Math. Soc.* **A60** (1996), 31–41.

[BRW2] Submaps of maps. IV: Degree restricted nonplanar maps, *Ars Combin.* **35** (1993), 135–142.

Bender, E.A. and L.B. Richmond

[BeR1] A survey of asymptotic behavior of maps, *J.Comb. Theory* **B40** (1986), 297–329.

[BeR2] Submaps of maps III, k-connected nonplanar maps, *J. Combin. Theory* **B55** (1992), 125–132.

Bender, E.A. and N.C. Wormald

[BeW1] Almost all convex polyhedra are asymmetric, *Canad. J. Math.* **37** (1985), 854–871.

[BeW2] The number of rooted convex polyhedra, *Canad. Math. Bull.* **31** (1986), 99–102.

[BeW3] The number of loopless planar maps, *Discrete Math.* **54** (1985), 526–545.

[BeW4] The asymptotic number of rooted nonseparable maps on a surface, *J. Combin. Theory* **A49** (1988), 370–380.

Berman, G. and W.T. Tutte

[BerT1] The golden root of a chromatic polynomial, *J.Comb. Theory* **6** (1969), 301–302.

Biggs, N.

[Big1] Algebraic Graph Theory, Cambridge Uni. Press, 1974.

Birkhoff, G.D. and D.C. Lewis

[BiL1] Chromatic polynomials, *Trans. AMS* **60** (1946), 355–451.

Brahana, H.R.

[Bra1] Systems of circuits in two- dimensional manifolds, *Ann. Math.* **32** (1921), 144–168.

Brooks, R.L., C.A.B. Smith, A.H. Stone and W.T. Tutte

[BSST1] The dissection of rectangles into squares, *Duke Math. J.* **7** (1940), 312–140.

Brown, W.G.

[Bro1] Enumeration of quadrangular dissections of the disc, *Canad. J. Math.* **17** (1965), 302–317.

[Bro2] Enumeration of triangulations of the disc, *Proc.London Math. Soc.* **14** (1964), 746–768.

[Bro3] An algebraic technique for solving certain problems in the theory of graphs, in Theory of Graphs: Proc. Colloq. Tihany, Hungary, 1966, Acad. Press, 1968, 57–60.

[Bro4] On the existence of square roots in certain rings of power series, *Math. Ann.* **158** (1965), 82–89.

Brown, W.G. and W.T. Tutte

[BrT1] On the enumeration of rooted nonseparable planar maps, *Canad. J. Math.* **16** (1964), 572–577.

Bryant, R.P. and D. Singerman,

[Bry1] Foundations of the theory of maps on surfaces with boundary, *Quart. J. Math.* Oxford Ser.(2) **36** (1985), 17–41.

Conder, M.

[Con1] Asymmetric combinatorially-regular maps, *J. Algebraic Combin.* **5** (1996), 323–328.

Cori, R.

[Cor1] Un code pour les graphes planaire et ses applications, Asterisque **27**, 1975.

[Cor2] Graphes planaires et systemes de parentheses, Center National de la Rocherche Scientifique, Instiut Blaise Pascal, 1969.

Cori, R. and A. Mach

[CoM1] Maps, hypermaps and their automorphisms: a survey. I, II, III, *Exposition. Math.* **10** (1992), 403–427, 429–447, 449–467.

Cori, R., S. Dulusq and G. Viennot

[CDV1] Shuffle of parenthesis systems and Vaxter permutations, *J. Comb. Theory* **A43** (1986), 1–22.

Cori, R. and J. Richard

[CoR1] Enumeration des graphes planaires a l'aide des series formelles en variables non commutatives, *Discrete Math.* **2** (1972), 115–162.

Cori, R. and B. Vauquelin

[CoV1] Planar maps are well labeled trees, *Canad. J. Math.* **33** (1981), 1023–1042.

Dong, F. M. and Y. P. Liu

[DoL1] Counting rooted near- triangulations on the sphere, Preprint, Institute of Applied Math. Academia Sinica, 1990. Also in *Discrete Math.* **123** (1993), 35–45.

Dong, F. M. and J. Y. Yan

[DoY1] On the enumeration of rooted outerplanar maps (Chinese), *Acta Math. Sinica* **32** (1989), 501–511.

Drmota, M.

[Drm1] Asymptotic distributions and a multivariate Darboux method in enumeration problems, *J. Comb. Theory*, **A67** (1994), 169–184.

Edmonds, J.R.

[Edm1] A combinatorial representation for polyhedral surfaces, *Notices Amer. Math. Soc.* **7** (1960), 646.

Ferri, M

[Fer1] Crystallizations of 2– fold branched coverings of S^3, *Proc. AMS* **73** (1979), 271–276.

Fiderico, P. J.

[Fid1] The number of polyhedra, *Philips Res. Report* **30** (1975), 220–231.

Gagliardi, C.

[Gag1] Surface maps and n-dimensional manifolds, Fourth Conference on Topology (Italian) (Sorrento, 1988). *Rend. Circ. Mat. Palermo* (2) Suppl. No. 24

(1990), 97–126.

Gao, Z. C.

[Gao1] The asymptotic number of rooted triangular maps on a surface, *J. Combin. Theory* **B52** (1991), 236–249.

[Gao2] The asymptotic number of rooted 2- connected triangular maps on a surface, *J. Combin. Theory* **B54** (1992), 102–112.

[Gao3] The number of rooted 2- connected triangular maps in the projective plane, *J. Comb. Theory* **B53** (1991), 130–142.

[Gao4] A pattern for the asymptotic number of rooted maps on surfaces, *J. Combin. Theory Ser.* **A64** (1993), 246–264.

[Gao5] The number of degree restricted maps on general surfaces, *Discrete Math.* **123** (1993), 47– 63.

Gao, Z.C. and L. B. Richmond

[GaR1] Root vertex valency distributions of rooted maps and rooted triangulations, *European J. Combin.* **15** (1994), 483–490.

Gessel, I. and G. Viennot

[GeV1] Binomial determinants, paths, and hook length formulae, *Advances Math.* **58** (1985), 300–321.

Good, I. J.

[Goo1] Generalizations to several variables of Lagrange's expansion with applications to stochastic processes, *Proc. Cambridge Phil. Soc.* **56** (1960), 367–380.

Gould, H. W.

[Gou1] Coefficient identities for powers of Taylor and Dirichlet series, *Amer. Math. Monthly* **81** (1974), 3–14.

Goulden, I. P. and D. M. Jackson

[GoJ1] Combinatorial Enumerations, Wiley, 1983.

Hall, D. W. and D. C. Lewis

[HaL1] Coloring six-rings, *Trans. AMS* **64** (1984), 184–191.

Harary, F. and W. T. Tutte

[HaT1] The number of plane trees with given partition, *Mathematika* **11** (1964), 99–101.

[HaT2] On the order of the group of a planar map, *J. Comb. Theory* **1** (1966), 394–395.

Harary, F., G. Prins and W. T. Tutte

[HPT1] The number of plane trees, *Akad. van W., Proc.* **67** (1964), 319–329.

Heath, L. S.

[Heal] Graph embeddings and simplicial maps, ACM Symposium on Parallel Algorithms and Architectures (Velen, 1993), *Theory Comput. Syst.* **30** (1997), 51–65.

Heffter, L.

[Hef1] Ueber das Problem der Nachbargebiete, *Math. Ann.*, **38** (1891), 477–508.

Hering, F., R. C. Read and G. C. Shephard

[HeRs1] The enumeration of stack polytopes and simplicial clusters, *Discrete Math.* **40** (1982), 303–217.

Jackson, D. M.

[Jac1] Some combinatorial problems associated with products of conjugate classes of the symmetric group, *J. Comb. Theory* **A49** (1988), 363–369.

[Jac2] On an integral representation for the genus series for 2- cell embeddings, *Trans. Amer. Math. Soc.*, **334** (1994), 755–772.

[Jac3] Algebraic and analytic approaches for the genus series for 2- cell embeddings on orientable and non- orientable surfaces, in *Formal Power Series and Algebraic Combinatorics*, DIMACS **24**, 1994, 115–132.

[Jac4] The genus series for maps. *J. Pure Appl. Algebra* **105** (1995), 293–297.

Jackson, D. M. and R. C. Read

[JaR1] A note on permutation without runs of given length, *Aequationes Math.* **17** (1978), 336–343.

Jackson, D. M. and T. I. Visentin

[JaV1] A character theoretic approach to embeddings of rooted maps in an orientable surface of given genus, Comb. Optim. CORR88- 45, Uni. Waterloo, 1988. Also see *Trans. Amer. Math. Soc.*, **322** (1990), 343–363.

[JaV2] Character theory and rooted maps in an orientable surface of given genus, *Trans. Amer. Math. Soc.*, **322** (1990), 365–376.

[JaV3] A formulation for the genus series for regular maps, *J. Combin. Theory* **A74** (1996), 14–32.

Jacques, A

[Jaq1] Constellations et graphes topologiques, in Combinatorial Theory and its Applications II, Budapest, 1970.

[Jaq2] Sur legenre d'unc paire de substitutions, *C. R. Acad. Sci. Paris* **267** (1968), 625–627.

James, G. and A. Kerber

[JamK1] The Representation Theory of the Symmetric Group, Encycl. Math. **16**, Addison–Wesley, 1981.

Jones, V. F. R.

[Jon1] A polynomial invariant for links via Von Neumann algebra, *Bull. AMS* **12** (1985), 103–112.

Kauffman, L. H.

[Kau1] New invariants in knot theory, *Amer. Math. Monthly* **95** (1988), 195–243.

Kozyrev, V. P. and S. V. Maisrii

[KoM1] Enumeration and generation of plane rooted trees with given parameters (Russian), Proc. All-Union Sem. Disc. Math. Appls. (Russian), Moskov, Gos. Uni., Mekl.-Mat.Fak., 1986, 92–99.

Levine, H. I.

[Lev1] Homotopic curves on surfaces, *Proc. AMS* **14** (1963), 986–990.

Lienhardt, P.

[Lie1] n-dimensional generalized combinatorial maps and cellular quasi- manifolds, *Internat. J. Comput. Geom. Appl.* **4** (1994), 275–324.

Lins, S.

[Lin1] Graph–encoded maps, *J. Comb. Theory* **B32** (1982), 171–181.

Liskovets, V. A.

[Lis1] Enumeration of nonisomorphic planar maps I, Problems in Group Theory and Homological Algebra, 1981, 103–115.

[Lis2] Enumeration of nonisomorphic planar maps II, Geometric Methods in Problems in Analysis and Algebra, 1981, 106–117.

[Lis3] Enumeration of nonisomorphic planar maps, *Sel. Math. Sov.* **4**:4(1985), 303–323.

[Lis4] A census of nonisomorphic planar maps, *Colloq. Math. Soc. J. Bolyai* **25** *Algeb. Methods in Graph Theory*, 1978, 479–494.

[Lis5] A reductive technique for enumerating non-isomorphic planar maps, *Discrete Math.* **156** (1996), 197–217.

[Lis6] Some asymptotical estimates for planar Eulerian maps, *Combin. Probab. Comput.* **5** (1996), 131–138.

Liskovets, V. A. and T. T. S. Walsh

[LiW1] The enumeration of non- isomorphic 2- connected planar maps, *Canad. J. Math.* **35** (1983), 417–435.

[LiW2] Ten steps to counting planar graphs, *Cong. Numer.* **60** (1987), 269–277.

Little, C. H. C.

[Lit1] Cubic combinatorial maps, *J. Comb. Theory* **B44** (1988), 44–63.

Liu, Y. P.

[Liu1] Extending Kotzig's theorem into the non–orientable cases, Comb. Optim. CORR82- 36, Uni. Waterloo, 1982.

[Liu2] Planar graph theory and the four color problem (I)– planarity and duality (in Chinese with English abstract) *J. Math. Res. Expos.* **3** (1983), 3: 123–136.

[Liu3] K- valent maps on the surfaces, Comb. Optim. CORR82- 35, Uni. Waterloo, 1982; Also in *Acta Math. Appl. Sinica* Eng. Series **1** (1984), 57–62.

[Liu4] On the number of rooted c- nets, Comb. Optim. CORR83- 35, Uni. Waterloo, 1983; Also in *J. Comb. Theory* **B36** (1984), 118–123.

[Liu5] A note on the number of rooted planar triangulations (Chinese with English abstract), *Acta Math. Appl. Sinica* **11**:2 (1986), 146–148.

[Liu6] The maximum non- orientable genus of a graph (Chinese with English abstract), *Scientia Sinica* Special Issue on Math. **I** (1979), 191–201.

[Liu7] The maximum orientable genus of a graph, *Scientia Sinica* Special Issue on Math. **II** (1979), 41–55.

[Liu8] The maximum orientable genus of some kinds of graphs (Chinese), *Acta Math. Sinica* **24** (1981), 817–132.

[Liu9] A note on Edmonds' surface duality theorem of graphs (Chinese), *Chinese J. Operations Research* **2**:1 (1983), 62–63.

[Liu10] Map color theorem and surface embeddings of graphs I (Chinese), *Math. Theory Prac.* No.1 (1981), 65–78.

[Liu11] Map color theorem and surface embeddings of graphs V (Chinese), *Math. Theory Prac.* No.1 (1982), 34–44.

[Liu12] The orientable genus of $K_n - K_3$, *KEXUE TONGBAO* (Chinese Ed.) **25** (1980), 959; Or see *KEXUE TONGBAO*(Sci. Bull., Eng. Ed) **26** (1981), 188.

[Liu13] Enumeration of rooted separable planar maps, Comb. Optim. CORR82- 4, Uni. Waterloo, 1982; Also in *Utilitas Math.* **25** (1984), 77–94.

[Liu14] Planar graph theory and the four color problem (II)– the five color theorem and formations of the four color problem (in Chinese with English abstract), *J. Math. Res. Expos.* **4** (1984), 1: 121–136.

[Liu15] Enumerating rooted loopless planar maps, Comb. Optim. CORR 83- 4, Uni. Waterloo, 1983; Also in *Acta Math. Appl. Sinica* Eng. Series **2** (1985), 14–26.

[Liu16] Enumerating rooted simple planar maps, Comb. Optim. CORR 83- 9, Uni. Waterloo, 1983; Also in *Acta Math. Appl. Sinica* Eng. Series **2** (1985), 101–111.

[Liu17] Counting rooted planar maps, Comb. Optim. CORR83 26, Uni. Waterloo, 1983; Also in *J. Math. Res. Expos* **4**:3(1984), 37–46.

[Liu18] Planar graph theory and the four color problem (III)– theoretical results derived from the four color problem (in Chinese with English abstract), *J. Math. Res. Expos.* **5** (1985), 1: 125–144.

[Liu19] Enumeration of rooted vertex non- separable planar maps, Comb. Optim. CORR83- 14, Uni. Waterloo, 1983; Also in *Chinese Ann. Math.* **9B** (1988), 390–403.

[Liu20] A functional equation for enumerating non- separable planar maps with vertex partition, *KEXUE TONGBAO* (Chinese Ed.) **10** (1985), 646–649; Or see

KEXUE TONGBAO (Sci. Bull., Eng. Ed.) **31** (1986), 73–77.

[Liu21] Some enumerating problems of maps with vertex partition, *KEXUE TONGBAO* (Chinese Ed.) **30** (1985), 1521–1525; Or see *KEXUE TONGBAO* (Sci. Bull., Eng. ED.) **31** (1986), 1009–1014.

[Liu22] Planar graph theory and the four color problem (IV)– three colorings and the geometric characterization of the four color problem (in Chinese with English abstract), *J. Math. Res. Expos.* **5** (1985), 3; 123–136.

[Liu23] A functional equation for enumerating nonseparable Eulerian planar maps with vertex partition, *KEXUE TONGBAO* (Chinese Ed.) **31** (1986), 81–84. Or see *KEXUE TONGBAO* (Sci. Bull., English Ed.) **31** (1986), 941–945.

[Liu24] Combinatorial map theory, A Plenary Talk to the 2-nd National Conf. Comb. Graph Theory, Guangzhou, 1985.

[Liu25] Enumerationof rooted outerplanar maps withvertex partition,*KEXUE TONGBAO* (Chinese Ed.) **31** (1986), 1045–1048; Or see *KEXUE TONGBAO* (Sci. Bull., Eng. Ed.) **32** (1987), 295–299.

[Liu26] Planar graph theory and the four color problem (V)– an approach for the four color problem by computers (in Chinese with English abstract), *J. Math. Res. Expos.* **6** (1986), 2: 175–188.

[Liu27] On the enumeration of planar maps with vertex partition (Chinese), A Plenary Talk to the 4-th National Conf. Graph Theory, Jinan, 1985.

[Liu28] On the functional equations for enumerating planar maps, A Report on the National Conf. for the 50-th Ann. of the Founding of the Chinese Math. Soc., Shanghai, 1985.

[Liu29] On the enumeration of simple planar maps (in Chinese), *Acta Math. Sinica* **31** (1988), 279–282.

[Liu30] Enumeration of rooted nonseparable outerplanar maps,*KEXUE TONGBAO* (Chinese Ed.) **33** (1988), 473; Or see *Chinese Sci. Bull.*(English Ed.) **33** (1988) 1140–1141.

[Liu31] The enumeration of simple planar maps, RUTCOR Research Report RRR37-87, Rutgers Uni., 1987; Also in *Utilitas Math.* **34** (1988), 97–104.

[Liu32] On face partition of rooted outerplanar maps, *KEXUE TONGBAO* (Chinese Ed.) **33** (1988), 1441–1444; Or see *Chinese Sci. Bull.* (Eng. Ed) **34** (1989), 365–369.

[Liu33] On the vertex partition equation of rooted loopless planar maps, RUTCOR Research Report, RRR17- 89, Rutgers Uni., 1987; Also in *Acta Math. Scientia* **10** (1990), 167–172.

[Liu34] An enumerating equation of simple planar maps with face partition, RUTCOR Research Report, RRR36- 87, Rutgers Uni., 1987; Main part also in *Acta Math. Appl. Sinica* **12**(1989), 292–295.

[Liu35] On the enumerating equations of planar maps (Chinese with English abstract), *Chinese Advances Math.* **18** (1989), 446–460.

[Liu36] Enumeration of rooted non- separable outerplanar maps, *Acta Math. Appl. Sinica* Eng. Series **5** (1989), 169–175.

[Liu37] An enumerating equation of rooted loopless planar maps with vertex partition (Chinese with English abstract), *Acta Math. Appl. Sinica* **12** (1989), 210–

214.

[Liu38] On the number of Eulerian planar maps, Research Report Series A, No. 24, Department of Statistics, Uni. Rome "La Sapienza", 1989. Also in *Acta Math. Scientia* **12** (1992), 418–423.

[Liu39] On an enumeration equation of simple planar maps with face partition (Chinese with English abstract), *Acta Math. Appl. Sinica* **12** (1989), 292–295.

[Liu40] Chromatic sum equations for rooted planar maps, *Cong. Numer.* **45** (1984), 275–280.

[Liu41] Enumeration of rooted bipartite planar maps (Chinese), *Acta Math. Scientia* **9** (1989), 21–28.

[Liu42] On chromatic sum equations for rooted cubic planar maps, *KEXUE TONGBAO* (Chinese Ed) **31** (1986), 1285–1289;, Or see *Chinese Sci. Bull.* (English Ed) **32** (1987), 1230–1235.

[Liu43] On chromatic sum equations, An Hourly Talk, 1-st China- USA Conf. Graph Theory Appls., Jinan, 1986.

[Liu44] Chromatic sum equations for rooted cubic planar maps, *Acta Math. Appl. Sinica* Eng. Series **3** (1987), 136–167.

[Liu45] Correction to "Chromatic sum equations for rooted cubic planar maps", *Acta Math. Appl. Sinica* Eng. Series **4** (1986), 95–96.

[Liu46] Chromatic sum equations for rooted nonseparable planar maps (in Chinese with English Abstract), *Comm. Appl. Math. Comput.* **2** (1988), 1–11.

[Liu47] On chromatic equation for rooted outerplanar maps, *KEXUE TONGBAO* (Chinese Ed.) **33** (1988), 1261–1265;, Or see *Chinese Sci. Bull.* (English Ed.) **34** (1989), 812–817.

[Liu48] On chromatic and dichromatic sum equations for planar maps, *Discrete Math.* **84** (1990), 167–179.

[Liu49] Recent progress on chromatic and dichromatic sum equations for planar maps, Research Report Series A No.4, Department of Statistics,Uni. Rome "La Sapienza", 1990. Also in *Seris Formelles et Comb. Algeb.* (Eds. M. Delest, G. Jacob and P. Leroux), 1991, 305–318.

[Liu50] Chromatic enumeration for rooted outerplanar maps, *Chinese Ann. Math.* **11B** (1990), 491–502.

[Liu51] A note on the number of cubic planar maps, *Acta Math. Scientia* **12** (1992), 282–285.

[Liu52] A note on the number of loopless Eulerian planar maps, *J. Math. Res. Expos.* **12** (1992), 165–179.

[Liu53] On functional equations arising from map enumerations, *Discrete Math.* **123** (1993), 93–109.

[Liu54] On the enumeration of rooted planar maps with vertex partition (in Chinese with English abstract), *Comm. Appl. Math. Comput.* **5** (1991), 1–20.

[Liu55] On the vertex partition equation of loopless Eulerian planar maps, *Acta Math. Appl. Sinica* English Series **8** (1992), 45–85.

[Liu56] Dichromatic sum equations for outer planar maps, *Appl. Math.* (A J. Chinese Unis.) **8** (1993), 64–68.

[Liu57] A polyhedral theory on graphs, *Acta Math. Sinica* **10**(1994), 136–142.

[Liu58] Embeddability in Graphs, Kluwer, 1995.

[Liu59] Combinatorial invariants on graphs, *Acta Math. Sinica* **11** (1995), 211–220.

[Liu60] Rectilinear Embeddings: Theory and Methods (in Chinese), Science Press, 1994.

[Liu61] A note on the number of bipartite planar maps, *Acta Math. Scientia* **12** (1992), 85–88.

Macdonald, I.G.

[Macl] Symmetric Functions and Hall Polynomials, Clarendon Press, Oxford, 1979.

MacLane, S.

[MaL1] A structural characterization of planar combinatorial graphs, *Duke Math. J.*, **3** (1937), 416–472.

Meir, A. and J.W. Moon

[MeM1] On the asymptotic method in combinatorics, *J. Comb. Theory* **A51** (1989), 77–89.

Mohar, B.

[Moh1] Convex representations of maps on the torus and other flat surfaces, *Discrete Comput. Geom.* **11** (1994), 83–95.

Mullin, R. C.

[Mul1] On the enumeration of tree– rooted maps, *Canad. J. Math.* **19**(1967), 174–183.

[Mul2] On the average activity of a spanning tree of a rooted map, *J.Comb. Theory* **3**(1967), 103–121.

[Mul3] On the average number of trees in certain maps, *Canad. J. Math.* **18**(1966), 33–41.

[Mul4] On counting rooted triangular maps, *Canad. J. Math.* **7**(1965), 373–382.

[Mul5] The enumeration of Hamiltonian polygons in triangular maps, *Pacific J. Math.* **16** (1966), 139–145.

[Mul6] The enumeration of rooted triangular maps, *Amer. Math. Monthly* **71**(1964), 1007–1010.

Mullin, R. C., E. Nemeth and P.J.Schellenberg

[MNS1] The enumeration of almost cubic maps, in Proc. S-E Conf. Theory Comput. Baton Rouge, 1970, 281–295.

Mullin, R.C. and P.J. Schellenberg

[MuSc1] The enumeration of c- nets via quadrangulations, *J. Comb. Theory* **4** (1968), 259–276.

Mullin, R. C. and R. G. Stanton
[MuSt1] A map–theoretic approach to Davenport–Schinzzel sequences, *Pacific J. Math.* **40**(1972), 167–172.

Nedela, R. and M. Škoviera
[NeS1] Exponents of orientable maps, *Proc. London Math. Soc.* (3) **75** (1997), 1–31.

Polya, G. and R. C. Read
[PoR1] Combinatorial Enumeration of Groups, Graphs, and Chemical Components, Springer, 1987.

Read, R.C.
[Rea1] An introduction to chromatic polynomials, *J. Comb. Theory* **4**(1968), 52–71.

Richmond, L.B., R. W. Robinson and N. C. Wormald
[RRW1] On Hamilton cycles in 3-connected cubic maps, *Ann. Discrete Math.* **27** (1985), 141–150.

Richmond, L.B. and N. C. Wormald
[RiW1] Almost all maps are asymmetric, *J. Comb. Theory* **B63** (1995), 1–7.

Riordan, J.
[Rio1] Combinatorial Identities, Wiley, 1968.

Riskin, A.
[Ris1] On type 2 diminimal maps on the torus, *J. Combin. Inform. System Sci.* **17** (1992), 288–301.

Robertson, N. and P.D. Seymour
[RoS1] Graph minors VII: Disjoint paths on a surface, *J.Comb. Theory* **B45** (1988), 212–254.

Robinson, R.W. and N.C. Wormald
[RoW1] Almost all bipartite cubic maps are Hamiltonian, in Proc. Waterloo Silver Jubilee Conf., Vol.3, 1982.

Rota, G. C.
[Rot1] The number of partitions of a set, *Amer. Math. Monthly* **71** (1964), 498–504.

Serre, J. P.
[Ser1] Linear Representations of Finite Groups, Springer, 1977.

Servatius, B. and H. Servatius
[SeS1] Self-dual maps on the sphere. Algebraic and Topological Methods in Graph Theory (Lake Bled, 1991), *Discrete Math.* **134** (1994), 139–150.

Širň, J. and M. Škoviera and H-J. Voss

[SiS1] Sachs triangulations and regular maps. Algebraic and Topological Methods in Graph Theory (Lake Bled, 1991), *Discrete Math.* **134** (1994), 161–175.

Stanley, R.P.

[Sta1] Enumerative Combinatorics, Vol.1, Wadsworth and Brooks, 1986.

Tutte, W. T.

[Tut1] A ring in graph theory, *Proc. London Math. Soc.* **21**(1946), 98–101.

[Tut2] A contribution to the theory of chromatic polynomials, *Canad. J. Math.* **6**(1954), 80–91.

[Tut3] A Census of planar triangulations, *Canad. J. Math.* **14**(1962), 21–38.

[Tut4] A census of Hamiltonian polygons, *Canad. J. Math. Soc.* **68** (1962), 402–417.

[Tut5] A census of slicings, *Canad. J. Math.* **14**(1962), 708–722.

[Tut6] A new branch of enumerative graph theory, *Bull. Amer. Math. Soc.* **68** (1962), 500–504.

[Tut7] A census of planar maps, *Canad. J.Math.* **15** (1963), 249–271.

[Tut8] The number of planted plane trees with a given partition, *Amer. Math. Monthly* **71** (1964), 272–277.

[Tut9] On the enumeration of almost bicubic rooted maps, Rand Corp. Memorandum RM- 5887- PR, 1969.

[Tut10] Even and odd 4- colourings, in Proof Techniques in Graph Theory (ed. by F. Harary), Acad. Press, 1969, 161–169.

[Tut11] On the enumeration of four–coloured maps, *SIAM J. Appl. Math* **17**(1969), 454–460.

[Tut12] Counting planar maps, *J.Recreational Math.* **1**(1968), 19–27.

[Tut13] On the enumeration of planar maps, *Bull. Amer. Math. Soc.* **74**(1968), 64–74.

[Tut14] On dichromatic polynomials, *J. Comb. Theory* **2** (1967), 301–320.

[Tut15] On the enumeration of two- coloured rooted and weighted plane trees, *Aequationes Math.* **4**(1970), 143–156.

[Tut16] On chromatic polynomials and the golden ratio, *J.Comb. Theory* **9** (1970), 289–296.

[Tut17] The golden ratio in the theory of chromatic polynomials, *Ann. New York Acad. Sci.* **175** (1970), 391–402.

[Tut18] More about chromatic polynomials and the golden ratio, in Combinatorial Structures and their Applications, Gordon and Breach, 1970, 439–453.

[Tut19] Dichromatic sums for rooted planar maps, *Proc. Symp. in Pure Math.* **XIX** (combinatorics) (1971), 235–245.

[Tut20] The use of numerical computations in the enumerative theory of planar maps, Jeffery–Williams Lectures, 1968–1972, Canad. Math. Congress, 1972, 73– 89.

[Tut21] Chromatic sums for rooted planar triangulations: the cases $\lambda = 1$ and $\lambda = 2$, *Canad. J. Math.* **25** (1973), 426–447.

[Tut22] Chromatic sums for rooted planar triangulations II: the case $\lambda = \tau + 1$, *Canad. J. Math.* **25** (1973), 657–671.

[Tut23] Chromatic sums for rooted planar triangulations III: the case $\lambda = 3$, *Canad. J. Math.* **25** (1973), 780–790.

[Tut24] Chromatic sums for rooted planar triangulations IV: the case $\lambda = \infty$, *Canad. J. Math.* **25**(1973), 309–325.

[Tut25] What is a map?, in New Directions in the Theory of Graphs (ed. by F. Harary), Acad. Press, 1973, 309–325.

[Tut26] Some polynomials associated with graphs, in Combinatorics(Proc. the 1973 British Combinatorial Conference), 161–167.

[Tut27] Chromatic sums for rooted planar triangulations V : special functions, *Canad. J. Math.* **26** (1974), 893–907.

[Tut28] Codichromatic graphs, *J. Comb. Theory* **B16** (1974), 168–174.

[Tut29] Map–colouring problems and chromatic polynomials, *Amer. Scientist* **62** (1974), 702–705.

[Tut30] On elementary calculus and the Good formula, *J. Comb. Theory* **B18** (1975), 97–137.

[Tut31] The dichromatic polynomial, Proc. 5-th British Comb. Conf., Util. Math., 1975, 605–635.

[Tut32] Chromatic sums, Proc. Kalamazoo Conf., 1977, 590– 598.

[Tut33] On a pair of functional equations of mathematical interest, *Aequationes Math.* **17** (1978), 121–140.

[Tut34] On the enumeration of convex polyhedra, *J. Comb. Theory* **B28** (1980), 105–126.

[Tut35] Combinatorial oriented maps, *Canad. J. Math.* **31**(1979), 986–1004.

[Tut36] Chromatic solutions, *Canad. J. Math.* **34**(1982), 741–758.

[Tut37] Chromatic solutions II, *Canad. J. Math.* **34** (1982), 952–960.

[Tut38] On the spanning trees of self- dual maps, *Ann. New York Acad. Sci.* **319** (1970), 540–548.

[Tut39] Graph Theory, Addison–Wesley, 1984.

[Tut40] The dissection of equilateral triangles into equilateral triangles, *Proc. Cambridge Philos. Soc.* **44** (1948), 463–482.

[Tut41] Chromatic sums revisited, *Aequatione Math.*, **50** (1995), 95–134.

Vince, A.

[Vin1] Combinatorial maps, *J. Comb. Theory* **34** (1983), 1–21.

[Vin2] Map duality and generalizations, *Ars Combin.* **39** (1995), 211–229.

van de Waerden, B. L.

[Wae1] Einfuhrung in die algebraische Geometrie, Springer, 1939.

Walsh, T. T. S.

[Wal1] Hypermaps versus bipartite maps, *J. Comb. Theory* **B18** (1975), 155– 163.

[Wal2] Counting non- isomorphic three- connected planar maps, *J. Comb. Theory* **B32** (1982), 33– 44.

[Wal3] Generating non- isomorphic maps without storing them, *SIAM J. Discrete Math.* **4** (1983), 161- 178.

Walsh, T. R. S. and A. B. Lehman

[WaL1] Counting rooted maps by genus I, *J. Comb. Theory* **B13** (1972), 192–218.

[WaL2] Counting rooted maps by genus II, *J Comb. Theory* **B13** (1972), 122–144.

[WaL3] Counting rooted maps by genus III, *J. Comb. Theory* **B18** (1975), 222–259.

Wawrzynczyk, A.

[Waw1] Group Representations and Special Functions, D. Reidel, 1984.

Whittaker, E. T. and G. N. Watson

[WhW1] A Course of Modern Analysis, Cambridge Uni. Press, 1940.

Witten, E.

[Wit1] Quantum field theory and the Jones polynomial, *Comm. Math. Phys.* **121** (1989), 351–399.

Wormald, N. C.

[Wor1] Counting unrooted planar maps, *Discrete Math.* **36**(1981), 205–225.

[Wor2] On the frequency of 3-connected subgraphs of planar graphs, *Bull. Austral. Math. Soc.* **34** (1986), 309–317.

[Wor3] The asymptotic number of nonsingular and nonseparable maps on the projective plane, Fourteenth Australasian Conference on Combinatorial Mathematics and Computing (Dunedin, 1986). *Ars Combin.* **23** (1987), 133–137.

Yan, J. Y. and Y. P. Liu

[YaL1] Asymptotic behavior of the enumerations of rooted planar and outerplanar maps, *KEXUE TONGBAO*(Chinese Ed.) **35** (1990), 88–91; Or see *Chinese Sci. Bull.* (English Ed.) **36** (1991), 188–192.

Yang, K. W.

[Yng1] Integration in the umbral calculus, *J. Math. Appl.* **74** (1980), 200–211.

[Yng2] Isomorphisms of group extensions, *Pacific J. Math.* **50** (1974), 299–304.

Index

Symbol

Number

A

B

U

V

W

MIX
Papier aus verantwortungsvollen Quellen
Paper from responsible sources
FSC® C105338

If you have any concerns about our products,
you can contact us on
ProductSafety@springernature.com

In case Publisher is established outside the EU,
the EU authorized representative is:
Springer Nature Customer Service Center GmbH
Europaplatz 3, 69115 Heidelberg, Germany

Printed by Libri Plureos GmbH
in Hamburg, Germany